AF761529

Sustainability in Plant and Crop Protection

Volume 13

Series Editor
Aurelio Ciancio, Sezione di Bari, Consiglio Nazionale delle Ricerche Istituto per la Protezione delle Piante, Bari, Italy

The series describes new perspectives recently developed in pest and disease management, based on innovative tools emerging from basic and applied research. The volumes will aim at interested readers involved in plant protection and crop management, for whom soil biodiversity, crop sustainability and, in general, organic approaches are fundamental. Different cropping systems will be treated by researchers involved in cutting edge studies worldwide. A number of basic issues including sustainability, life-cycle assessment, evolution, plant nutrition and organic cropping technologies will provide a common framework, within which different components of the crop production cycle will be focused on. These will range from roots and endophytes to pest and disease control, through the management of soil microbiome and fertility. These issues will be examined at the field and crop levels, including the effects of invasive species and climate changes on agroecosystems. Recent advancements in massive sequencing will represent the basis of dedicated volumes, dealing with transcriptomics and related approaches. They will illustrate the potentials and benefits of extensive DNA and RNA data analyses and studies, for practical purposes in crop protection, disease management and food production.

Contributions on any of the above cited topics are welcome. Potential Editors proposing a new volume are requested to contact the Series Responsible Editor and provide a short CV (400 words max) listing a selection of their most significant publications. In order to broaden the base of contributors and avoid redundancies, only one volume per Editor is allowed. Exceptionally, in case of many contributed chapters, a two-issues volume can eventually be considered.

More information about this series at http://www.springer.com/series/13031

Imran Ul Haq • Siddra Ijaz
Editors

Plant Disease Management Strategies for Sustainable Agriculture through Traditional and Modern Approaches

Springer

Editors
Imran Ul Haq
Department of Plant Pathology
University of Agriculture Faisalabad
Faisalabad, Pakistan

Siddra Ijaz
Centre of Agricultural Biochemistry
and Biotechnology (CABB)
University of Agriculture Faisalabad
Faisalabad, Pakistan

ISSN 2567-9805 ISSN 2567-9821 (electronic)
Sustainability in Plant and Crop Protection
ISBN 978-3-030-35954-6 ISBN 978-3-030-35955-3 (eBook)
https://doi.org/10.1007/978-3-030-35955-3

© Springer Nature Switzerland AG 2020
This work is subject to copyright. All rights are reserved by the Publisher, whether the whole or part of the material is concerned, specifically the rights of translation, reprinting, reuse of illustrations, recitation, broadcasting, reproduction on microfilms or in any other physical way, and transmission or information storage and retrieval, electronic adaptation, computer software, or by similar or dissimilar methodology now known or hereafter developed.
The use of general descriptive names, registered names, trademarks, service marks, etc. in this publication does not imply, even in the absence of a specific statement, that such names are exempt from the relevant protective laws and regulations and therefore free for general use.
The publisher, the authors, and the editors are safe to assume that the advice and information in this book are believed to be true and accurate at the date of publication. Neither the publisher nor the authors or the editors give a warranty, expressed or implied, with respect to the material contained herein or for any errors or omissions that may have been made. The publisher remains neutral with regard to jurisdictional claims in published maps and institutional affiliations.

This Springer imprint is published by the registered company Springer Nature Switzerland AG.
The registered company address is: Gewerbestrasse 11, 6330 Cham, Switzerland

Preface

Plant scientists, particularly pathologists, always are in a continuous struggle exploring new ways for plants protection, facing challenges such as adverse environmental and soil conditions and various kinds of diseases. The field of plant pathology is always challenging as it is of great importance for researchers to achieve the Nobel goal of food security and safety for an ever-increasing world population.

George N. Agrios mentioned in his book *Plant Pathology* that with a careful estimation, up to 42% losses in crops may be attributed to biotic stresses world over annually and these estimated losses are different from those caused by abiotic stresses. The situation is worse in developing countries where people face the challenge of food security, suffering from malnutrition and starvation. Plant scientists, therefore, always try to explore and develop more advanced, efficient, economic, and balanced ways to get maximum food yields. This may be achieved by protecting crops from diseases and insects, keeping the environment clean and healthy, and reducing any adverse effects on human and animal populations.

Moreover, advancement and evolution in biological science disciplines, such as microbiology, biotechnology, bioinformatics, and information and communication technology, offered new dimensions to plant pathology for the development of new disease management strategies. By keeping this perspective in view, we are making this attempt to keep plant scientists updated with latest developments in plant disease management strategies, aiming at the best integration of conventional and innovative methods.

We tried our best to collect and compile useful, practical, and recent information on plant disease management from diverse groups of authors and countries associated with well-reputed teaching and research organizations. We hope we reached the objective of updating and equipping readers with the most comprehensive and latest knowledge available today.

This book considers traditional and modern approaches for plant disease management. For a sustainable agriculture, methods based on sustainable management of phytopathogens are indeed an indispensable factor. In a nutshell, we tried to cover competitive areas of plant disease management, assembling best classical and modern strategies, most suitable for a sustainable agriculture.

Faisalabad, Pakistan Imran Ul Haq
Faisalabad, Pakistan Siddra Ijaz

Contents

About the Editors

Dr. Imran Ul Haq is a Plant Pathologist and is specialized in fungal molecular taxonomy. Currently, he is Associate Professor and in-charge of Fungal Molecular Biology Laboratory as well as Fungal Molecular Biology Laboratory Culture Collection (FMB-CC-UAF) at the Department of Plant Pathology, University of Agriculture, Faisalabad, Pakistan. He is running different research projects funded by national and international agencies. His research interests are fungal molecular taxonomy, integration of nanotechnology with other control strategies for sustainable plant disease management. He has authored 3 Laboratory manuals as well as book chapters and has more than 50 research publications in well-reputed, peer-reviewed, high-impact national and international journals. He is also author of a book *Recombinant DNA Technology*.

Dr. Siddra Ijaz is a Molecular Biologist and is currently serving as Assistant Professor at the Centre of Agricultural Biochemistry and Biotechnology (CABB), University of Agriculture, Faisalabad, Pakistan. She has several competitive research grants funded by national as well as international agencies, and established molecular biology laboratory. Her research focus includes plant genome engineering using transgenic technologies, genome editing through CRISPR/Cas9, and nanobiotechnology. She has authored two books *Molecular Basis of Life* and *Recombinant DNA Technology*, as well as book chapters. She has more than 40 research publications in high-impact, peer-reviewed national and international scientific journals.

Contributors

Maqshoof Ahmad Department of Soil Science, University College of Agriculture and Environmental Sciences, The Islamia University of Bahawalpur, Bahawalpur, Pakistan

Zohaib Ahmad Discipline of Plant Pathology, University College of Agriculture and Environmental Sciences, The Islamia University of Bahawalpur, Bahawalpur, Pakistan

Muhammad Tahir Akram Institute of Horticultural Sciences, University of Agriculture Faisalabad, Faisalabad, Punjab, Pakistan

Saleem Ashraf In-Service Agricultural Training Institute, Rahim Yar Khan, Punjab, Pakistan

Muhammad Azam Institute of Horticultural Sciences, University of Agriculture Faisalabad, Faisalabad, Punjab, Pakistan

Maria Babar Centre of Agricultural Biochemistry and Biotechnology (CABB), University of Agriculture Faisalabad, Faisalabad, Pakistan

Siva K. Balasundram Department of Agriculture Technology, Universiti Putra Malaysia (UPM), Serdang, Selangor, Malaysia

Shaghef Ejaz Department of Horticulture, Bahauddin Zakariya University, Multan, Pakistan

Anjum Faraz Department of Plant Pathology, University of Agriculture Faisalabad, Faisalabad, Pakistan

Kamlesh Golhani Department of Agriculture Technology, Universiti Putra Malaysia (UPM), Serdang, Selangor, Malaysia

Zia Ul Haq Department of Plant Pathology, Faculty of Agriculture Sciences and Technology, Bahauddin Zakariya University, Multan, Pakistan

Mariam Hassan Ayub Agricultural Research Institute (AARI), Faisalabad, Pakistan

Yasir Iftikhar Department of Plant Pathology, College of Agriculture, University of Sargodha, Sargodha, Pakistan

Siddra Ijaz Centre of Agricultural Biochemistry and Biotechnology (CABB), University of Agriculture Faisalabad, Faisalabad, Pakistan

Moazzam Jamil Department of Soil Science, University College of Agriculture and Environmental Sciences, The Islamia University of Bahawalpur, Bahawalpur, Pakistan

Kiranjot Kaur Khalsa College, Amritsar, India

Ali Hassan Khan Ayub Agricultural Research Institute (AARI), Faisalabad, Pakistan

Imran Khan Institute of Soil and Environmental Sciences, University of Agriculture Faisalabad, Faisalabad, Pakistan

M. Arslan Khan Institute of Horticultural Sciences, University of Agriculture Faisalabad, Faisalabad, Punjab, Pakistan

Muhammad Naeem Khan Ayub Agricultural Research Institute (AARI), Faisalabad, Pakistan

Muhammad Yahya Khan Sub Campus Burewala, University of Agriculture Faisalabad, Burewala, Pakistan

Nabeeha Aslam Khan Department of Plant Pathology, University of Agriculture Faisalabad, Faisalabad, Pakistan

Lih Ling Kong Institute of Tropical Agriculture, Universiti Putra Malaysia, Seri Kembangan, Serdang, Selangor, Malaysia

Muhammad Zunair Latif Department of Plant Pathology, University of Agriculture Faisalabad, Faisalabad, Pakistan

Guoqing Li State Key Laboratory of Agricultural Microbiology and Hubei Key Laboratory of Plant Pathology, Huazhong Agricultural University, Wuhan, China

Yang Long State Key Laboratory of Agricultural Microbiology and Hubei Key Laboratory of Plant Pathology, Huazhong Agricultural University, Wuhan, China

Muhammad Luqman College of Agriculture, University of Sargodha, Sargodha, Pakistan

Najm-ul-Seher Department of Soil Science, University College of Agriculture and Environmental Sciences, The Islamia University of Bahawalpur, Bahawalpur, Pakistan

Sajid Mahmood Nadeem Sub Campus Burewala, University of Agriculture Faisalabad, Burewala, Pakistan

Bukhtawer Nasir Centre of Agricultural Biochemistry and Biotechnology (CABB), University of Agriculture Faisalabad, Faisalabad, Pakistan

Farheen Nazli Department of Soil Science, University College of Agriculture and Environmental Sciences, The Islamia University of Bahawalpur, Bahawalpur, Pakistan

Pesticide Quality Control Laboratory Bahawalpur, Agriculture Department, Government of the Punjab, Bahawalpur, Pakistan

Anita Puyam Rani Lakshmi Bai Central Agricultural University, Jhansi, Uttar Pradesh, India

Rashad Qadri Institute of Horticultural Sciences, University of Agriculture Faisalabad, Faisalabad, Punjab, Pakistan

Hafiza Arooj Razzaq Centre of Agricultural Biochemistry and Biotechnology (CABB), University of Agriculture Faisalabad, Faisalabad, Pakistan

Ashara Sajid Department of Plant Pathology, College of Agriculture, University of Sargodha, Sargodha, Pakistan

Muhammad Kaleem Sarwar Department of Plant Pathology, University of Agriculture Faisalabad, Faisalabad, Pakistan

Qaiser Shakeel Discipline of Plant Pathology, University College of Agriculture and Environmental Sciences, The Islamia University of Bahawalpur, Bahawalpur, Pakistan

Redmond R. Shamshiri Department of Agriculture Technology, Universiti Putra Malaysia (UPM), Serdang, Selangor, Malaysia

Hafiz Abdul Samad Tahir Pakistan Tobacco Board, Tobacco Research Station, Okara, Ministry of Commerce, Government of Pakistan, Islamabad, Pakistan

Imran Ul Haq Department of Plant Pathology, University of Agriculture Faisalabad, Faisalabad, Pakistan

Ganesan Vadamalai Department of Plant Protection, Universiti Putra Malaysia (UPM), Seri Kembangan, Malaysia

Yaodong Yang Hainan Key Laboratory of Tropical Oil Crops Biology, Coconut Research Institute, Chinese Academy of Tropical Agricultural Sciences, Wenchang, Hainan, China

Muhammad Yaseen College of Agriculture, University of Sargodha, Sargodha, Pakistan

Chapter 1
History and Recent Trends in Plant Disease Control: An Overview

Imran Ul Haq and Siddra Ijaz

Abstract Plants are continuously exposed to certain biotic and a biotic stresses, causing serious crop losses every year. Prevailing situation is representing today a serious threat to global food security and safety. Any professional plant pathologist needs to have theoretical as well as practical knowledge and a clear understanding of plant diseases and of the factors involved, knowing how to discover effective control means. This chapter has been designed to provide the reader a brief overview regarding the concept of plant diseases, their diagnosis and the threats they pose to crop production and protection. Here we discuss and focus on basic principles including: plant disease management, conventional and advanced methods of controlling diseases and integration of various control measures, historical perspectives, disease management in the current era, future directions and challenges.

Keywords Plant pathology · Historical perspectives · Principles of plant disease control · Recent trends in plant pathology

1.1 Introduction

Plant pathology is the science concerned with a detailed study of plant diseases (caused by biotic and abiotic factors), mechanisms of inducing diseases in plants and efforts for their survival by overcoming diseases and achieving plants full genetic potential. The field of plant pathology is dynamic. It is worth studying all practical efforts needed to achieve the noble goal of providing safe and diverse food

I. Ul Haq (✉)
Department of Plant Pathology, University of Agriculture Faisalabad, Faisalabad, Pakistan

S. Ijaz
Centre of Agricultural Biochemistry and Biotechnology (CABB),
University of Agriculture Faisalabad, Faisalabad, Pakistan

© Springer Nature Switzerland AG 2020

I. Ul Haq, S. Ijaz (eds.), *Plant Disease Management Strategies for Sustainable Agriculture through Traditional and Modern Approaches*, Sustainability in Plant and Crop Protection 13, https://doi.org/10.1007/978-3-030-35955-3_1

for our ever increasing global population. Plant diseases affect not only crop yields but also their quality, and reduce farmers resource-use efficiency. Plant health management strategies preventing crop losses (yield and quality) enhance production and significantly contribute towards food security and safety (Strange and Scott 2005). With an increasing world population and its food needs, the agricultural research in twentieth century remained focused on increasing crops productivity (Evans 1998; Smil 2000; Nellemann et al. 2009).

Plant pathogens mostly include microorganisms, such as fungi, bacteria, viruses as well as abiotic stresses such as adverse environments, lack or excess of nutrients, extreme temperature ranges, high moisture, dry condition, soil pH, pollution, light intensity and chemical injuries. Unwanted plants (weeds) are also a big growth constraint, causing heavy crop losses other than pathogens and insects. Damages caused by insects or animals to plants are not included in plant pathology.

Plant disease management strategies practiced as long as agriculture itself. Despite of all the scientific and technological advancements and their contributions in controlling diseases and significantly reduce the occurrence and severity of epidemics to date, plant protection is still a big challenge for agricultural scientists, and it is more complex than ever before (FAO 2011; Brown 2011). Plant pathogens (fungi, bacteria, viruses) not only interact one each other during the infection process, but also with other abiotic factors. Crop health management hence requires a multidisciplinary approach (Teng et al. 1984). To achieve the goal of sustainable plant disease management, some research areas need to be focused on. They are: host resistance development, pathogens avoidance or evasion, reduction of inoculum and remediation strategies, integration, set up of environmental conditions affecting pathogens reproduction and growth, evolution of new pathogenic races (He et al. 2016).

1.2 Plant Pathology: Food Safety and Security

Preventing the infestation and contamination of food from disease-causing microorganisms is important and represents a major concern for food safety. Food security is defined as the nutritious, healthy and uninterrupted food supply to all people around the world, for a healthy life style. Undernourished people suffering from food security issue in 2010 were 925 million worldwide, a highly unacceptable number (FAO 2010; Clapp 2014). By causing various diseases, plant pathogens affect crop plants ranging from merely mild symptoms to calamities, which may turn out to be famine or may intensify current shortage of food for million people. Plant pathogens have much variability in their population and can easily overcome resistance, thus ruining the epic work of plant breeders. Hence, conventional plant breeding equipped with latest plant disease management technologies, GMOs and marker-aided selection, have a crucial role to play in food security, powered by sufficient resources (Strange and Scott 2005). Oerke (2006) explained that food safety and security issues due to plant stresses, especially plant diseases, are severe in countries with less resources, which become critical when postharvest losses are included. Hundred of billion USD are wasted due to plant diseases every year, in terms of money. If global harvest and postharvest losses are combined these exceed 16–28% of global production, with higher percentages in developing countries lacking facilities and infrastructure to control spreading of plant pathogens (Agrios 2005). Recently, *Puccinia graminis tritici*, one of the most severe pathogen of wheat causing black stem rust, has damaged wheat crops in low income areas of Africa, Middle East and Asia, causing serious issues of food security (Flood 2010). Similarly, another example of trans-boundary impact of plant pathogens is *Fusarium xylarioides,* which has caused loss of almost 1 billion USD to coffee producers in central and Eastern Africa (Flood 2009).

1.3 Brief History of Plant Pathology

In the early stages of nomadism when leaves, fruits, seeds, and meat were the main source for survival, plant diseases were present and named as mildews, blights and blasts (Russell 2005). With the advancement in living style and a shift from nomadism to domestication, individual families began to grow crops, mainly cereals (wheat, barley, oat and rye), legumes (chickpeas, lentils and fava beans) and fruit trees (figs, apples, olives, peaches and citrus).Among other fruit crops grapes there were squash and melons and other plants also cultivated for survival of people and their animals as well. Around 470 BC the Greek philosopher Democritus mentioned the ways to control mildews, blasts and blights in his writings (Agrios 2005). Plant diseases were commonly considered at that time as divine punishments. Therefore, in fourth century BC, when Romans faced heavy crop losses by rust diseases, a deity called “Robigus”. They began to worship and offer sacrifices with the belief that this deity would prevent the dreaded rusts and other diseases (Littlefield 1981). Democritus suggested that sprinkling of olive grounds were helpful in controlling plant blights. Homer, described around 1000 BC the therapeutic properties of sulfur against plant diseases. However, most ancient reports dealt with pseudo-beliefs. Very little information was written since then anywhere for almost 2000 years.

In 1200 AD it was observed that mistletoe parasitizes the host plant and makes it sick. The host plant could be saved by the removal of infected plant parts. In the mid of sixteenth century, it was noted by French farmers that wheat rust infections were always more frequent and growing near barberry plants. Farmers believed that these plants played a key role as source of primary inoculum which on later stage infected the cultivated wheat fields. They hence requested the Government to pass out a legislation aimed at eradication of wild bushes of barberry to protect wheat (Agrios 2005). Meanwhile, it was observed that a species or variety was more resistant than others. Probably, in late 1600s southern England farmers used sodium chloride solution (brine) as seed treatment to control wheat bunt. Later on in mid-1700s the brine solution was substituted with copper sulfate which significantly improved the bunt disease control efficacy (Russell 2005). In 1670, Thoullier found that ergotism was caused by consuming contaminated grains and claimed that it did spread from person to person. In mid 1700s tree canker(s) were cured by excisions of the infected part(s) and grafting wax was used to cover the cut area (Agrios 2005).

In 1755, the French researcher Tillet proved through experimentation that smut infected plants could be increased by dusting the smut spores on wheat kernels before planting. Infection rates could be reduced by the application of copper sulfate to the wheat kernels before sowing (Tillet 1755). In 1761, copper sulfate was used to control wheat bunt (Schulthess 1761).

In the early 1800s, mildew of fruit trees was controlled by the application of either lime sulfur or aqueous sulfur solutions. In 1807, the French researcher Prevost properly concluded that wheat smut could be controlled with copper sulfate that could inhibit the germination of smutted spores (Prevost 1807). Late blight epidemic throughout Europe (particularly in Ireland) required scientists' efforts in controlling the diseases. In 1846, John Lindley observed the effectiveness of copper on potato plants but his report did not get popularity. By 1854, in England, powdery mildew infection on leaves and grapes was controlled by applying a mixture of lime and sulfur to the infected parts. The same practice was adopted by French vineyard industry for management of the powdery mildew of grapes, which caused heavy losses of up to 80% to their vine production industry (Spencer 1978; Kenrick 1833). In 1860s, another disease attacked the French vine industry. It was noticed that the Phylloxera aphid *Daktulosphaira vitifoliae* Fitch 1856 (syn. *Phylloxera vastatrix* Planchon, 1868), probably introduced with vines imported from the USA as powdery mildew resistant, was associated with this disease. However, grapevines in North America were resistant against these aphids so they were imported and used as rootstocks for the grafting of European vines. Some of these grafted vines showed excellent degree of resistance against the aphids.

In 1870s the German scientist Kuhn studied control measure strategies, to specifically cope with seed borne infections. In 1878, European grapevines were attacked by downy mildew. The attack was so severe that within 5 years of its appearance it spread throughout French and adjacent vine producing countries. Scientists were trying hard to cope with downy mildew and for this purpose they applied different substances to the soil or even dusted the vines, but nothing worked significantly. In October 1882, the French botanist Pierre Alexis Millardet was strolling through a powdery mildew affected grape vine orchard. He was surprised to see that vines alongside road were still green and healthy. By close observation of leaves he found that the vines were treated with some kind of chemical. Later on, by investigating from the orchard manager he understood that it was a common practice to treat the vines with a poisonous mix of chemicals (copper sulphate and lime) to protect them from passers. In 1885, he find out the best combination (8-8-100) to control downy mildew of grapes which is known today as Bordeaux mixture (Millardet 1885). The discovery of the Bordeaux mixture started the chemical control of plant diseases. The mixture was an excellent fungicides as well as bactericide. It was successfully used over a century to control several diseases (leaf spots, late blight of potato, leaf blights and downy mildews) throughout the world. This discovery improved the way of thinking that plant diseases could be successfully controlled by using chemicals (Agrios 2005).

In 1913, the concept of seed treatments with organic mercury compounds flourished which dominated until the 1960s when mercury was banned due to its high levels of toxicity. In 1928 Alexander Fleming, with the discovery of penicillin, came up with a different idea of controlling plant diseases. In 1934, discovery of the dithiocarbamate fungicide thiram led to the development of ferbem, zineb and maneb along with the development of many other protective fungicides. In 1965, carboxin was discovered followed by other systemic fungicides (Russell 2005). In 1950s, streptomycin was used as first antibiotic against bacterial diseases. Soon after, cycloheximide provided some promising results for management of fungal pathogens. In 1967, plant diseases caused by mollicutes as well as by fastidious bacteria were effectively controlled by tetracycline. In 1954 and 1963, a few bacterial and fungal strains were observed that developed resistance against a bactericide and fungicide.

The appearance of pathogenic race(s)/strains resistant to specific chemicals entirely changed the strategies of plant disease management. Strategies such as use of fungicides in combinations, alteration of compounds and application of systemic compound(s) at earlier stages followed by broad spectrum compound(s) on later stages of the diseases were extensively suggested (McManus et al. 2002). By 1980s, most of pesticides (85–90%) related to public concerns were banned either by US government, EU authorities or manufacturers. This initiative enforced the scientists' communities to re-examine and improve alternative disease control measures that were used by ancient times, which now represent the basis of integrated disease management (IDM, see Sect. 1.6).

In the early twentieth century it was reported that some microorganisms have the capability to harbor or suppress soil borne diseases caused by soil borne pathogen(s). Furthermore, the discovery of penicillin enforced the researchers to find out quite similar non-pathogenic microorganisms that could be able to antagonize or inhibit the pathogenic one(s). In 1963, the first biocontrol was obtained against root and butt rot of pines caused by *Heterobasidion annosum*, controlled by inoculation of *Phleviopsis gigantean* (a non-pathogenic fungus) to the stumps of freshly cut pines. In 1972, the crown gall bacterium of stone fruit was controlled by pre-inoculating seeds and root transplants with a related but non-pathogenic bacterium. Tobacco mosaic virus was controlled in tomato fields by pre-inoculation of seedlings with non-pathogenic strains obtained by artificial virus mutation. Control of citrus tristeza was obtained through cross protection. In late 1980s, control of viral diseases was obtained through genetic engineering techniques. Another recent development in plant disease control includes the systemic acquired resistance by creating necrotic lesions when pathogenic microorganisms are applied to the plants. In 1960s, plant defense activators were synthesized and market tested in 1996, with reasonable success (Agrios 2005).

1.4 Basic Principles of Plant Disease Management

Plant pathologist must have all kinds of information regarding the host plant, the pathogen attacking that particular host, the data about disease occurrence and history, agronomy, environmental conditions, cost of production etc., while planning and implementing the most appropriate disease management strategies.

Principles of plant disease management may be summarized as follows:

1. ***Exclusion***: this is probably the first defense line in IDM aimed at prevention of introduction and dispersal of inoculum into the area where it was not present before. ***Regulatory control methods***, usually adopted with the objective to exclude the inoculum from host plant, or from a certain area. These measures are applied by means of ***Quarantine and Inspections***, based on a number of regulatory practices. It is the responsibility of the national or state regulatory agencies to ensure the prevention of introduction and dispersal of the pathogens between and within the country or states, by implementing the quarantine laws (Fegan et al. 2004; Nutter and Madden 2005). Quarantine usually imposes complete restriction on import of any agricultural product for specific pathogenic threats, or may impose partial sanctions depending on the results of the material inspection. The import restrictions under the quarantine regulation might be imposed on all countries, provinces or regions (Fry 1982; Palti 1981; Sill 1982). Mayths and Baker (1980) suggested principles to make the quarantine regulations more effective. According them (i) the organisms suspected to cause damage to the crops should be restricted only through these regulations and (ii) these regulations should not affect the trade but only restrict the organisms capable of causing diseases. They should (iii) also be operated under the amended or improved law, according to the situation prevailing at the time.
2. ***Avoidance** or **Evasion***. Sometimes farmers select sites for crops cultivation usually based on a low risk of disease occurrence, due to certain factors such as unfavorable climatic conditions for disease development, absence of vectors, etc. This is a form of exclusion based on time. In time, avoidance of pathogen inoculum (by selecting proper planting date, site, seed or propagation material as well as agronomic practices and plant protection measures) is one of the effective strategy for controlling plant diseases. The philosophy involved in this strategy is to avoid the critical time period (favorable environmental and growth conditions) during which that pathogen may get established, causing infection and thus inducing crop losses. (Nutter and Guan 2001; Savary et al. 2006).
3. ***Eradication*** involves cultural (horticultural, agronomic practices), physical as well as biological and chemical methods aimed at reducing the initial inoculum of a certain pathogen from its host plant, plant part or from certain geographical area (site/field). Among the cultural practices, physical eradication of infected host, removal of infected plant parts, burial or burning of debris, eradication of alternate hosts and crop rotation are the methods most applied for eradication. Other practices such as soil solarization and mulching are more recent approaches, which are now widely practiced in different parts of the USA, Australia, and

many other countries, adding to sterilization, hot air treatment, hot water treatment, sun drying, Light wavelengths and refrigeration are physical methods used for inoculum reduction. Use of antagonists, trap crops, suppressive soil, cross protection are the biological approaches to reduce the pathogen population (Palti 1981; Coelho et al. 1999; Du Toit and Hernandez-Perez 2005).

Sanitation practices aim at reducing or eliminating a pathogen population from a diseased field, through disinfection of warehouses, tools and equipment that also may contribute in reducing the initial inoculum, using chemicals (Fry 1982; Lipps 1985; Palti 1981; Sharvell 1979).

4. ***Resistance*** is the ability of the host plant to defend itself against biotic or abiotic stresses. Some of plant pathologists keep host resistance under the "direct protection", one of the principles of plant disease management. However, the complex and scientifically interesting mechanism of host resistance to pathogens and the effects the disease development rates should be recognized as separate principles of plant disease management. The mechanisms of host resistance towards diseases may involve one or more mechanisms reducing the inoculum and the rate of infection and disease progress (Van der Plank 1963; Zadoks Schein 1979). There are two major types of resistance: (1) Resistance opposing the establishment of infection by reducing the amount of initial inoculum. It is also named vertical resistance, complete resistance, race-specific or monogenic resistance, and (2) other type of resistance opposing infection and disease severity on plants, called horizontal resistance, quantitative resistance, or partial resistance. Cultivation of resistant varieties is an easy to adopt, eco-friendly, safer and effective way to control plant diseases. Resistance breaks down, however, is a big challenge as new virulent races of pathogens evolve. Once a resistant variety is developed its useful "life span" may be enhanced by adopting proper agronomic, cultural, as well as plant protection measures.
5. ***Direct protection*** using chemicals for plant disease management is a significant component of disease management. Chemicals (fungicides) have been used to protect crop from pathogens since the 1940s. Since then, the application of fungicides contributed significantly in controlling plant diseases (Leadbeater and Gisi 2015). Replacement of older non-systemic fungicides with systemic fungicides (more effective and specific) was the major development in the field of chemical management in the '60s. For example among the systemic fungicides the triazole group gained a 24% business share of the total fungicide market (Hewitt 1998). On other hand, some non-systemic fungicides also had a significant business volume in developing countries, especially due to a lower cost Development of new classes of fungicides posed significant effects on disease management. However, the resistance of pathogens to many of the newly developed products is still a big challenge for plant pathologists. Use of fungicides is more effective and efficient when combined with other control methods in an IDM program (De Waard et al. 1993). Actually, low toxicity, low residues in edible parts, and ecofriendly availability of agrochemicals to meet international health standards is a public demand (Knight et al. 1997). Since the development of first fungicidal formulations to date, a large number (likely several hundred)

formulations have been developed and made available commercially, worldwide. These fungicide formulations are applied in different ways, depending on the nature of targeted pathogens such as dusts and wetable powders, liquids, granular, and fumigants.

Several attempts have been made so far to improve the efficacy of fungicide formulations, by means of additional chemicals. Emulsifiers contribute in the formation of soluble suspensions. Foam suppressors increase the contact of the spray suspension maximally, on the plant surface. Penetrants enhance absorption capability of systemic formulation. Stickers enhance chemical persistence on plant surface. Surfactants reduce surface tension, and also enhance the penetration of the chemical to the plant subsurface. Wetting and dispersing agents improve particle suspension during application. UV filters enhance formulation photolytic stability on the plant surface (Agrios 2005).

The chemical industry branch for plant disease management started in nineteenth century with the discovery of simple inorganic copper and sulphur products. It is increasing its list of chemicals with complex, novel groups and various mode of action, effective against groups of pathogens (Hewitt 2000; Leadbeater and Gisi 2015). Thiram and captan are early broad spectrum, contact organic fungicides whereas streptomycin discovered in 1955 is still an effective and most common (when permitted) antibiotic, with gentamycin allowed in Latin America and oxolinic acid in Israel (Shtienberg et al. 2001; McManus et al. 2002).

Since 1930, new fungicides are providing a promising role to restrict and limit pathogens boundaries and for effective management of economically important diseases. A wide range of formulations and products for fungi and bacterial diseases are offered today on the market, whereas the nematicide industry showed minimum growth. This was most probably due to less awareness as indicated by few early reports (Hague and Gowen 1987). In the second half of the nineteenth century, carbon di-sulphide was discovered, a pioneer effective nematicide, followed by chloropicrin used successfully in early nineteenth century in England against nematodes and other soil pathogens (Schacht 1859; Kuhn 1881; Mathews 1919).

Fungicides target various known and unknown mechanisms of actions essential for fungal growth and development. Aniline-pyrimidines and streptomycin disturb protein synthesis and enzymatic activity; benzimidazole carbamates inhibit microtubule formation; hydroxyl-pyrimidine group fungicides attack metabolic processes involving nucleotides; chlorothalonil blocks glutathione conversion into its various forms (Hollomon and Chamberlain 1981; Chen et al. 2001; Halling et al. 2002; Gupta et al. 2004; Carr et al. 2005; Mueller et al. 2008; Yang et al. 2011; Koo et al. 2009).

Fungicides also have various mode of applications through fumigation, flooding, injector, dusters or as foliar, etc. according to the target pathogen. Least effective fungicides are continuously being replaced by chemicals with active redistribution and translocation ability, having novel, new chemistry and complex action sites such as mancozeb, propineb, captan etc. (Maude 1996; Klittich 2008; Ivic 2010; Milenkovski et al. 2010). For research and development of an effective new fungicide with distinct characteristics, huge amounts of funding and time is required, for its long term survival in market and safety to humans (Mcdougall 2010; Leadbeater and Gisi 2015). Pesticide rules and regulations are becoming strict with passage of time due to human and environmental health concerns, not hindering agricultural development, with USA, Brazil and the EU as leading examples (Pelaez et al. 2013).

Awareness regarding resistance against fungicides is important as it can waste resources utilized on products development and marketing. Various fungus and fungicide factors are responsible for resistance development (Brent and Hollomon 1995). Risk of resistance development vary in various groups of fungicides with strobilurin and benzimidazole at high risk, starting 2 years after the product introduction on the market (Brent and Hollomon 2007).

1.5 Biological Control

The exploitation of the antagonistic potential of microorganisms to control plant pathogens has gained popularity in recent years. Biological control is considered as the best alternative method to reduce the use of pesticides. It is a general consideration that biological control keeps environment healthy, eco-friendly, self-sustainable and has long lasting impacts. Several microorganism antagonistic to plant pathogens are available in the market, globally. The commercial formulations of these antagonists are named "microbial pesticides" or "biopesticides". Their possible modes of action includes parasitism, competition, antibiosis, induced resistance and inactivation of pathogen enzymes (Agrios 2005).

1.6 Integrated Plant Disease Management

In order to attain the target of a sustainable disease control in an economic and sustainable way, establishment and implementation an IDM system is the most appropriate option. While designing an IDM plan it must be kept in consideration that the system must include almost all possible control methods. Usually, an IDM system is designed with the objective of controlling all diseases of a single crop. However, it may also target a specific disease (major threat to a crop or occurring in epidemic form) e.g., potato late blight.

The major aims and components of an IDM program are: (i) eliminating or reducing the initial inoculum, (ii) keeping its effectiveness in time, (iii) enhancing the host resistance, (iv) slowing down the infection process and (v) the pathogen secondary cycles (Agrios 2005).

1.7 Recent Advances in Plant Disease Management

Modern plant pathology has been greatly accelerated with the aid of molecular tools and advancements in plant disease control strategies. Since the last few decades, molecular plant pathology has been proved very helpful by introducing several new ways and providing the better opportunities for disease diagnosis and control. In this regard, biotechnology and genetic engineering played a key role. Molecular techniques such as DNA-based identification of plant pathogen(s), rapid sequencing, quantitative real time PCR (qPCR), diagnostic assays, biomarkers, and whole genome sequencing greatly improved the way of pathogen(s) detection, disease diagnosis and management.

Field ready serological assays are very helpful in the decision making process and in pre-screening against targeted diseases. Biomarkers such as volatile chemicals are in practice for the detection of pest outbreaks. Biosensors coupled with information technology networks can provide real-time surveillance or monitoring of emerging problems caused by pests and diseases (Lucas 2011). Molecular tools such as microsatellites, remote sensing, image analysis, global information system, geo-statistics, and geographic information systems are very helpful in the monitoring and surveillance of plant diseases spread and risk assessment.

Development of different disease forecasting models and computer simulation-based methods for surveillance of plant disease epidemics are very important for their mapping (Agrios 2005). The adoption of the E-Phyto (electronic phytosanitary certificate) provides a great deal in safe trade of plants and plant-based products, by introducing innovative technologies. In recent years the potential of nanotechnology in plant disease management has been greatly put in practice. Surface response resonance and other nano-based sensors are very helpful in the detection of pesticides residues, seed borne mycoflora, seed certification and quarantine. Nano-based pesticides significantly reduced environment and health-associated risks.

Automatic purification of nucleic acids and specific proteins from pathogen(s) increased the possibilities of early diagnosis. The knowledge for identification of genetic basis, signaling molecules, network and pathways that control plant defense could be very helpful for the development of a new generation of genetic modulators. Additionally, it will provide more opportunities for the development of genetically modified organisms that could respond well to biotic stresses (Lucas 2011).

Plants have intrinsic networks to respond to phytopathogens upon their intrusion. They use a vast array of proteome and metabolome resources for their defense. Plant biologists are continuously struggling to explore the world of these biomolecules involved in plant-pathogen interaction and disease development. The resistance-associated genes have been tracking to be identified since a long epoch and are being used in the development of transgenic resistant plants. These genes are being used in the perspective of introducing resistance, triggering the endogenous resistance mechanism of plants by their overexpression against phytopathogens (Rommens and Kishore 2000).

However, advancements in the field of molecular biology facilitated to exploration of the molecular basis of plant pathogen interactions. Plant innate responses could be engineered to get durable resistance including systemic acquired resistance and hypersensitive response (Strittmatter et al. 1995). It could be fascinating to hypothesize that the introduction of resistance gene(s) would mediate incompatible reactions with annexing phytopathogens and lead to hypersensitive response ensuing localized cell death.

References

Agrios, G. N. (2005). *Introduction to plant pathology*. New York: Elsevier Academic Press Publication.

Brent, K. J., & Hollomon, D. W. (1995). *Fungicide resistance in crop pathogens: How can it be managed*. FRAC Monograph No. 1 (2nd, Rev. ed.). Fungicide Resistance Action Committee. Monograph 1 GCPF, FRAC, Brussels, pp. 1–48.

Brent, K. J., & Hollomon, D. W. (2007). *Fungicide resistance in crop pathogens: How can it be managed?* (2nd Rev. ed.) Online. Fungicide Resistance Action Committee (FRAC). CropLife International, Brussels, Belgium.

Brown, L. R. (2011). *World on the edge. How to prevent environmental and economic collapse*. New York/London: W. W. Norton Company.

Carr, J. F., Gregory, S. T., & Dahlberg, A. E. (2005). Severity of the streptomycin resistance and streptomycin dependence phenotypes of ribosomal protein S12 of *Thermus thermophilus* depends on the identity of highly conserved amino acid residues. *Journal of Bacteriology, 187*, 3548–3550.

Chen, S. K., Edwards, C. A., & Subler, S. (2001). Effects of the fungicides benomyl, captan and chlorothalonil on soil microbial activity and nitrogen dynamics in laboratory incubations. *Soil Biology and Biochemistry, 33*, 1971–1980.

Clapp, J. (2014). Food security and food sovereignty: Getting past the binary. *Dialogues in Human Geography, 4*, 206–211.

Coelho, L., Chellemi, D. O., & Mitchell, D. J. (1999). Efficacy of solarization and cabbage amendment for the control of *Phytophthora* spp. in North Florida. *Plant Disease, 83*, 293–299.

De Waard, M. A., Georgopoulos, S., Hollomon, G. D. W., Ishii, H., Leroux, P., Ragsdale, N. N., & Schwinn, F. J. (1993). Chemical control of plant diseases: Problems and prospects. *Annual Review of Phytopathology, 31*, 403–421.

Du Toit, L. J., & Hernandez-Perez, P. (2005). Efficacy of hot water and chlorine for eradication of *Cladosporium variable, Stemphylium botryosum,* and *Verticillium dahliae* from spinach seed. *Plant Disease, 89*, 1305–1312.

Evans, L. T. (1998). *Feeding the ten billion. Plants and population growth*. Cambridge: Cambridge University Press.

FAO. (2010). http://faostat.fao.org

FAO. (2011). http://faostat.fao.org. Accessed 17 Oct 2011.

Fegan, R. M., Olexa, M. T., & McGovern, R. J. (2004). Protecting agriculture: The legal basis of regulatory action in Florida. *Plant Disease, 88*, 1040–1043.

Flood, J. (2009). *Coffee wilt disease*. Wallingford: CAB International. 200 pp. ISBN 978-1-84593-641-9.

Flood, J. (2010). The importance of plant health to food security. *Food Security, 2*(3), 215–231.

Fry, W. E. (1982). *Principles of plant disease management*. London: Academic.

Gupta, K., Bishop, J., Peck, A., Brown, J., Wilson, L., & Panda, D. (2004). Antimitotic antifungal compound benomyl inhibits brain microtubule polymerization and dynamics and cancer cell proliferation at mitosis, by binding to a novel site in tubulin. *The Biochemist, 43*, 6645–6655.

Hague, N. G. M., & Gowen, S. R. (1987). Chemical control of nematodes. In R. H. Brown & B. R. Kerry (Eds.), *Principles and practice of nematode control in crops* (pp. 131–178). Sydney: Academic.

Halling-Sørensen, B., Sengeløv, G., & Tjørnelund, J. (2002). Toxicity of tetracyclines and tetracycline degradation products to environmentally relevant bacteria, including selected tetracycline-resistant bacteria. *Archives of Environmental Contamination and Toxicology, 42*, 263–271.

He, D. C., Zhan, J. S., & Xie, L. H. (2016). Problems, challenges and future of plant disease management: From an ecological point of view. *Journal of Integrative Agriculture, 15*, 705–715.

Hewitt, H. G. (1998). *Fungicides in crop protection*. Wallingford: CAB International.

Hewitt, G. New modes of action of fungicides. *The Royal Society of Chemistry, 2000, 2000*, 28–32.

Hollomon, D. W., & Chamberlain, K. (1981). Hydroxypyrimidine fungicides inhibit adenosine deaminase in barley powdery mildew. *Pesticide Biochemistry and Physiology, 16*(2), 158–169.

Ivic, D. (2010). Curative and eradicative effects of fungicides. In O. Carisee (Ed.), *Fungicides* (pp. 3–22). Rijeka: In Tech.

Kenrick, W. (1833). *The new American orchardist*. Boston: Carter, Hendee Co. /Russell, Ordiorne Co.

Klittich, C. J. (2008). Milestones in fungicide discovery: Chemistry that changed agriculture. *Plant Health Progress, 9*(1), 31.

Knight, S. C., Anthony, V. M., Brady, A. M., Greenland, A. J., Heaney, S. P., Murray, D. C., et al. (1997). Rationale and perspectives on the development of fungicides. *Annual Review of Phytopathology, 35*, 349–372.

Koo, B. S., Park, H., Kalme, S., Park, H. Y., Han, J. W., Yeo, Y. S., et al. (2009). α-and β-tubulin from *Phytophthora capsici* KACC 40483: Molecular cloning, biochemical characterization, and antimicrotubule screening. *Applied Microbiology and Biotechnology, 82*, 513–524.

Kuhn, J. (1881). *The results of the experiments to determine the cause of beet exhaustion and to study the nature of nematodes* (Vol. 3). G. Schönfeld.

Leadbeater, A., & Gisi, U. (2015). Chapter 1: The challenges of chemical control of plant diseases. In *Recent developments in management of plant diseases* (pp. 3–18). New Delhi: Springer.

Lipps, P. E. (1985). Influence of inoculum from buried and surface corn residues on the incidence and severity of corn anthracnose. *Phytopathology, 75*, 1212–1216.

Littlefield, L. J. (1981). *Biology of the plant rusts: An introduction*. Ames: Iowa State University Press.

Lucas, J. A. (2011). Advances in plant disease and pest management. *The Journal of Agricultural Science, 149*(S1), 91–114.

Mathews, D. J. (1919). *Report of the work of the W. B. Randall research assistant*. Nursery and Market Garden Industry Development Society, Ltd. Experiment and Research Station, Cheshunt, Herts, UK, Annual report 5, pp. 18–21.

Maude, R. B. (1996). *Seedborne diseases and their control: Principles and practice*. Wallingford: CAB International.

Mayths, G., & Baker, E. A. (1980). An appraisal of the effectiveness of quarantines. *Annual Review of Phytopathology, 18*, 85–101.

McDougall, P. (2010). *The cost of new agrochemical product discovery, development and registration in 1995, 2000 and 2005–8*. Available at http://www.croplife.org/view_document.aspx?docId=2478

McManus, P. S., Stockwell, V. O., Sundin, G. W., & Jones, A. L. (2002). Antibiotic use in plant agriculture. *Annual Review of Phytopathology, 40*, 443–465.

Milenkovski, S., Baath, E., Lindgren, P. E., & Berglund, O. (2010). Toxicity of fungicides to natural bacterial communities in wetland water and sediment measured using leucine incorporation and potential denitrification. *Ecotoxicology, 19*, 285–294.

Millardet, P. M. A. (1885). (1) Traitement du mildiou et du rot. (2) Traitement du mildiou par le melange de sulphate de cuivre et dechaux. (3) Sur l'histoire du traitment du mildiou par le sulphate de cuivre. J. Agric. Prat. 2, 513–516, 707–719, 801–805; Engl. trans. by F. L. Schneiderhan in Phytopathol. Classics 3 (1933).

Mueller, D., Bradley, C. A., & Nielsen, J. (2008). *Field crop fungicides for the north central United States*. Agricultural Experiment Station, Iowa State University.

Nellemann, C., MacDevett, M. Manders, T., Eickhout, B., Svilhus, B. Prins, A. G., & Kaltenborn, B. P. (Eds.). (2009). *The environmental food crisis – The environment's role in averting future food crises*. A UNEP rapid response assessment. United Nations Environment.
Nutter, F. W., Jr., & Guan, J. (2001). Disease losses. In O. C. Maloy & T. D. Murray (Eds.), *Encyclopedia of plant pathology* (pp. 340–351). New York: Wiley.
Nutter, F. W., Jr., & Madden, L. V. (2005). Plant disease as a possible consequence of biological attacks. In R. A. Greenfield & M. S. Bronze (Eds.), *Biological terrorism* (pp. 793–818). Norfolk: Horizon Scientific Press/Caister Scientific Press.
Oerke, E. C. (2006). Crop losses to pests. *The Journal of Agricultural Science, 144*, 31–43.
Palti, J. (1981). *Cultural practices and infectious crop diseases*. Heidelberg/New York: Springer-Verlag Berlin.
Pelaez, V., da Silva, L. R., & Araújo, E. B. (2013). Regulation of pesticides: A comparative analysis. *Science and Public Policy, 40*, 644–656.
Prevost, B. (1807). *Memoir on the immediate cause of caries or coal of the bles, and several. other diseases of plants, and on the condoms of caries*. Paris: Bernard Quai des Augustins. [Translation by Keitt GW, 1939. Memoir on the immediate cause of bunt or smut of wheat, and of several other diseases of plants, and on preventives of bunt. Menasha, WI, USA: American Phytopathological Society, Phytopathological Classics No. 6.].
Rommens, M. C., & Kishore, G. M. (2000). Exploiting the full potential of disease-resistance genes for agricultural use. *Current Opinion in Biotechnology, 11*, 120–125.
Russell, P. E. (2005). A century of fungicide evolution. *The Journal of Agricultural Science, 143*, 11–25.
Savary, S., Teng, P. S., Willocquet, L., & Nutter, F. W., Jr. (2006). Quantification and modeling of crop losses: A review of purposes. *Annual Review of Phytopathology, 44*, 89–112.
Schacht, H. (1859). About some enemies and diseases of the sugar beet. *Journal of the Association for the Beet Sugar Industry in the Zoll, 9*, 239–250.
Schulthess, H. (1761). Proposal of some remedies saved by experience against the fire in the grain. *Abhandl Naturf Gesell Zurich, I*, 498–506.
Sharvell, E. G. (1979). *Plant disease control*. Westport: The AVI Publishing Company.
Shtienberg, D., Zilberstaine, M., Oppenheim, D., Herzog, Z., Manulis, S., Shwartz, H., & Kritzman, G. (2001). Efficacy of oxolinic acid and other bactericides in suppression of *Erwinia amylovora* in pear orchards in Israel. *Phytoparasitica, 29*, 143–154.
Sill, W. H., Jr. (1982). *Plant protection: An integrated interdisciplinary approach*. Ames: The Iowa State University Press.
Smil, V. (2000). *Feeding the world: A challenge for the twenty-first century*. Cambridge: The Massachusetts Institute of Technology Press.
Spencer, D. M. (1978). Powdery mildew of strawberries. In *The powdery mildews* (pp. 355–358). New York: Academic.
Strange, R. N., & Scott, P. R. (2005). Plant disease: A threat to global food security. *Annual Review of Phytopathology, 43*, 83–116.
Strittmatter, G., Janssens, J., Opsomer, C., & Botterman, J. (1995). Inhibition of fungal disease development in plants by engineering controlled cell death. *Bio/Technology, 13*, 1085–1089.
Teng, P. S., Shane, W. W., & MacKenzie, D. R. (1984). Crop losses due to plant pathogens. *Critical Reviews in Plant Sciences, 2*, 21–47.
Tillet, M. (1755). *Dissertation sur la cause qui corrompt et noircit les grans de ble dans les epis; et sur les moyens de prevenir ces accidents*. Bordeaux; Engl. transl. by H. B. Humphrey in Phytopathological Classics No. 5 (1937).
Van der Plank, J. E. (1963). *Plant disease: Epidemics and control*. New York: Academic.
Yang, C., Hamel, C., Vujanovic, V., & Gan, Y. (2011). Fungicide: Modes of action and possible impact on nontarget microorganisms. *ISRN Ecology, 2011*, 1–8.
Zadoks, J. C., & Schein, R. D. (1979). *Epidemiology and plant disease management* (p. 427). New York: Oxford University Press.

Chapter 2
Plant Genetics and Physiology in Disease Prognosis

Ganesan Vadamalai, Lih Ling Kong, and Yasir Iftikhar

Abstract The dynamics of plant physiology and protein expression may largely contribute in disease diagnosis. Biochemical changes and secondary metabolites cross talk while pathogen and plant interact through cellular defense mechanisms. Plants genetics in relation to resistance levels vs pathogens helps in categorizing varieties and also the pathogen, on the basis of symptoms development. Although symptomology is the basic criterion for identification of plant diseases, other serological, biochemical and molecular assays are highly sensitive and useful for correct diagnosis of plant diseases. Advances in plant physiology and genetics, under varying spatio-temporal scales, are used for the detection and management of diseases. Thus, biochemical characterization of diseased plants opens new trends in disease diagnosis to formulate management strategies. In this chapter we focused on the comparison between genetics and physiology of diseased and healthy plants. Moreover, effect of biochemical changes due to certain pathogens on host plants are also discussed as concerns detection. The use of proteome in disease diagnosis is also described. Genetics of resistance and susceptible varieties vs diseases was highlighted for disease diagnosis. As different plant pathogens such as fungi, bacteria, nematodes, viruses and virus-like pathogens have different expression profiles during disease progression, physiology and genetics of diseased plants appear as useful tools for diagnosis.

Keywords Disease prognosis · Plant physiology · Plant genetics · Induced resistance · Biochemical detection

---

G. Vadamalai
Department of Plant Protection, Universiti Putra Malaysia (UPM), Serdang, Selangor, Malaysia

L. L. Kong
Institute of Tropical Agriculture, Universiti Putra Malaysia, Seri Kembangan, Serdang, Selangor, Malaysia

Y. Iftikhar (✉)
Department of Plant Pathology, College of Agriculture, University of Sargodha, Sargodha, Pakistan
e-mail: yasir.iftikhar@uos.edu.pk

© Springer Nature Switzerland AG 2020

I. Ul Haq, S. Ijaz (eds.), *Plant Disease Management Strategies for Sustainable Agriculture through Traditional and Modern Approaches*, Sustainability in Plant and Crop Protection 13, https://doi.org/10.1007/978-3-030-35955-3_2

2.1 Introduction

Any deviation from normal functioning in physiology, morphology and genetics of plant is referred as a disease. Therefore, certain biochemical changes and their expression also differentiate the diseased plants from healthy ones. Although symptomology is the first, basic criterion for the identification of plant diseases, other serological, biochemical and molecular assays are highly sensitive and useful for the correct and real-time diagnosis of plant diseases. Therefore, advances in the physiology and genetics of plant disease diagnosis are being used for the detection and management of diseases.

Plant pathogens are detected and characterized using different basis to manage the diseases in plants. Observation of symptomology and microscopic morphology are conventional methods for detection. Host-pathogen interactions are involved in physiological and genetic alterations present in diseased plants. Pathogens influence the physiological pathways and expression of many genes. Biochemical, serological and molecular assays have been recognized as recent trends in diagnosis. Biochemistry of diseased plants offers quick and reliable detection methods. Similarly, amplification of gene expression through different immunological and nucleic acid based assays are fast approaches that offer advantages over conventional methods.

Biochemical and molecular techniques are very useful against obligate fungal pathogens which are slow growing on the medium such as powdery mildew, downy mildew, and rusts etc. During a pathogen attack on the there are two types of host resistance/susceptibility reactions, such as compatibility and in-compatibility. The in-compatible reaction is involved in the hypersensitive response and production of some metabolites which help the plant to combat the invading pest or pathogens. In a compatible reaction the host is susceptible and conducive to the pathogen growth. Biochemical and molecular assays are important for a reliable disease diagnosis. Physiological characterization always involves many biochemical assays, instead of a single test as applied in morphology based diagnosis.

2.2 Recent Trends in Biochemical and Molecular Detection of Plant Pathogens

Plant pathogens are detectable on the basis of their specific biochemical activities in the host, and are detected by biochemical and molecular techniques such as electrophoresis, Polymerase chain reaction (PCR), restriction fragment length polymorphism (RFLP), dot blot hybridization, DNA finger printing etc. Study of DNA is fundamental for molecular detection of any pathogen. Analyzing DNA of plant and pathogens may show the alteration induced in physiological traits of diseased tissue samples. Electrophoresis is helpful in separating the complex mixture of DNA into fragments of different sizes. Polyacrylamide or agarose medium are required for electrophoresis, depending on the DNA or protein profile analysis. Protein profiling is performed through the reaction of different restriction enzymes, and is separated through electrophoresis. The electrophoresis analysis has been found very effective to detect and characterize the disease caused by many pathogens, over the last decades. A derivative of electrophoresis named "Isozyme electrophoresis" is being used effectively to differentiate species and strains. Grouping within the species of true fungi and oomycetes can be achieved through iso-enzyme electrophoresis. Using this assay different species of *Phytophthora* (*P. cinnamomi P. cambivora* and *P. cactorum*) can be separated. A phylogenetic tree may be constructed using iso-enzyme analysis. For example, virulence and non-virulence in black leg of canola was investigated to this analytical approach. *Fusarium* species distributed around the world have also been differentiated through cellulose-acetate electrophoresis (Oudemans and Coffey 1991a, b).

Viroids are the smallest pathogens infecting plants. Though they are short length RNAs, they have genetic variations. Different viroids have been detected and characterized through polyacrylamide gel electrophoresis (PAGE) and hybridization. Potato spindle tuber viroid, isolates of *Coconut cadang-cadang viroid* (CCCvd), citrus viroids, isolates of citrus bent leaf viroids have been characterized through PAGE (Morris and Wright 1975; Randles 1985; Hodgson et al. 1998; Barbosa et al. 2005; Cao et al. 2009; Khoo et al. 2017). Similarly, viruses and their isolated may be separated through PAGE and blotting (Narayanasamy and Doraiswamy 2003). Citrus viruses and different isolates of citrus tristeza virus have also been characterized through electrophoresis (Narayanasamy 2008).

2.3 Physiology of Diseased Plants

Plants pathogens exist in different strains, isolates, pathovar, races and biotypes, depending on their genetics. Gene for gene hypothesis (Flor 1946) opened different perspectives in genetics to be studied. Using conventional methods for strain identification is a time taking procedure. Pathogens are able to cause diseases in plants when one or more pathogenicity genes are present. With the passage of time different races, strains and biotypes evolved due to factors such as environment, competition for the host and point mutation in genetic makeup. Screening to find out the resistance source against different pathogens is useful to formulate management strategies. Genetic variability can also be determined through analyses of differential hosts.

Cultivars having different resistance genes have been selected and used to identify pathogen races physiologically different. This is the best way to define races in host plants, to avoid confusion. There are many examples of interaction between hosts and pathogens. Potato (*Solanum* tuberosum) interacts with *Phytophthora* infestans when the conditions are favorable for both of them. R1 resistance genes are present in the varieties of potato against virulent strain of *P. infestans* race 1. However, some fungal strains are complex and virulent to varieties having more than one resistance genes. For example, *P. infestans* virulent races named 1, 2 and 3 show virulence to specific resistance genes present in the potato that are indicated as R1, R2 and R3 resistance genes. Some cultivars having no resistance genes are susceptible to all *P. infestans* races that cause infection and lead to the minimal quality and yield losses. Another example is the interaction of the fungus *Cladosporium fulvum* and tomato *Lycopersicon esculentum* (Higgins and de Wit 1985). This interaction shows the specificity of the resistance genes in tomato against the virulent genes present in the pathogen genome. In case of plant pathogenic bacteria, the races are defined on the basis of avirulence genes, against the resistance genes present in the host plant. *Pseudomonas syringae* race 6 has the 6 avirulence genes acting against the resistance genes present in soybean (*Glycine max*) plants (Staskawicz et al. 1984). The races of *Xanthomonas campestris* pv. *malvacearum* are similarly defined (Gabriel et al. 1986).

There are different levels of resistance in different plant parts, according to age, and in genetically identical plants as well as in different parts and tissues (Innes 1974; Jones and Hayes 1971; Mares 1979). Rates of resistance increase or decrease in different plant parts as in case of stem and roots resistance, that increases in first 2 weeks as compared to fruits and leaves where resistance level decreases with age (Hunter 1978; Hunter et al. 1978; Jones and Hayes 1971; Nilsen et al. 1979; Wheeler 1977). Leaves show more resistance at developing stage, usually when flowering on plants occurs. These changes are referred to as ontogenic changes in plants, against the pathogen attack (Bell 1980). When a bacterial pathogen enters into host plant the host defense mechanism is activated, showing enhanced levels of resistance to infection by causing necrosis or local lesions. Such type of resistance is referred to as induced resistance, which is sometime induced by different chemicals in plants (Matta 1971). Various reviews are available in the literature on the resistance mechanisms active in plants after bacteria and nematodes attacks (Kuc and Hammerschmidt 1978; O'Brien and Fisher 1978).

Induced resistance develops similarly in different plant species with minor variations due to the effects of host species and inducing agents involved. A typical example of induced resistance is that developed against the bacterial pathogen *Pseudomonas lachrymans* (Caruso and Kuc 1979). The initial density of the inducing pathogen is very important for the extent and duration of the subsequent induced resistance. First appearance of systemic induced resistance in true leaves after inoculation occurs after 72–96 h. It reaches the highest level within in 7 days and continues for 4–7 weeks. Another inoculation made on a higher leaf gradually expands and persists for longer period until fruiting, and is called booster inoculation. After 96 h of inoculation the leaves removed show few lesions and induce resistance stands for a longer time, even after removal from plant (Jenns and Kuc 1979).

Bacterial multiplication is restricted by the high level of general resistance in the hosts, without involving necrosis of plants, and race-specific resistance occurs when high level of necrosis is present. Multiplication of virulent strains occurs exponentially until necrosis occurs and then populations drop. Population decline indicates the accumulation of antibiotics and other reactive chemical compounds in intercellular spaces between the lesions formed by the pathogen (Webster and Sequeira

1977; Sequeira and Hill 1974; Essenber et al. 1979). Avirulent bacteria strains were found to increase the electrolyte leakage of pepper 16 h before necrosis occurred in infected plants. Magnitude and velocity of electrolyte leakage increased greatly by the infiltration of bacteria or water into the tissues (Ook 1975). Structural changes causing necrotic resistance reactions in tobacco and cotton plants were explained in detail (Goodman et al. 1976; Sequeira et al. 1977; Cason et al. 1978; Essenber et al. 1979). Mechanisms of resistance initiates only after the attachment of bacteria to the cell wall of the host Many bacteria then spread into the system as saprotrophic as well as avirulent microorganisms. As a result the bacteria are engulfed by the fibrillar matrix similar to the cutin or suberin. Saprotrophic bacteria burst and cause toxicity to the cell, which leads to its degeneration and internal necrosis, thus depriving the intruder of its food source (Sequeira et al. 1977; Sing and Schroth 1977). A virulent bacterium, after inoculation into the host plant, leads instead to a different process, with erosion of the outer cell wall, most commonly the cuticle, within 20 min to 4 h. From the cell wall some extrusions are protruded from fibrillar layers which attach to the bacterium that penetrates into the host. Subsequently, the plasmalemma starts to produce convoluted and membrane bounded vesicles that accumulate in the plasmalemma and the cell wall spaces. As a result, the plant cuticle and cell wall become thicker and denser. At the same time, the cytoplasm starts vesiculation and accumulation of droplets of osmophilic compounds on the plasmalemma, which is closer to the bacteria. At the end, all the membrane systems as well as all organelles are extensively degenerated and the corresponding host cell collapse and necrotizes. Necrotic changes may also occur in adjacent cells while some non-necrotic adjoining cells show a normal response. Bacterial envelopment, attachment and erosion of the cell wall do not occur in susceptible reactions, as compared to the disruption of the cell, membrane and organelles which occur 7 days after infection.

In a different process and pathosystem, root-knot nematode multiplication is directly proportional to the formation of giant cells in the roots, as the nematodes depend on cell proliferation as well as hypertrophy for its survival. The juvenile stages penetrate the host radical tissues and induce the formation of a multinucleate giant cell in which they feed. This structure takes the place of normally occurring vascular tissues. The nematode juvenile stages in susceptible roots become sedentary and develop into adult females after the initiation of feeding. Once feeding starts, the adult females laid egg masses after 20–35 days, depending on temperature. The giant cell maintenance and establishment in the roots is hence responsible for the parasite success. Nematode attack, penetration and prevalence in roots follow the same path both in resistant and susceptible cultivars. For example, in alfalfa resistant cultivars, the rate of nematode larvae declines after 3–4 days of infection due to the failure of establishment of a suitable host-pathogen relationship. Due to resistance to the nematode attack, very few larvae can be found, 1 week after penetration, in the roots, and no gall formation occurs (Reynolds et al. 1970; Griffin and Elgin 1979). A little gall formation may be found in some other resistant crops but the development of larvae is slow. Roots may still contains eggs and females develop to form the few giant cells which lead to some gall formation (Dropkin 1969; Jena and Rao 1977). Necrotic cells around the nematodes may be found after 3–4 days of infection, especially in resistant cultivars (Veech and Endo 1970; Waterman et al. 1978). All the cellular events occurring during the process of necrosis are similar to the host-specific necrogenic resistance towards viruses, bacteria and fungi.

During viral infection different changes and reactions occur within the plants, such as: anatomical and histological changes, necrosis, hyperplasia, hypoplasia, whereas mesophyll cells become smaller and less differentiated with few intercellular spaces. Moreover, cytological changes can also be observed. There is disturbance in the internal cell organization, with modification and disintegration of cell organelles. There is no detectable cytological effects in nuclei during viral infection, but some viruses induce the formation of nuclear inclusion bodies, with reduction in chromatin. Geminiviruses induce the swelling, hypertrophism and disintegration of nuclei and chloroplast. Starch accumulation is another biochemical change observed during viral infections. There is an aggregation and regeneration of mitochondria and deposition of vesicles, the cell wall becomes thickened, and the callus with other electron dense material are deposited. Disturbance in ribosomes along with proliferation in plasmodesmata lead to restriction of virus movement during infection in plants. Inclusion bodies are one of the major cytological effect associated to a viral infection.

Biochemical changes induced by viral infections are mainly reflected in the contents of nitrogen fractions, carbohydrates and sugars, phenolic compounds, alkaloids, growth regulators/hormones, nucleic acids and low molecular metabolites. These effect are expressed on unit basis such as: fresh dry weight of tissue, protein/nitrogen determination, leaf area, RNA/DNA contents, yield/out puts, major and minor nutrients, chlorophyll contents, symptom expression and other specific and non-specific changes.

2.4 Rhizobacteria and Host Plants

Root exudates and lysates, which are secreted by roots, provide nutrition and niche for soil bacteria proliferation. In the rhizosphere the number of bacteria is 10 times greater than in bulk soil. A huge variety of different taxa cover the 15% of the leaf surface area by making micro colonies on these plant parts. During infection, these bacteria use the nutrients which are secreted by the host cell for their growth. Some metabolites may also be secreted by the host cell into the rhizosphere, utilized as part of a defense mechanism. Several of these metabolites can act as signalling compounds that are perceived by neighbouring cells within the same micro-colony, by cells of other bacteria that are present in the rhizosphere, or by root cells of the host plant (Van Loon and Bakker 2003; Bais et al. 2004; Gray and Smith 2005; Kiely et al. 2006).

The symbiotic relationship between rhizobia and legumes is the best example of signal exchange. In this mechanism, flavonoid compounds are released by the host plants which act as signaling molecules for the secretion of nod factor by the rhizobia. Root hairs perceive the nod factors. Function of the nod factors is similar to the function of hormones, inducing nodules in those roots where the rhizobia fix the atmospheric nitrogen. Bacteria get established within the host plant and use the carbohydrates it provides, providing in return the nitrogen fixed for plant amino acids biosynthesis (Brencic and Winans 2005; Gray and Smith 2005). This is the prime example of a symbiotic relationship between soil bacteria and plants.

Starting from this relationship, several Plant Growth Promoting Rhizobacteria (PGPR) have been found, which promote symbiosis. Rhizobacteria also help in growth as well as development of different leguminous crops when the environment is deficient in nitrogen. Different hormone modulations are involved in the regulation mechanisms involved in growth and development of plants (Frankenberger and Arshad 1995). Some bacterial species produce additional growth hormones including cytokinins, ethylene, gibberellins and auxins (Pieterse and Van Loon 1999). *Arabidopsis* seedlings vertically grown under genobiotic conditions with Hoagland nutrient medium showed maximum growth under these conditions (Tanimoto 2005). In wheat and pearl millet growth is promoted by auxins released by *Azospirillum brasilense*. Bacterial mutants that lost production of indole acetic acid up to 70%, also showed a largely affected growth promotion capacity (Barbieri and Galli 1993).

There are also non-pathogenic bacteria which antagonize pathogens by competition for nutrients available in soil. Sometime competition occurs by secretion of the lytic enzymes as well as antibiotics (Handelsman and Stabb 1996; Van Loon and Bakker 2003). Rhizobacteria are involved in activating the defense mechanism of plants against pathogenic species. This defense mechanism is known as induced systemic resistance (ISR), as initially described by the Van Peer et al. (1991). Rhizobacteria were found to protect the leaves of cucumber from the attack of different fungi, such as anthracnose and *Colletotrichum orbiculare* (Wei et al. 1991). It was concluded that ISR acts as an enhanced defense capacity in plants (Van Loon and Bakker 2005). In some cases, when a pathogen proliferates in the host plant, the defense mechanism is activated thus reducing the symptom expression on some plant parts. Such type of tolerance is associated with physiological races. Some physiological races are present, for example, in *Erwinia carotovora*, the causal agent of the potato soft rot.

A reduced disease incidence can also be the outcome of alterations in the microbial populations present in the rhizosphere, due to an altered host physiology (Mark et al. 2005). Exudates are secreted from roots upon colonization by pathogens. Secretion of exudates vary in quantity and composition according to the different stages of root development (Phillips et al. 2004; Whitehead et al. 2001; Bauer and Mathesius 2004).

2.5 Induced Resistance Against Pathogen Infection

Arabidopsis is a test plant used for scientific experiments. The dependence of tomato, tobacco and *Arabidopsis* on salicylic acid (SA), jasmonic acid (JA) and/or ethylene (ET) has been determined. Systemic induced resistance (ISR) is elicited greatly by PGPR along with the ET pathway. In strains of bacteria such as *Serratia marcescens* 90–66, *P. fluorescens* CHAO, SA is produced with additional siderophores (Van Loon and Bakker 2005). *Pseudomonas* aeruginosa strain 7NSK2 showed ISR as it directly produces SA (De Meyer et al. 1999). In the case of tomato, ISR was triggered by the SA produced by its own siderophores, antibiotics pyocyanin and pyochelin that induce oxygen free radicals, and not by the inoculated 7NSK2 strain. The oxygen free radicals activate the production of SA which helps to induce ISR (Audenaert et al. 2002). In case of *Bacillus* spp. signaling pathways are also activated in the same mechanism occurring in the *Pseudomonas*, and directly act on the NPR1 gene, activating resistance in plants (Kloepper et al. 2004). PGPR are also host-specific as some rhizobacteria elicit ISR in some plants but are not active in other hosts, indicating a specificity in the host-pathogen interaction. It was a common perception that resistant reactions triggered by the rhizobacteria are specific to the different host plants, but that the bacteria do not colonize the roots of all hosts (Bakker et al. 2003; Meziane et al. 2005; Van Loon and Bakker 2005). Specific plant receptors are involved in the recognition of different bacterial components. Therefore, levels of resistance and susceptibility play an important role in the biology of plant pathogens, as well as in the planning and formulation of any related integrated disease management.

References

Audenaert, K., Pattery, T., Comelis, P., & Hofte, M. (2002). Induction of systemic resistance to *Botrytis cinerea* in tomato by *Pseudomonas aeruginosa* 7NSK2: Role of salicylic acid, pyochelin, and pyocyanin. *Molecular Plant-Microbe Interactions, 15*, 1147–1156.

Bais, H. P., Park, S. W., Weir, T. L., Callaway, R. M., & Vivanco, J. M. (2004). How plants communicate using the underground information superhighway. *Trends in Plant Science, 9*, 26–32.

Bakker, P. A. H. M., Ran, L. X., Pieterse, C. M. J., & Van Loon, L. C. (2003). Understanding the involvement of rhizobacteria-mediated induction of systemic resistance in biocontrol of plant diseases. *Canadian Journal of Plant Pathology, 25*, 5–9.

Barbieri, P., & Galli, E. (1993). Effect on wheat root development of inoculation with an *Azospirillum brasilense* mutant with altered indole-3-acetic acid production. *Research in Microbiology, 144*, 69–75.

Barbosa, C. J., Pina, J. A., Perez-Panades, J., Bernad, L., Serra, P., Navarro, L., & Duran-Vila, N. (2005). Mechanical transmission of citrus viroids. *Plant Disease, 89*, 749–754.

Bauer, W. D., & Mathesius, U. (2004). Plant responses to bacterial quorum sensing signals. *Current Opinion in Plant Biology, 7*, 429–433.

Bell, A. A. (1980). The time sequence of defense. In J. G. Horsfall & E. B. Cowling (Eds.), *Plant disease: An advanced treatise* (Vol. 5, pp. 53–73). New York: Academic.

Brencic, A., & Winans, S. C. (2005). Detection of and response to signals involved in host-microbe interactions by plant-associated bacteria. *Microbiology and Molecular Biology Reviews, 69*, 155–194.

Cao, M. J., Atta, S., Liu, Y. Q., Wang, X. F., Zhou, C. Y., Mustafa, A., & Iftikhar, Y. (2009). First report of citrus bent leaf viroid and citrus dwarfing viroid from citrus in Punjab, Pakistan. *Plant Disease, 93*(8), 840.

Caruso, F. L., & Kuc, J. (1979). Induced resistance of cucumber to anthracnose and angular leaf spot by *Pseudomonas lachrymans* and *Colletotrichum lagenarium*. *Physiological Plant Pathology, 14*, 191–201.

Cason, E. T., Richardson, P. E., Essenberg, M. K., Brinkerhoff, L. A., Johnson, W. M., & Venere, R. J. (1978). Ultrastructural cell wall alterations in immune cotton leaves inoculated with *Xanthomonas malvacearum*. *Phytopathology, 68*, 1015–1021.

De Meyer, G., Audenaert, K., & Hofte, M. (1999). *Pseudomonas aeruginosa* 7NSK2-induced systemic resistance in tobacco depends on in plant salicylic acid accumulation but is not associated with PR1a expression. *European Journal of Plant Pathology, 105*, 513–517.

Dropkin, V. H. (1969). The necrotic reaction of tomatoes and other hosts resistant to *Meloidogyne*: Reversal by temperature. *Phytopathology, 59*, 1632–1637.

Essenber, M., Cason, E. T., Hamilton, B., Brinkerhoff, L. A., Gholson, R. K., & Richardson, P. E. (1979). Single cell colonies of *Xanthomonas malvacearum* in susceptible and immune cotton leaves and the local resistant response to colonies in immune leaves. *Physiological Plant Pathology, 15*, 53–56.

Flor, H. H. (1946). Genetics of pathogenicity in Melampsora lini. *Journal of Agricultural Research, 73*, 335–357.

Frankenberger, W. T., & Arshad, M. (1995). *Phytohormones in soils-microbial production and function*. New York: Marcel Dekker.

Gabriel, D. W., Burges, A., & Lazo, G. R. (1986). Gene-for-gene interactions of five cloned avirulence genes from *Xanthomonas campestris* pv. *malvacearum* with specific resistance genes in cotton. *Proceedings of the National Academy of Sciences of the United States of America, 83*, 6415–6419.

Goodman, R. N., Huang, P., & White, J. A. (1976). Ultrastructural evidence for immobilization of an incompatible bacterium, *Pseudomonas pisi*, in tobacco leaf tissue. *Phytopathology, 66*, 754–764.

Gray, E. J., & Smith, D. L. (2005). Intracellular and extracellular PGPR: Commonalities and distinctions in the plant bacterium signaling process. *Soil Biology and Biochemistry, 37*, 395–412.

Griffin, G. D., & Elgin, J. H. (1979). Penetration and development of *Meloidogyne hapla* in resistant and susceptible alfalfa under differing temperatures. *Journal of Nematology, 9*, 51–56.

Handelsman, J., & Stabb, E. V. (1996). Biocontrol of soilborne plant pathogens. *The Plant Cell, 8*, 1855–1869.

Higgins, V. J., & de Wit, P. J. G. M. (1985). Use of race- and cultivar-specific elicitors from intercellular fluids for characterising races of *Cladosporium fulvum* and resistant tomato cultivars. *Phytopathology, 75*, 695–699.

Hodgson, R. A. J., Wall, G. C., & Randles, J. W. (1998). Specific identification of Coconut tinangaja viroid for differential field diagnosis of viroids in coconut palm. *Phytopathology, 88*, 774–781.

Hunter, R. E. (1978). Effects of catechin in culture and in cotton seedlings on the growth and polygalacturonase activity of *Rhizoctonia solani*. *Phytopathology, 68*, 1032–1036.

Hunter, R. E., Halloin, J. M., Veech, J. A., & Carter, W. W. (1978). Terpenoid accumulation in hypocotyls of cotton seedlings during aging and after infection by *Rhizoctonia solani*. *Phytopathology, 68*, 347–350.

Innes, N. L. (1974). Resistance to bacterial blight of cotton varieties homozygous for combinations of B resistance genes. *The Annals of Applied Biology, 78*, 89–98.

Jena, R. N., & Rao, Y. S. (1977). Nature of resistance in rice (*Oryza sativa* L.) to the root-knot nematode (*Meloidogyne graminicola* Golden and Birchfield) II. Histopathology of nematode infection in rice varieties. *Proceedings of the Indian Academy of Sciences – Section B, 86*, 87–89.
Jenns, A. E., & Kuc, J. (1979). Graft transmission of systemic resistance of cucumber to anthracnose induced by *Colletotrichum lagenarium* and tobacco necrosis virus. *Phytopathology, 69*, 753–756.
Jones, I. T., & Hayes, J. D. (1971). The effect of sowing date on adult plant resistance to *Erysiphe graminis* f. sp. *avenae* in oats. *The Annals of Applied Biology, 68*, 31–39.
Kiely, P. D., Haynes, J. M., Higgins, C. H., Franks, A., Mark, G. L., Morrissey, J. P., & O'Gara, F. (2006). Exploiting new systems-based strategies to elucidate plant-bacterial interactions in the rhizosphere. *Microbial Ecology, 51*, 257–266.
Khoo, Y. W., Iftikhar, Y., Murugan, T., Roslin, N. A., Adawiyah, R., Kong, L. L., & Vadamalai, G. (2017). First report of Citrus bent leaf viroid in Malaysia. *Journal of Plant Pathology, 99*(1), 287–304.
Kloepper, J. W., Ryu, C. M., & Zhang, S. A. (2004). Induced systemic resistance and promotion of growth by *Bacillus* spp. *Phytopathology, 94*, 1259–1266.
Kuc, J., & Hammerschmidt, R. (1978). Acquired resistance to bacterial and fungal infection. *Annals of Applied Biology, 89*, 313–317.
Mares, D. J. (1979). Microscopic study of the development of yellow rust (Puccinia striijormis) in a wheat cultivar showing adult plant resistance. *Physiological Plant Pathology, 15*, 289–296.
Mark, G. L., Dow, L. M., Kiely, P. D., Higgins, H., Haynes, J., Baysse, C., et al. (2005). Transcriptome profiling of bacterial responses to root exudates identifies genes involved in microbe-plant interactions. *Proceedings of the National Academy of Sciences of the United States of America, 102*, 17454–17459.
Matta, A. (1971). Microbial penetration and immunization of uncongenial host plants. *Annual Review of Phytopathology, 9*, 387–410.
Morris, T. J., & Wright, N. S. (1975). Detection by polyacrylamide gel of a diagnostic nucleic acid from tissue infected with Potato spindle tuber viroid. *American Potato Journal, 52*, 57–63.
Meziane, H., Van der Sluis, I., Van Loon, L. C., Hofte, M., & Bakker, P. A. H. M. (2005). Determinants of *Pseudomonas putida* WCS358 involved in inducing systemic resistance in plants. *Molecular Plant Pathology, 6*, 177–185.
Narayanasamy, P., & Doraiswamy, S. (2003). *Plant viruses and viral diseases*. Chennai: New Century Book House.
Narayanasamy, P. (2008). *Molecular biology in plant pathogenesis and disease management. Microbial plant pathogens Volume 1*. Dordrecht: Springer Science+Business Media B.V.
Nilsen, K. N., Hodges, C. F., & Madsen, J. P. (1979). Pathogenesis of *Drechslera sorokiniana* leaf spot on progressively older leaves of *Poa pratensis* as influenced by photoperiod and light quality. *Physiological Plant Pathology, 15*, 171–176.
O'Brien, P. C., & Fisher, J. M. (1978). Studies on the mechanism of resistance of wheat to Heterodera avenae. *Nematologica, 24*, 463–471.
Ook, A. A. (1975). Effect of low concentrations of *Xanthomonas vesicatoria* infiltrated into pepper leaves. *Phytopathology, 65*, 487–489.
Oudemans, P., & Coffey, M. D. (1991a). Isozyme comparison within and among worldwide sources of three morphologically distinct species of Phytophthora. *Mycological Research, 95*, 19–30.
Oudemans, P., & Coffey, M. D. (1991b). A revised systematics of twelve papillate Phytophthora species based on isozyme analysis. *Mycological Research, 95*, 1025–1046.
Phillips, D. A., Fox, T. C., King, M. D., Bhuvaneswari, T. V., & Teuber, L. R. (2004). Microbial products trigger amino acid exudation from plant roots. *Plant Physiology, 136*, 2887–2894.
Pieterse, C. M. J., & Van Loon, L. C. (1999). Salicylic acid-independent plant defence pathways. *Trends in Plant Science, 4*, 52–58.

Randles, J. W. (1985). Coconut cadang-cadang viroid. In K. Maramorosch & J. McKelvey (Eds.), *Viroids and prions-subviral pathogens of plants and animals* (pp. 39–74). New York: Academic Press.

Reynolds, H. W., Carter, W. W., & O'Bannon, J. H. (1970). Symptomless resistance of alfalfa to *Meloidogyne incognita acrita*. *Journal of Nematology, 2*, 131–134.

Sequeira, L., & Hill, L. M. (1974). Induced resistance in tobacco leaves: The growth of *Pseudomonas solanacearum* in protected tissues. *Physiological Plant Pathology, 4*, 447–455.

Sequeira, L., Gaard, G., & De Zoeten, G. A. (1977). Interaction of bacteria and host cell walls: Its relation to mechanisms of induced resistance. *Physiological Plant Pathology, 10*, 43–50.

Sing, V. O., & Schroth, M. N. (1977). Bacteria-plant cell surface interactions: Active immobilization of saprophytic bacteria in plant leaves. *Science, 197*, 759–761.

Staskawicz, B. J., Dahlbeck, D., & Keen, N. T. (1984). Cloned avirulence gene of *Pseudomonas syringae* pv. *glycinea* determines race incompatibility on *Glycine max* (L.) Merr. *Proceedings of the National Academy of Sciences of the United States of America, 81*, 6024–6028.

Tanimoto, E. (2005). Regulation of root growth by plant hormones: Roles for auxin and gibberellin. *Critical Reviews in Plant Sciences, 24*, 249–265.

Van Loon, L. C., & Bakker, P. A. H. M. (2003). Signalling in rhizobacteria-plant interactions. In H. De Kroon & E. J. W. Visser (Eds.), *Root ecology* (Ecological Studies) (Vol. 168, pp. 297–330). Berlin/Heidelberg: Springer.

Van Loon, L. C., & Bakker, P. A. H. M. (2005). Induced systemic resistance as a mechanism of disease suppression by rhizobacteria. In Z. A. Siddiqui (Ed.), *PGPR: Biocontrol and biofertilization* (pp. 39–66). Dordrecht: Springer.

Van Peer, R., Niemann, G. J., & Schippers, B. (1991). Induced resistance and phytoalexin accumulation in biological control of Fusarium wilt of carnation by *Pseudomonas* sp. strain WCS417r. *Phytopathology, 81*, 728–734.

Veech, J. A., & Endo, B. Y. (1970). Comparative morphology and enzyme histochemistry in root-knot resistant and susceptible soybeans. *Phytopathology, 60*, 896–902.

Waterman, M. A., Aist, J. R., & Israel, H. W. (1978). Centrifugation studies help clarify the role of papilla formation in compatible barley powdery mildew interactions. *Phytopathology, 68*, 797–802.

Webster, D. M., & Sequeira, L. (1977). Expression of resistance in bean pods to an incompatible isolate of *Pseudomonas syringae*. *Canadian Journal of Botany, 55*, 2043–2052.

Wei, G., Kloepper, J. W., & Tuzun, S. (1991). Induction of systemic resistance of cucumber to *Colletotrichum orbiculare* by select strains of plant growth-promoting rhizobacteria. *Phytopathology, 81*, 1508–1512.

Wheeler, H. (1977). Increase with age in sensitivity of oat leaves to victorin. *Phytopathology, 67*, 859–861.

Whitehead, N. A., Barnard, A. M., Slater, H., Simpson, N. J., & Salmond, G. P. (2001). Quorum-sensing in Gram-negative bacteria. *FEMS Microbiology Reviews, 25*, 365–404.

Chapter 3
Conventional Plant Breeding Program for Disease Resistance

Ali Hassan Khan, Mariam Hassan, and Muhammad Naeem Khan

Abstract Disease resistance is of great concern for plant breeding programs. Diseases are a major yield-limiting factor, caused by many air born, soil born or waterborne microorganisms, which in fact are a risk for food security. Improving efficacy of management practices can increase yields, but only to a limited extent, whereas plant breeding as a technology increases yields to large extents. Advancements in new science and technology allow the development of tools whereas old ones are also refined. Most cost-effective and environment-friendly methods applied in disease resistance programs include adoption of conventional breeding approaches. There are two type of resistance, namely vertical (controlled by major genes) and horizontal (controlled by minor genes). Breeding programs change with respect to crops, diseases and pathogens. In spite of this, main objective is the accumulation of favorable gene(s) into cultivars, to deal with a given scenario. Selection, introduction, hybridization and screening are the main steps of a successful breeding program. Landraces, related species, mutations and wild relatives are the sources of resistance. They can be utilized for resistance introduction in commercial cultivars. Selection of resistant cultivar is the most robust and cheap method, allowing thereby introduction of resistant cultivar into a new region. Moreover, resistant cultivars are used to cross with local cultivars for introduction of resistance genes into them. The rapid evolution of phytopathogens and crops susceptibility pose severe issues, therefore disease resistance represents a complex aspect of any program. Being also affected by the environment it still represents a big challenge for breeders.

Keywords Backcross · Polycross · Recurrent selection · Conventional breeding · Synthetic varieties

A. H. Khan (✉) · M. Hassan · M. N. Khan
Ayub Agriculture Research Institute (AARI), Faisalabad, Pakistan

© Springer Nature Switzerland AG 2020

I. Ul Haq, S. Ijaz (eds.), *Plant Disease Management Strategies for Sustainable Agriculture through Traditional and Modern Approaches*, Sustainability in Plant and Crop Protection 13, https://doi.org/10.1007/978-3-030-35955-3_3

3.1 Introduction

Since the beginning of human civilization, agriculture is the backbone of the world's economy. People depend on agriculture either directly or indirectly for food, feed, shelter and clothes. Therefore crop protection is of great concern for the world food security. The increase in the world population with a rapid pace will boost the demand of food and other raw materials (Miedaner 2016). Crop protection is therefore important for food security.

In nature, plants face different biotic and abiotic challenges and are affected by a variety of microorganisms, including bacteria, viruses, fungi etc. Diseases cause severe damage to crop plants and result in biomass reduction, stunting growth and ultimately plant death. However, the damage depends upon pathogen prevalence at infection by. The biggest challenge faced for food security by twenty-first century scientists is to improve yield stability through the development of disease resistant crops. Breeding for disease resistance is not only important to avoid crop damages but also to protect the ecosystem by chemicals usually applied for disease management (Hogenboom 1993; Strange 2013). The importance of resistance as well as its stability for plant production provide an ultimate reason for disease resistance breeding (Clifford and Lester 1988).

Conventional breeding for resistance is an excellent technique to shield crops from damage, both from the ecological and economic point of view (Wolfe and Gessler 1992). Breeding strategies depend on the disease, the pathogen and the crop. The basic necessities of breeding for disease resistance are the genetic sources of resistance and the methodology for its introduction into economically important and commercially acceptable cultivars (Roane 1973). Conventional plant breeding consists of mainly three steps: (a) germplasm collection, (b) recognition of desired phenotypes and their (c) hybridization to get better cultivars (Fehr 1987; Stoskopf et al. 1993).

3.2 Disease Economic Impact

Crop epidemic diseases have caused huge economic losses and even famine. It has also been shown that most plant diseases are the result of human activities. A plant disease can be defined as any change that interrupts the plant normal development and decreases its economic value (Lucas et al. 1992). A disease interferes with the regular function of several plant parts and results in decreased yields or reduced quality. Visible reactions in plants are called "symptoms", including wilting, stunting, yellowing, death, and whole or partial abnormal growth. Three components are necessary for a disease to occur, in any plant system:

1. A susceptible host plant.
2. A virulent pathogen.
3. A favourable environment.

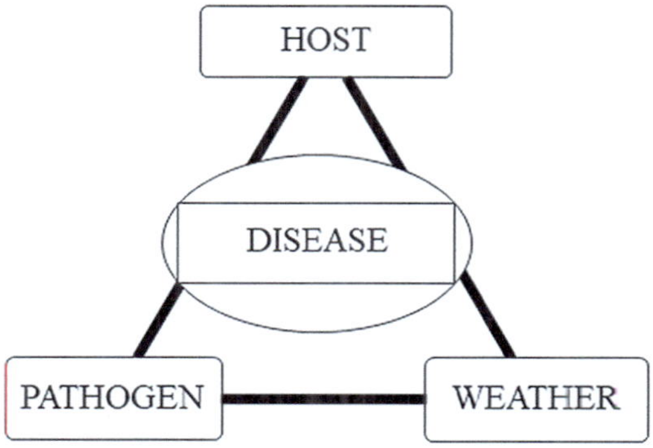

Fig. 3.1 A schematic representation of the three components model of disease establishment in plants

When these three components are present at the same time (Fig. 3.1), a disease will occur (Zadoks 2001).

3.2.1 *Examples of Plant Diseases*

In 1840, an epidemic of potato late blight caused by an Oomycete (*Phytophthora infestans)* resulted in the Irish famine. This is still one of the most significant diseases of potato (Strange 2013.) In early nineteenth century social issues occurred in Ireland between English land lords, who had major concern on revenues, and farmers who depended on potatoes as main source of diet for their families (Large 1940). The disease was first observed in Belgium in 1845 (Bourke 1964), and later spread to other countries i.e. England, Scotland, Ireland, France, Germany and Scandinavia. It caused million deaths and forced people to migrate to North America.

The causal agent of the brown spot of rice is the fungus *Helminthosporium oryzae.* In favorable conditions, this disease can result in severe damages to rice crops. In 1942, it caused disastrous consequences in Bengal, with a dramatic impact on people. In rural areas farmers left their places and migrated to other cities searching for food and employment. They faced starvation and many people died (Padmanabhan 1973). The affected population number was two million.

In USA (1970–1971), corn leaf blight disease became epidemic with losses that were dramatic. However, the USA agricultural industry was extremely diversified, and human distress was much less than in the previously cited epidemics. Total losses were officially predictable at 1 billion USD, over the nation (Ullstrup 1972).

Cassava *(Manihot esculenta)* was cultivated approximately 4000–6000 years ago (Fauquette and Fargette 1990). This plant originated in South America, where it is the third carbohydrate source for importance. Its per annum production is about 136 million tonnes. In Africa it is an important crop and total yield reaches 57 million tonnes. Epidemics of the African Cassava Mosaic Virus in the continent are frequent, and prevalence may reach 80–100% of plants, with projected losses around 50% of yield (Fauquette and Fargette 1990).

Bayoud is a fungal disease caused by *Fusarium oxysporum.* In Morocco about ten million date palms are affected, with three million trees also killed in Algeria. This disease not only causes production losses, but also speeds up the process of desertification (Assef et al. 1986).

Cloves are used in Indonesia in the production of kretek cigarettes (tobacco mix 40% ragged clove bud) (Bennet et al. 1985). A disease known as "Sumatra disease", caused by *Pseudomonas syzygii*, let down Indonesia's plan of self dependence in cloves production. The name of this disease derived from the island where the crop is cultivated. Losses are around 10–15% of yields, about 50 million USD (Strange 2013).

A cocoa *(Theobroma cacao)* disease, known as swollen shoot, shows shoots that become swelled with a severe dieback. Millions of cocoa trees have been infected and died in West Africa and about 190 million trees have been eradicated to control its spread (Thresh and Owusu 1986).

3.3 Pathogens Targeted by Plant Breeders

Depending on their nature, pathogens can be seen microscopically or, in some cases, with naked eyes. They can be soilborne or airborne. The groups of disease causing agents are fungi, oomycetes, bacteria, viruses, viroids, phytoplasms, and nematodes. Plant breeders, with different degrees of success, allocated various amounts of resources to select for resistance in these categories. Plant species and germplasm differ according to their susceptibility to diseases resulting from pathogens from each group. Cereal products tend to have significant problems with airborne fungal diseases whereas, in contrast, soybean is mostly attacked by viruses.

3.3.1 Fungi

The fungi are divided into four classes that are Ascomycotina, Basidiomycotina, Mastigomycotina, and Zygomycotina, according to the morphology of sexual constitution and the sporulating organs produced subsequent to sexual reproduction (Isaac 1991). Another category, Deuteromycotina, is set aside for fungi where no sexual phase is recognized. Organisms that are detrimental to plants are grouped in five classes. Mastigomycotina includes genus *Phytophthora* meaning "plant destroyer", an appropriate word for that genus. Further important disease caused by members of this genus include the black pod disease of cocoa and blight of pigeon pea. Members of *Rhizopus*, a genus of Zygomycotina, cause major losses in several plants i.e. cassava, peanuts, sorghum and cucurbits. It has also importance as causing postharvest diseases in soft fruits (Michailides and Spott 1990). One of the worst pathogenic strains of known plants is *Claviceps*, a member of Ascomycotina. The danger of this organism is not related to damaged grain crops, but to its fungal sclerotia, which have extremely toxic compounds such as alkaloids.

Rusts and smuts are included in Basidiomycotina. Rusts and smuts are extremely specialized obligate parasites, and represent a constant danger to crops (Barnett and Binder 1973). Additionally, Basidiomycotina contains several of plant largest parasites, such as bracket fungi that damage tree species (Jonsson et al. 2005). Some fungi can also act as vectors for viruses while giving them little harm. For example, *Olpidium brassicae* lettuce can transmit large vascular virus (LBVV) and tobacco necrosis virus (TNV), whereas *Polymyxa graminis* transmits various viruses that induce significant diseases on host plants.

3.3.2 Nematodes

They can be used not only as a direct loss of crop but also as source/vectors of viruses of plant. Only two of the 17 orders of nematodes cause damages to plants (Tylenchida and Dorylaimida), either inducing direct losses or transmitting plant viruses (orders Dorylaimida and Triplonchida) (Wyss 1981). In addition, nematode feeding damages the root tissues providing entrance into the root for numerous parasites, in particular fungi. Some loss estimates can be obtained for nematodes by comparing nematicide-treated with control soils. For example, Ingham and Detling (1990) observed that, treating mixed grass prairie with carbofuran, the nematicide reduced the nematode population by about 82%, increasing production of up to 52%.

3.3.3 Protozoa and Algae

The likelihood of a protozoid origin for some severe plant diseases was not entirely documented until 1976, when they were linked to two major defects in coconut palms. McCoy and Martinez-Lopez (1982) observed nine cases of dwarf coconut palms in the USA, deadly wilted by *Phytomonas staheli*. Phloem necrosis in palms is currently recognized to be due to a *Phytomonas* (Trypanosomatidae) (Douet 1984).

Cephaleuros virescens is the causal agent of the Algal leaf spot disease. It has been associated with many disease warning sign in more than 50 high plants. In another study Tahiti lime (*Citrus latifolia*) showed up to 98% of leaves contaminated by an alga of this genus (Marlatt and Pohronezny 1983; Strange 1993). Other algae of genera *Chlorochytrium, Rhodochytrium* and *Phyllosiphon* are also involved in plant diseases (Strange 2006).

3.3.4 Bacteria

The bacteria causing diseases in plants were formerly classified into genera of Gram positive (i.e. *Corynebacterium*) and Gram negative (i.e. *Agrobacterium*, *Erwinia*, *Pseudomonas*, *Xanthomonas* and *Xylella*). Recently, this classification has been thoroughly revisited, with recommendations for classifying coryneform bacteria in *Curtobacterium* spp.

Other genera include *Arthrobacter*, *Rhodococcus* and *Clavibacter* (Davis 1986). members of the *Corynebacterium-Clavibacter* clade can cause diseases in a number of plants. One of the most severe one is *Corynebacterium sepedonicum* (Deboer and McCan 1989). Pathogenic species of *Clavibacter* (Raju and Wells 1986) include *Clavibacter xyli* subssp. *Xylene*, causing ratoon stunt of sugarcane. Grisham (1991) observed that this organism caused losses around 14% of cane in the 1st year of sowing, elevated to 27% in the 3rd year.

In around 200 species of dicotyledonous plants the crown gall and root gall diseases are caused by infections by *Agrobacterium tumefaciens*.

3.3.5 Actinomycetes

Some members of Streptomycetes cause potato warts. The weight decrease of potatoes is minute. However, the financial failure of the grower is noteworthy because potatoes showing malicious warts are not preferred by the consumers. A similar disease in carrot was also documented (Janse 1988).

3.3.6 Mycoplasmas and Spiroplasms

This group of bacteria is characterized by the absence of a cell wall. Mycoplasmas are spherical whereas spiroplasms, as the name suggests, are spiral-shaped. They need vectors for transmission into susceptible organisms, and are responsible for diseases such as aster yellow and corn stunt. In citrus, fruit production can be decreased by 50–100% due to the 'stubborn disease', caused by *Spiroplasma citri*. This disease may be observed in 5–10% of citrus trees in California, and higher frequencies in Mediterranean countries (Smith and Banks 1986).

3.3.7 Viruses and Viroids

Plant viruses are classified into 38 groups (Boswell et al. 1986) based on morphology, RNA or DNA (single or double stranded). Serological method and nucleic acid probes are used to determine the characteristics or relatedness of plant viruses. There are more than 700 recognized viruses of plants, numerous of which have a wide host range and result in catastrophic diseases.

Viroids consist only of a circular single-strand RNA. There are at least 12 recognized root diseases induced by viroids. These comprise of economically important pathogens attacking potato, citrus, and coconut palm trees. In Philippines, the latter destroyed more than 30 million palm trees, in spite of its simple structure that consists of around 300 nucleotides (Hanold and Randles 1991). Neither viruses nor viroids are able to proliferate in the absence of hosts.

3.4 Management of Plant Diseases

Once the causal factor of a disease is determined correctly, it is possible to develop plans for its management and control. Over the last century much research has been carried out on pathogens, diseases and management methods. Today we can take advantage of this vast amount of knowledge to sustain control programmes. Smart management of plant diseases is an economic necessity. It helps to prevent epidemics and disastrous famines. There are three basic approaches for plant diseases management i.e., chemotherapy, prevention and genetic resistance (Fig. 3.2).

Chemical control against phytopathogens is an important approach. The use and misuse of chemicals (fungicides and pesticides) are known since the '60s, in which the dangers of pesticides were highlighted.

Prevention is based on the consideration that the most effective disease control strategy is to keep the host and pathogens far from each other. This type of management can be taken in several forms. A government unit (county, state or nation) may establish prohibition and prevention rules. Such quarantines are practiced in parallel with inspections.

The use of cultivars resistant to diseases is one of the less expensive, safest and most practical solution. The use of resistant cultivars is attractive to those who must rely on expensive pesticides to protect large tracts of low-income crops, such as wheat. Expert scientists, time and money are needed to grow resistant varieties. In crop plants, resistance is a foundation step in any disease management program.

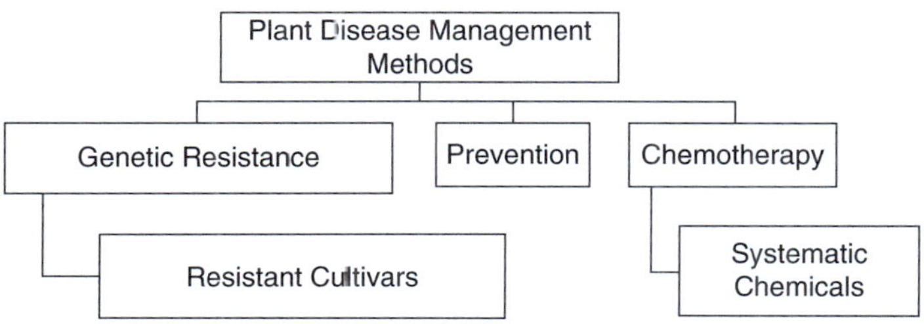

Fig. 3.2 Approaches for plant disease management strategies

3.5 Genetics of Disease Resistance

Van der Plank (1963) was the first to classify resistance as vertical or horizontal. Vertical resistance (VR) is also known as race specific. Horizontal resistance (HR), also called polygenic resistance, relies on genetically different and physiologically different species. Other terms used for such resistance are either quantitative or partial resistance.

3.5.1 Vertical Resistance

VR is conditioned by oligogenes and is successful against some races of a pathogen, but not all. HR, which is polygenic in inheritance, appears efficient against all races of that pathogen. It is now clear that achieving a population resistance, where a given pathogen population cannot increase and damage the host population possessing VR or HR, or a combination of both, is a challenging task that needs much attention. In cases of outbreaks and epidemics, VR is of sufficient value in effective control measures, but it has been found unsatisfactory against widespread epidemics (Severns et al. 2014).

A variety with HR, showing most resistance to all pathogen races, does not affect the pathogen population growth from the initial inoculum level but it does reduce the rate at which such an increase normally takes place (Garrett 1999). As a multiline cultivar possesses many genes for VR, the initial inoculum of the pathogen, in course of time, becomes small. The oligogenes give VR ease of manipulation in greenhouse and field trials and were found superior in yield assessments.

Plant pathologists accustomed to work with populations. Breeders pursue a yield boom, quite unaware of the nature of adversity. In fact, the disease importance increases not only with the degree of inbreeding required for its containment, but also with the extensive use of cultivars having the same germplasm. The medial method suggested was the use of non-uniform crop varieties.

3.5.2 Horizontal Resistance

For many plant diseases the value of VR in ongoing breeding program was analyzed and found clearly inappropriate. The basic point in favor of HR breeding techniques is that the resistance effectiveness as it does not "break down". VR instead is subject to this effect. In synthesis, while VR confers complete but non-permanent protection, HR confers incomplete but permanent protection. It is hence useful to recognize some terms introduced in HR breeding programs. A "pathodeme" is a host population in which all individuals have a given resistance in common. A "pathotype" is a pathogenic group with the same pathogenicity reaction on a particular host. When a variety pathodemes is inoculated with a variety of pathotypes, and the disease incidence displays a differential interface between pathodemes and pathotypes, the resistance and pathogenicity are called vertical. When there is no differential interaction, they are called horizontal. The pathodemes and pathotypes can also be described as vertical or horizontal. VR involves mechanisms which are within the pathogen's capacity for change. HR, on the other hand, involves mechanisms which are beyond the pathogen's capacity for change. The term "capacity for change" means that every pathogen can change as it has an in-built capacity within the well understood term "natural variability ". There are, however, limits to that variability and HR involves mechanisms beyond those limits. It should be understood that change here means population dynamics and not evolution. Furthermore, VR is inherited oligogenically (i.e., controlled by a few genes for major heritable changes by looking for applied characters). HR is almost always a polygenically inherited resistance (i.e., controlled by a number of genes). However, oligogenic HR does occur in rare cases as not all the HR components are inherited polygenically. *Vice versa,* not all oligogenic resistance is VR. The most important point is that oligogenic HR is a qualitative inheritance, a mechanism of functioning with related final effects. As opposed to these rare cases, the general run of universal HR is quantitative. The influence of breeding techniques comparing oats and rye may well illustrate this point.

It may be difficult in some diseases to obtain sufficient HR to control a disease in natural growing conditions. It would seem best to initiate breeding with HR first and then to reinforce it with VR, should HR prove inadequate to meet the situation. For instance, a good horizontal pathodeme can be used as the basis of a multiline of several different vertical pathodemes. The result then would be a slowing down of an epidemic, as the epidemiological effects of a multiline are similar to those of HR. This approach has much merit as it is nearer natural conditions where HR is essential and VR is a supplementary protection occurring as natural multilines.

In absence of a pathogen, erosion of HR can take place in nature. There are two types of erosions, one called "phenotypic erosion" and the other known as "genotypic erosion". Generally, in most crops, a high degree of susceptibility (the suscept-pathogen relationship) may be ascribed to an erosion of HR but it can be restored by breeding within existing local cultivars, i.e. by restoring lost genes. However, search for these local materials will have to go beyond locally available cultivars, including wild progenitors. Any factor(s) which masks HR will reduce pressure on selection for it. These factors can be fungicides and similar artificial disease control measures, or VR itself. VR can be eliminated making certain that the population does not possess genes for it. This has been found possible in the case of potato bred against *Phytophthora infestans* but it was not possible in wheat, where complete absence of VR to *Puccinia graminis* is unknown. Individuals showing hypersensitivity reactions such as ' flecking' flecking" are evidences of VR and such plants can be eliminated. Similarly, if VR confers complete resistance against non-matching vertical pathotypes, screening of the host populations can be done for a "slight disease" rather than a "no disease" condition. Also VR can be eliminated by ensuring that all individuals in the host population are susceptible to a single vertical pathotype. VR provides a complete and lasting control of a disease only when the host population flexibility is maximal, and the pathogen population flexibility is minimal.

In crops where breeding for HR is undertaken, the basic assumption should be that the existing levels of HR are due to phenotypic erosion, in which case, breeding could be confined to existing cultivars.

However, if this assumption is not warranted a search may be made beyond existing cultivars, and efforts channeled for getting together a wider genetic base. Genetic heterogeneity can be achieved by random polycross with the assistance of a male gametocide. By suitably increasing the intensity of the epidemic conditions, a sufficient selection pressure for HR can be mounted. VR has several attributes such as hypersensitivity reaction, race- or pathotype-specific effect, inheritance based on major genes resistance and non-durable resistance.

3.6 Resistance Breeding Strategies

3.6.1 Prioritize the Importance of Diseases

A significant concern in resistance breeding is to establish the suitable priority level. Breeding efforts for resistance may be unsatisfactory if the consequences are extremely damaging to other traits, e.g. yield. These special effects may occur because of genetic linkages, and are unfortunately common (Johnson 1978).

In addition, resistance breeding needs resources. Highest efficacy will be attained by placing it at a level of priority as low as reliable, by producing suitable genotypes. It can be determined by the importance of a disease and the likelihood of incorporation of a worthwhile resistance level, while conducting a breeding program considering the choices for other control schemes. Some diseases are widespread, being as such a noticeable goal to control through breeding. For other diseases, the resolution is not much clear and accurate data on their prevalence in crops is not sufficient (Wolfe and Barrett 1979). There may be, therefore, a biased component in disease targeting. Main objective of resistance breeding is to identify diseases of importance at current genotypes have enough resistance sources. Thus breeders may need to spend energy on selection for resistance, or against high susceptibility for in fact an insignificant diseases.

The occurrence of wheat yellow rust in UK during 1970–1988 was typically low (Polley and Thomas 1991), as yellow rust was not a vital disease. Commonly less occurrence of this disease in the UK wheat crop is due to struggles by the UK wheat breeders for yield, and the National Institute of Agricultural Botany suggesting only genotypes resistant at more than a given, least possible level. On the other hand, this cannot completely exclude the threat of epidemics.

With more recent fungicides it is hard to control yellow rust in extremely susceptible genotypes. This was demonstrate by the development of an epidemic when the cultivar Slejpner became susceptible when it was commercially used (Bayles et al. 1989). Incidence of yellow rust on extremely susceptible genotypes happens regardless of availability of active fungicides, because of meteorological conditions or human factors. The potential threat of yellow rust has been famous for numerous years and resistance breeding has remained a main concern in most UK programs.

In other countries this was not the case. For example, frequent outbreaks of yellow rust on bread wheat in Italy are credited to the cultivation of extremely sensitive genotypes, lacking selection for resistance in breeding programs, in combination with much disease suitable climatic conditions, during several years (Chilosi and Corazza 1990).

Sometimes, a disease that has not even been considered important can develop in a crop. For example, due to a humid 1982 summer in the UK, the Avalon wheat genotype was considered as susceptible to ear blight, after its commercial presentation (Lupton 1983). During the development of Avalon, no selection against ear blight was made. On the other hand, a technique was later established for evaluation of susceptibility against Fusarium species by spraying the emerging spike with spore suspensions and retaining high humidity by mist irrigation (Jenkins 1984). By means of this method, bases of resistance were recognized and exploited.

3.6.2 Steps in Breeding for Disease Resistance

Collecting and maintaining genetic sources of resistance genes is the first step in breeding for pathogens resistance. The resources comprise of commercial and/or local varieties, wild related progenitors, species and related genera, mutagenesis.

3.6.2.1 Sources of Resistance Genes

A decisive factor which influences the plant breeder in the choice of breeding outlines for resistance is the knowledge about the accessibility of resistance sources. These are commonly available for many, generic diseases. On the other hand much less sources of resistance are available for specific diseases. In wheat resistance sources against rust are easily accessible, while for some diseases, such as the eyelid, it is difficult to find suitable sources. However, many sources of diseases resistance have been identified and successfully utilized by plant breeders. However, in some cases it is difficult or practically impossible to select a resistance source, because it cannot be found. This difficulty faced by the breeder and all the source of wheat disease resistance are not sufficiently documented (Scott 1981).

There are resistance sources which have not been observed in present cultivars but can be found in related species, from which they can be transferred successfully (Knott and Dvorak 1976). Availability of resistance source is indeed the most important starting point when initiating a breeding scheme. The most suitable situation is when sufficient genetic variation/sources are available in breeding populations or cultivars. Resistance sources should be tested for other important agronomical traits and can be directly introduced in other cultivars by crossing. Some source of resistance cultivars need acclimatization when the material is imported from neighboring regions having a similar environment. Much effort is needed when the source is a non-adapted cultivar.

3.6.2.2 Utilization of Genetic Resource

Plant breeders around the globe try to collect as much as possible of broad genetic bases, and many of their work is dedicated to collecting genetic sources which likely are beneficial for them. For example, in cotton breeding for resistance to bacterial blight (*Xanthomonas campestris*) R.L. Knight (1957) had to observe more than 1000 diverse accessions of wild and cultivated *Gossypium* spp., to find the source of resistance against the disease. Innes (1983) listed tetraploids and 2n *Gossypium* spp. and 2n wild species having 19 major genes, which were used to resist to blight. A total of 13 genes were identified in tetraploid *G. hirsutum*, only one gene in tetraploid G. *barbadense,* two in diploid *G. arboretum* and diploid G. *herbaceum*, and one gene in 2n *G. anomalum*.

If it is important to collect useful germplasm, another important aspect is being sure that such germplasm is conserved, well documented and easily accessible to breeders throughout the world, when needed. Techniques have been made available to conserve seeds without viability losses, keeping them in viable conditions for a long time (Roberts 1975). Tissue culture for long-term preservation is one of them, allowing preservation of vegetatively propagated species (Withers 1989). Developed and developing countries give importance to germplasm preservation and make more efforts to provide funding and services for germplasm conservation (Hawkes 1991). However, the protection of the world genetic resources is a difficult job that goes ahead of the budget of many governments.

The International Plant Genetic Resources Board (IBPGR) was established in 1974 with the directive to encourage and synchronize the breeding work, at the international level. Collection, preservation and documentation are fundamental functions of a gene bank (Brown et al. 1989). Most of the world's conserved genetic resources is in public sectors and was formed as the result of international agreements. Most agreements allow easy access, and without cost, to breeders. On the other hand, the Keystone Center (1991) has given emphasis to the global initiative for protection and sustainable utilization of plant genetic resources. It required joint efforts, and contribution by all members, trusting parties and institutions from all over the globe, including those that provide germplasm, information and technology, as well as financers and improvement organization.

3.7 Planned Deployment of Resistance Genes

3.7.1 Self-Pollinated Crops

In **self-pollinated** crops, backcross and pedigree breeding methods are used with modifications based upon breeder's convenience. In **mass selection**, plants from a population are bulked for resistance to make the foundation of a variety. Their heterozygosity may provide several merits over pure line varieties for disease resistance. In ***pure line selection*** pure line are derived from the progeny of selfed homozygous plants, selected from a variety. The progeny is evaluated for resistance in succeeding generations. If it is promising, then is the progeny is multiplied to develop a new variety (Allard 1960). In **line breeding** plants are selected (selfed or inter-pollinated), and the resulting progenies are tested or evaluated for resistance. **Hybridization** involves crossing of two lines for transferring disease resistance from donor parents, combining characteristics from each parent. It allows the plant breeder to combine diseases resistance into a single variety. The F_1 genetic constitution is identical though segregation in F_2 generation occurs as well as regain of homozygosity is attained in succeeding selfed generations. Selection followed by hybridization is based on following approaches.

Pedigree selection is mostly practiced in self-pollinated crops because it gives breeders the greatest opportunity to test their selection expertise (Allard 1960). The main limitation of this method is the high amount of material that a single breeder cannot handle alone. In pedigree selection, commonly practiced crosses are between a parent selected for his desired agronomic performance and another chosen parent, having a specific weaknesses, i.e. absence of disease resistance characters.

Disease resistant plants are selected in advanced generations and records are kept for all parental and offspring associations. Information from pedigree selection is helpful to avoid selection of narrowly related individuals with a close ancestry, whose likely value is almost similar. Usually in this method selection is started in the F_2 generation. Through F_3 and F_4 generations, selection is mostly tested in the best disease resistance plants of the best families, due to a desired level of heterozygosity maintained in these generations. Selection continues to generations F_5 and F_6, focussing on to family selection, often planted as individual rows or replicated in breeding nurseries. Ultimately, the lines that are consistent for the release of fresh cultivars are tested on sick field plots, and replicated in different ecological conditions (Allard 1960).

Pedigree selection also is used in cross-pollinated crops, i.e. in maize, for improvement of lines that have known desired traits and weakness for other particular traits (Agrawal et al. 1976). Pedigree selection may start with progenies developed in cross pollinated varieties, germplasm composites, synthetic or backcross populations, and also in F_2 populations. For maize, the objective is to select pure lines with a superior combining capacity in the production of high-performance F_1

hybrids capable of supporting diseases and other stresses characteristic of the area in which the hybrids will be grown. Crops in which quality standard are required in cultivars for quality demanding consumers, backcrossing breeding method mostly are used to transfer disease resistance from sources (donors) that may be agronomically inferior for yield, and quality cultivars that have been developed by typical pedigree selection (Fig. 3.3).

Bulk Selection is a desirable technique to combine the characteristics from both parents. Early segregating generations ($F_2 - F_6$) are bulked without selection. When homozygous plants are attained in later generation then selection is made for resistance and the plants are evaluated as in the pedigree method. Artificial epiphytotic conditions are established for selecting resistant plants (Singh 1986).

Backcross variety have been developed by crossing F1 hybrids with any of the known parent. This is useful when breeding for small grains crops. In this technique two plants are chosen and inter-mated. After regular backcrossing with one of the recurrent parent, the backcross progeny developed is almost identical to the recurrent one (Fig. 3.4). In this breeding method economically important variety, lacking a disease resistant character, is known as recurrent/recipient parent. At the same time a variety containing only a disease resistance character is known as donor parent. The élite varieties KDML105, Basmati and Manawthukha (Toojinda et al. 2005) were produced for resistances to rice disease blast in South and South East Asia by practicing this breeding technique (Sreewongchai et al. 2010). Advantages of using this method include i) the intervarietal transfer of characters (disease resistance, plant height, earliness, seed size, color, shape), ii) the interspecific transmission of disease resistance characters from associated/related species to cultivated ones, iii) the transfer of cytoplasmic material from one to another variety or species, (this requires, in case of cytoplasmic male sterility, the development of transgressive segregants and the production of isogenic lines). Some of the varieties which have been developed through this method is BD 8 of cotton (resistant to wilt), MS 521, MS 541A, MS 570 A of bajra, resistant to downy mildew, the transfer of wilt resistance to alfalfa (*Medicago sativa*) variety California common, from the variety Turkestan.

Varieties developed by this method require several back crossings to develop a new cultivar. In general, the newly developed variety cannot be better than the donor parent, except only for the character to transmit. Hybridization must be performed for every back cross, a factor that is time consuming, expansive for handling and by the time required until the backcross is over. Some line produced by this method show higher resistant to rust (leaf or ray rust). But they also had a yield potential 5–15% higher than the original variety (Singh et al. 2005). Certainly, a second round of crosses among lines with resistance derived from diverse donors might give advanced levels of resistance. This may be evaluated in case an additional resistance is desirable in several environments.

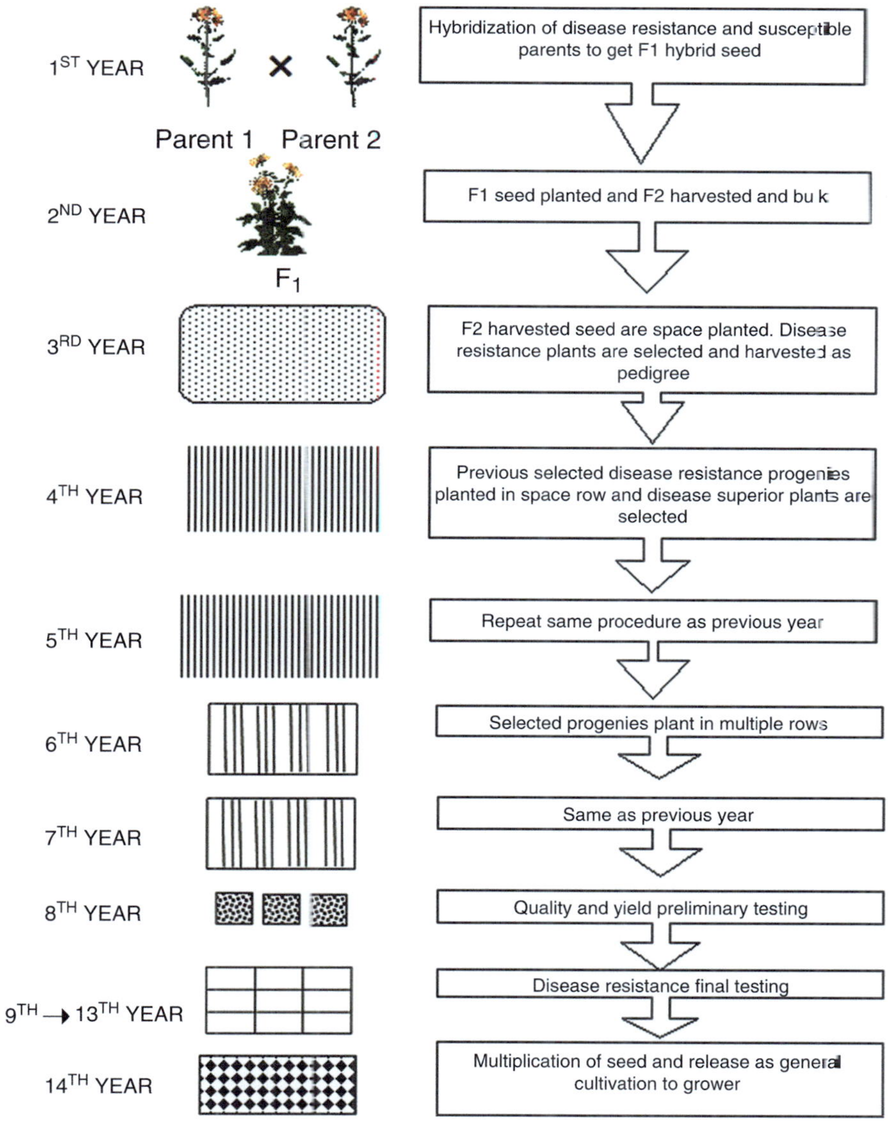

Fig. 3.3 Pedigree breeding method

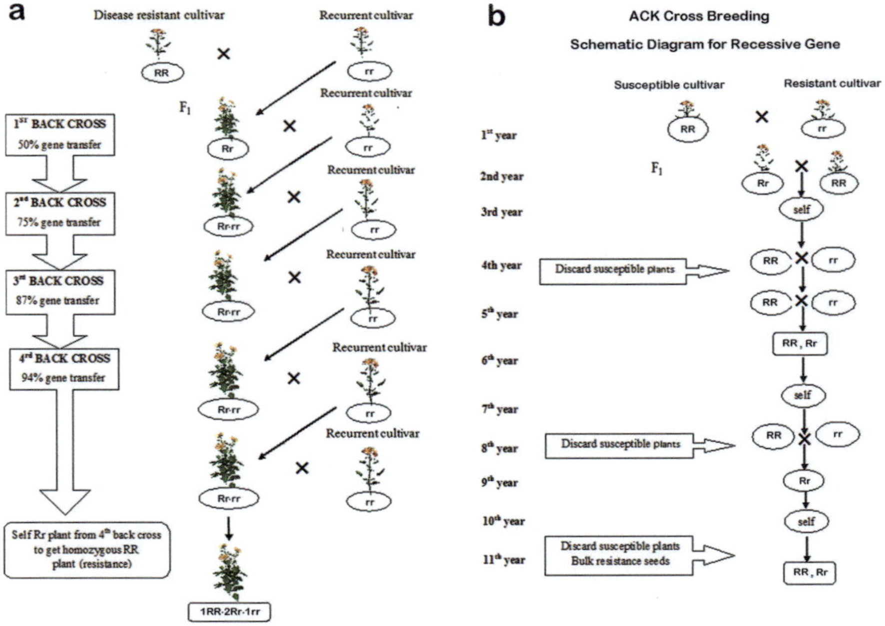

Fig. 3.4 Back cross breeding scheme for dominant (**a**) and recessive (**b**) genes

3.7.2 Cross-Pollinated Crops

Several methods can be employed in cross pollinated (allogamous) crops, to improve plant populations and develop disease resistant varieties.

In line breeding plants are selected (selfed or inter-pollinated), and resulting progenies are tested or evaluated for resistance (Singh 1986). **In polycross**, n resistant plants are selected from a heterozygous population and intercrossed, following the development of inbred lines. The progenies of a polycross are bulked, the resistant plants are selected and progenies of individual lines are tested independently (Singh 1986). **Synthetic/Hybrid** varieties: resistant lines, obtained through line breeding or recurrent selection, are used to produce hybrid or synthetic varieties. A synthetic variety is developed by intercrossing several selected plants that have been expected to be good combiners. Hybrid varieties are the product of controlled pollination between lines (Singh 1986). The parental lines must be maintained independently for reconstituting the synthetic or hybrid varieties.

In cross-pollinated crops, **recurrent selection** is also an effective method of selection for self-pollinated crops. For instance, it permits the accumulation of desirable genes to raise the level of polygenic resistance. Before commercial release of genotypes, recurrent selection provides information about partial resistance to estimate the genotypes potential. If a character, i.e. quantitative resistance, is governed by more than four genes, only a very small proportion of the progeny of a cross between a superior cultivar, susceptible, and a disease resistant donor parent, will have the required amount of resistance genes. Different cycles of recurrent selection for agronomic characters as well as required disease resistance will increase the selection efficiency. This is valid for genotypes from the population that combines better agronomic characters along with polygenic disease resistance. As observed by Eberhart (1990), "*For improvement in maize population, improvement program is the foundation which leads toward maximization of durable genetic gain every year*".

Selected genotypes of both populations are mated with genotypes of pure testers of the reciprocal population, assessing the performance of resultant F_1 crosses. Formerly, best selected genotypes are crossed in all possible combinations to rise next population cycle. Tester genotypes will change in advanced selection cycles, as development of pure genotypes (Eberhart 1990). The strength of selection can be improved by accumulating the quantity of investigated genotypes assessed at the test crosses, or by reducing the selected genotypes for re-combination in the next population cycle. As suggested by Eberhart (1990), genotypes 250–400 S_2 should be assessed for selection of 6–20 genotypes for recombination for the next cycle. This will attain a 4–8% strength of selection. In the beginning, it is required to select limited desirable characters that are essential to get stress resistance and high yield. Selection of more number of characters at the same time will result in slow selection gains for each single character. Eberhart (1990) endorsed the different selection stages to attain enhancements in different aspects. The S0 plants self-fertilized in all populations can be analyzed for eradication of unwanted traits established on highly hereditary characters, for example days to maturity and disease resistance. These selections should be performed prior to production, to decrease the number of plants that will self-pollinate. In the next period, selection was done on S_1 plants for fewer heritable characters, for example resistance to insects as well as lodging resistance. The selection between test crosses should be established mainly on yields, along with root rot and stem rot resistance. The recurrent selection programs for population improvement are logically appropriate for cross-pollinated crops (Fig. 3.5). However, several schemes of recurrent selection with self-pollinated crops have been tried.

The use of male sterility characters was suggested to assist the recurrent selection in barley sorghum and soybean (Gilmore 1964; Doggett and Eberhart 1968; Brim and Stuber 1973). However, Matzinger and Wernsman (1968) demonstrated that constant improvements in leaf yield can be gained by repetitive mass selection during random mating cycles between selected genotypes, in a heterogeneous synthetic tobacco cultivar that is usually self-fertilized. Jensen (1970) recommended a selective pairing technique of diallel with recurrent selection, for small grain populations to assist as a complement in conventional pedigree selection procedures. Díaz-Lago et al. (2002) revealed that programs with selection in early generation for partial crown rust resistance in oats, led towards a total increase of about 42% in resistance after four cycles of population improvement, to assist as an enhancement to conventional pedigree selection scheme. However, they concluded that

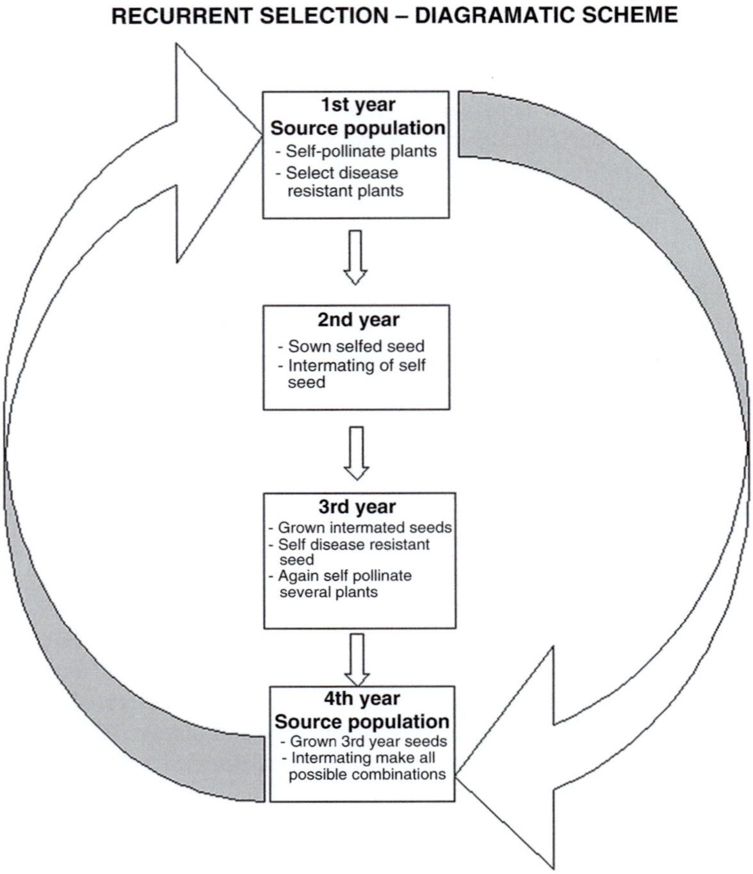

Fig. 3.5 Scheme of a recurrent selection breeding program

synchronized selection for the days to flowering would be essential as part of the supplementary resistance to crown rust was linked with late maturity.

Interspecific hybridization yields hybrids developed by crossing two species of the same genus, with the objective of transferring to a cultivated species one or few simply inherited characters such as disease resistance. Kufrijyoti is a potato variety developed through interspecific hybridization, which is resistant to late blight. Through interspecific hybridization, a resistance gene for cotton rust, due to *Puccinia cacabata*, has been transferred from *G. anomalum* and *G. arboretum* into *G. hirusutum* (Anjum et al. 1986).

Aminu (1940) reported the combination of genes for resistance to cotton leaf curl virus and other diseases between *G. hirusutum* and *G. arboreium.* Similarly, gene B6 present in the 'A' genome of *G. arboretum*, conferring resistance to bacterial blight by *Xanthomonas malvacearum*, was introgressed into *G. barbadense* (Knight 1957; Brinkerhoff 1970).

A source of common bunchy top plant resistance in *G. hirsutum* has been identified in "Delta Opal", which is used to transfer disease resistance (Ellis et al. 2016). Transmission of genes between species is a very preeminent technique that can result in a broad spectrum of resistance. In case of wheat, the *Lr34res* allele has been identified which gives resistance in maize against rust and the hemibiotrophic fungus responsible of northern corn leaf blight (NCLB) (Sucher et al. 2017). *Lr34res* has already been revealed to be operative for rice blast in rice (Krattinger et al. 2016) and vs different biotrophic fungi (Risk et al. 2012, 2013). Additionally, *Lr34res* has been linked with resistance for spot blotch (caused by the fungus *Bipolaris sorokiniana*) in wheat (Lillemo et al. 2013). Oligogenic and polygenic resistance have been recognized in corn, for race specific *Ht1* and *Htn1* genes, against widespread NCLB races (Welz and Geiger 2000; Hurni et al. 2015).

3.8 Evaluation

The imperative step in breeding for disease resistance is to evaluate the developed germplasm, which may be achieved in either greenhouse or field conditions. Both are laborious and expensive breeding steps. While executing a backcrossing program, selected breeding material must be stabilized under greenhouse conditions. In the greenhouse the material selected from successive backcross generations are tested for pathogens resistance, and the susceptible ones are eliminated. In field evaluation, selected material is grown in disease-free and disease-infested plots, to evaluate them for resistance against particular pathogens.

3.9 Release to Growers

The final breeding step is the release of lines, which have been evaluated as disease resistant, to the growers for cultivation.

3.10 Factors Affecting Expression of Disease Resistance

Some particular causes may complicate reproduction for resistance. These factors can be biotic or abiotic (Burdon 1987), as follows.

1. Abiotic factors must remain within a particular range to permit the development of the Pathogenic species (Sharma et al. 2003).
   (a) Temperature: expression of few resistant genes is restricted in case of excessively low or high temperature.

(b) Light: the intensity of light may disturb the efficacy of chemicals affecting as a result the pathogen resistance level.
(c) Soil fertility: extraordinary soil fertility produces extra succulent plants that are most vulnerable to infection. Other pathogens (opportunistic generalists) are also more effective in under-nourished plants.

2. Biotic factors

(d) Years: the response of a plant toward a pathogen can differ with time. Certain diseases are extremely effective at the beginning of plant growth than others (Burdon 1987).
(e) New races of pathogens: as mentioned, breeders should be aware that there is an efficacy of resistance for certain new races of pathogens, but not for others (Burdon 1987).
(f) Introduced resistance: infection caused by prior pathogenic infestations can induce a systemic resistant reaction towards a later infection by other pathogens (Kathiria et al. 2013). Such cases may also occur before infection.

3.11 Advantages of Breeding for Disease Resistance

Resistant varieties offer the cheapest means of disease control, with indirect benefits as they reduce the application of fungicides, thus reducing environmental contamination. The effectiveness of resistant varieties, however, is not affected by environmental conditions. Disease resistance breeding is an important goal for plant breeders around the globe. A combination of traits is desirable instead of just targeting one trait for releasing an attractive cultivar. Performance, superior quality and resistance to epidemic diseases are major considerations in improvement programs, with the first point aiming at the most significant progress.

1. Resistant varieties offer the cheapest means of disease control.
2. They obviate the use of fungicides, thus reducing environmental pollution
3. Effectiveness of resistant varieties is not affected by environmental conditions.
4. They safeguard from the inadvertent release of varieties that are most susceptible than earlier ones.

3.12 Problems in Breeding for Disease Resistance

The problems and inconveniences occurring during the development of a new variety may be summarized as follows.

1. Resistance breakdown (vertifolia effect, boom and bust cycle).
2. Horizontal resistance being durable, but difficulty may concern the accurate and reliable assessment of the resistance level.

3. Sometimes there is a negative correlation between yield and disease resistance, e.g. wheat leaf rust gene Lr34 causes a 5% reduction in grain yield (Draz et al. 2015).
4. Introgression of multiple resistance against several diseases requires meticulous planning and far greater efforts than that needed for a single resistance.

3.13 Breeding Challenges for Pathogen Resistance

Disease resistance breeding basically varies from that applied for other characters because the induced resistance may cause alteration in the evolution as well as population of the pathogens (Van Bueren et al. 2011).Identification of resistant genes is not possible but for the plant infected in disease conducive environmental conditions (Engering et al. 2013). The development of segregating populations is necessary for breeders. The challenge is in the identification and selection of desirable genotypes in such a way that they would remain genetically in force, even many years after first release . The breeder must use reliable methods to detect variations in the level of resistance among segregates. While naturally caused infection can be utilized, artificial inoculation is often much consistent. Main issue in parenting for disease resistance is that, with passage of time, conventional crops changes the environmental conditions (i.e., different agronomic practices) and races of pathogens (Walters et al. 2013). Plant breeders should develop new genotypes with desirable resistant genes by maintaining these changes, to guarantee constant crop productivity, avoiding development of destructive epiphytes and infections, and reduction in annual yield losses . Plant breeders should not develop highly resistant cultivars that are not economically valuable, as most convenient approach is to breed for medium resistance. For the horizontal quantitative resistance, breeding and selection for better performing genotypes is the most desirable approach, subsequently representing the highest average resistance.

References

Agrawal, R. A., Banerjee, S. K., Singh, M., & Katiyar, K. N. (1976). Resistance to insects in cotton. II. To pink bollworm, *Pectinophora gossypiella* (Saunders). *Coton et Fibres Tropicales, 31*, 217–221.

Allard, R. W. (1960). *Principles of plant breeding*. New York: Wiley/Toppan. 485 pp.

Aminu, K. C. (1940). Interspecific hybridization between Asiatic and new world cottons. *Indian Journal of Agricultural Sciences, 10*, 404–412.

Anjum, Z. I., Hayat, K., Chalkin, S., Azhar, T. M., Shehzad, U., Ashraf, F., et al. (1986). Occurrence of an antifungal principle in the root extracts of a Bayoud-resistant date palm cultivar. *Netherlands Journal of Plant Pathology, 92*, 43–47.

Assef, G. M., Assari, K., & Vincent, E. J. (1986). Occurrence of an antifungal principle in the root extracts of a Bayoud-resistant date palm cultivar. *Netherlands Journal of Plant Pathology, 92*, 43–47.

Barnett, H., & Binder, F. (1973). The fungal host-parasite relationship. *Annual Review of Phytopathology, 11*, 273–292.
Bayles, R. A., Channell, M. H., & Stigwood, P. L. (1989). New races of *Puccinia striiformis* in the United Kingdom in 1988. *Cereal Rusts Bulletin, 17*, 20–23.
Bennet, C. P. A., Hunt, P., & Asman, A. (1985). Association of a xylem-limited bacterium with Sumatra disease of cloves in Indonesia. *Plant Pathology, 34*, 487–494.
Boswell, K. F., Dallwitz, M. J., Gibbs, A. J., & Watson, L. (1986). The VIDE (virus identification data exchange) project: A data bank for plant viruses. *Review of Plant Pathology, 65*, 221–231.
Bourke, P. M. A. (1964). Emergence of potato blight, 1843-1846. *Nature, 203*, 805–808.
Brim, C. A., & Stuber, C. W. (1973). Application of genetic male sterility to recurrent selection schemes in soybean. *Crop Science, 13*, 528–530.
Brinkerhoff, L. A. (1970). Variation in *Xanthomonas malvacearum* and its relation to control. *Annual Review of Phytopathology, 8*, 85–110.
Brown, A. H. D., Frankel, O. H., Marshall, D. R., & Williams, J. T. (1989). *The use of plant genetic resources*. Cambridge: Cambridge University Press.
Burdon, J. J. (1987). *Diseases and plant population biology*. Cambridge: Cambridge University Press Archive, 208 pp.
Chilosi, G., & Corazza, L. (1990). Occurrence and epidemics of yellow rust on wheat in Italy. *Cereal Rusts and Powdery Mildews Bulletin, 18*, 1–19.
Clifford, B. C., & Lester, E. (1988). *Control of plant diseases, costs and benefits*. Oxford: Blackwell Science Publishing, 263 pp.
Davis, M. J. (1986). Taxonomy of plant-pathogenic coryneform bacteria. *Annual Review of Phytopathology, 24*, 115–140.
De Boer, S., & McCann, M. (1989). Determination of population densities of Corynebacteriumsepedonicum in potato stems during the growing season. *Phytopathology, 79*, 946–951.
Diaz-Lago, J. E., Stuthman, D. D., & Abadie, T. E. (2002). Recurrent selection for partial resistance to crown rust in oat. *Crop Science, 42*, 1475–1482.
Doggett, H., & Eberhart, S. A. (1968). Recurrent selection in sorghum. *Crop Science, 8*, 119–121.
Douet, M. (1984). *Effets structurant des consortiums maritimes sur le marche des lignes régulières: l'exemple des consortiums à participation française*. Thèse de Doctorat, Université Paris 1 Panthéon-Sorbonne, 635 pp.
Draz, I. S., Abou-Elseoud, M. S., Kamara, A. E. M., Alaa-Eldein, O. A. E., & El-Bebany, A. F. (2015). Screening of wheat genotypes for leaf rust resistance along with grain yield. *Annals of Agricultural Science, 60*, 29–39.
Eberhart, S. A. (1990). A comprehensive breeding system for developing improved maize hybrids. In B. Gebrekidan (Ed.), *Maize improvement, production and protection in Eastern and Southern Africa*. Nairobi: AMREF Printing Department.
Ellis, M. H., Stiller, W. N., Phongkham, T., Tate, W. A., Gillespie, V. J., Gapare, W. J., et al. (2016). Molecular mapping of bunchy top disease resistance in *Gossypium hirsutum* L. *Euphytica, 210*, 135–142.
Engering, A., Hogerwerf, L., & Slingenbergh, J. (2013). Pathogen-host-environment interplay and disease emergence. *Emerging Microbes & Infections, 2*, e5. https://doi.org/10.1038/emi.2013.5.
Fauquet, C., & Fargette, D. (1990). African cassava mosaic virus: Etiology, epidemiology and control. *Plant Disease, 74*, 404–411.
Fehr, W. R. (Ed.). (1987). *Principles of cultivar development*. New York: Macmillan.
Garrett, K. A. (1999). *Factors influencing the effects of host diversity on plant disease epidemics for wheat stripe rust and potato late blight.*
Gilmore, E. C. (1964). Suggested method of using reciprocal recurrent selection in some naturally self pollinated species. *Crop Science, 4*, 323–325.
Grisham, M. P. (1991). Effect of ratoon stunting disease on yield of sugarcane grown in multiple three-year plantings. *Phytopathology, 81*, 337–340.

Hanold, D., & Randles, J. W. (1991). Coconut cadang-cadang disease and its viroid agent. *Plant Disease, 75*, 330–335.
Hawkes, J. G. (1991). Genetic conservation of world crop plants. *Biological Journal of the Linnean Society, 43*, 1–80.
Hogenboom, N. G. (1993). Economic importance of breeding for disease resistance. In T. Jacobs & J. E. Parlevliet (Eds.), *Durability of disease resistance* (pp. 5–9). Dordrecht: Springer.
Hurni, S., Scheuermann, D., Krattinger, S. G., Kessel, B., Wicker, T., Herren, G., et al. (2015). The maize disease resistance gene Htn1 against northern corn leaf blight encodes a wall-associated receptor-like kinase. *Proceedings of the National Academy of Sciences of the United States of America, 112*, 8780–8785.
Ingham, R. E., & Detling, J. K. (1990). Effects of root-feeding nematodes on aboveground net primary production in a North American grassland. *Plant and Soil, 121*, 279–281.
Innes, N. L. (1983). Bacterial blight of cotton. *Biological Reviews, 58*, 157–176.
Isaac, S. (1991). *Fungal-plant interactions*. Dordrecht: Springer.
Janse, J. (1988). A *Streptomyces* species identified as the cause of carrot scab. *Netherlands Journal of Plant Pathology, 94*, 303–306.
Jenkins, G. (1984). *Winter and spring wheat*. Annual report of the Plant Breeding Institute. 1983, pp. 23–28.
Jensen, N. F. (1970). A diallel selective mating system of cereal breeding. *Crop Science, 10*, 629–635.
Johnson, R. (1978). Practical breeding for durable resistance to rust disease in self-pollinating cereals. *Euphytica, 27*, 529–540.
Jonsson, B. G., Kruys, N., & Ranius, T. (2005). Ecology of species living on dead wood. Lessons for dead wood management. *Silva Fennica, 39*, 289–309.
Kathiria, P., Sidler, C., Woycicki, R., Yao, Y., & Kovalchuk, I. (2013). Effect of external and internal factors on the expression of reporter genes driven by the N resistance gene promoter. *Plant Signaling Behavior, 8*, e24760.
Keystone Center. (1991). *Global initiative for the security and sustainable use of plant genetic resources*. Keystone: Keystone International.
Knight, R. L. (1957). Blackarm disease of cotton and its control. In *Plant protection conference* (Vol. 1956, pp. 53–59). London: Butterworths Scientific Publications.
Knott, D. R., & Dvorak, J. (1976). Alien germplasm as a source of resistance to disease. *Annual Review of Phytopathology, 14*, 211–235.
Krattinger, S. G., Sucher, J., Selter, L. L., Chauhan, H., Zhou, B., Tang, M., et al. (2016). The wheat durable, multipathogen resistance gene Lr34 confers partial blast resistance in rice. *Plant Biotechnology Journal, 14*, 1261–1268.
Large, E. C. (1940). *The advance of the fungi*. London: Jonathan Cape.
Lillemo, M., Joshi, A. K., Prasad, R., Chand, R., & Singh, R. P. (2013). QTL for spot blotch resistance in bread wheat line Saar co-locate to the biotrophic disease resistance loci Lr34 and Lr46. *Theoretical and Applied Genetics, 126*, 711–719.
Lucas, G. B., Lee Campbell, C., & Lucas, L. T. (1992). *Causes of plant diseases. Introduction to plant diseases: Identification and management* (pp. 9–14). Boston: Springer.
Lupton, F. G. H. (1983). *Winter wheat*. Annual report of the Plant Breeding Institute 1982, pp. 23–26.
Marlatt, R., & Pohronezny, K. (1983). Field survey of Tahiti lime, Citrus latifolia, for algal disease, melanose, and greasy spot in southern Florida. *Plant Disease, 67*, 946–949.
Matzinger, D. F., & Wernsman, E. A. (1968). Four cycles of mass selection in a syntheticvariety of an autogamous species *Nicotiana tabacum* L. *Crop Science, 8*, 239–243.
McCoy, R., & Martinez-Lopez, G. (1982). *Phytomonas staheli* associated with coconut and oil palm diseases in Colombia. *Plant Disease, 66*, 675–677.
Michailides, T., & Spotts, R. (1990). Postharvest diseases of pome and stone fruits caused by Mucorpiriformis in the Pacific Northwest and California. *Plant Disease, 74*, 537–543.

Miedaner, T. (2016). Breeding strategies for improving plant resistance to diseases. In J. M. Al-Khayri, S. M. Jain, & D. V. Johnson (Eds.), *Advances in plant breeding strategies: Agronomic, abiotic and biotic stress traits* (pp. 561–599). Cham: Springer International Publishing.

Padmanabhan, S. Y. (1973). The great Bengal famine. *Annual Review of Phytopathology, 11*, 11–26.

Plank, J. V. D. (1963). *Plant diseases: Epidemics and control. Plant protection conference, 1956* (p. 53). London: Butterworth.

Polley, R. W., & Thomas, M. R. (1991). Surveys of diseases of winter wheat in England and Wales, 1976–1988. *Annals of Applied Biology, 119*, 1–20.

Raju, B., & Wells, J. (1986). Diseases caused by fastidious xylem-limited bacteria and strategies for management. *Plant Disease, 70*, 182.

Risk, J. M., Selter, L. L., Krattinger, S. G., Viccars, L. A., Richardson, T. M., Buesing, G., & Herren, G. (2012). Functional variability of the Lr34 durable resistance gene in transgenic wheat. *Plant Biotechnology Journal, 10*, 477–487.

Risk, J. M., Selter, L. L., Chauhan, H., Krattinger, S. G., Kumlehn, J., Hensel, G., & Viccars, L. A. (2013). The wheat Lr34 gene provides resistance against multiple fungal pathogens in barley. *Plant Biotechnology Journal, 11*, 847–854.

Roane, C. W. (1973). Trends in breeding for disease resistance in crops. *Annual Review of Phytopathology, 11*, 463–486.

Roberts, E. H. (1975). Problems of long-term storage of seed and pollen for genetic resources conservation. In O. H. Frankel & J. G. Hawkes (Eds.), *Crop genetic resources for today and tomorrow* (pp. 269–295). Cambridge: Cambridge University Press.

Scott, P. R. (1981). Variation in host susceptibility. In M. J. Asher & P. J. Shipton (Eds.), *Biology and control of take-all* (pp. 219–236). London: Academic.

Severns, P. M., Estep, L. K., Sackett, K. E., & Mundt, C. C. (2014). Degree of host susceptibility in the initial disease outbreak influences subsequent epidemic spread. *Journal of Applied Ecology, 51*(6), 1622–1630.

Sharma, H. C., Venkateswarulu, G., & Sharma, A. (2003). Environmental factors influence the expression of resistance to sorghum midge, Stenodiplosis sorghicola. *Euphytica, 130*, 365–375.

Singh, D. P. (1986). *Breeding for resistance to diseases and insect pests*. Berlin/New York/London: Springer.

Singh, R. P., Huerta-Espino, J., & William, H. M. (2005). Genetics and breeding for durable resistance to leaf and stripe rusts in wheat. *Turkish Journal of Agriculture and Forestry, 29*, 121–127.

Smith, D. A., & Banks, S. W. (1986). Biosynthesis, elicitation and biological activity of isoftavonoid phytoalexins. *Phytochemistry, 25*, 979–995.

Sreewongchai, T., Toojinda, T., Thanintorn, N., Kosawang, C., Vanavichit, A., Tharreau, D., & Sirithunya, P. (2010). Development of elite indica rice lines with wide spectrum of resistance to Thaiblast isolates by pyramiding multiple resistance QTLs. *Plant Breeding, 129*, 176–180.

Stoskopf, N. C., Tomes, D. T., & Christie, B. R. (1993). *Plant breeding: Theory and practice*. Boulder: Westview Press.

Strange, R. N. (1993). Organisms that cause plant disease: Their detection, identification and proof of their role as pathogens. In *Plant disease control* (pp. 31–61). Boston: Springer.

Strange, R. N. (2006). *Introduction to plant pathology*. New Delhi: Wiley.

Strange, R. N. (2013). *Plant disease control: Towards environmentally acceptable methods*. Boston: Springer, US, 354 pp.

Sucher, J., Boni, R., Yang, P., Rogowsky, H., Büchner, H., Kastner, C., et al. (2017). The durable wheat disease resistance gene Lr34 confers common rust and northern corn leaf blight resistance in maize. *Plant Biotechnology, 15*, 489–496. https://doi.org/10.1111/pbi.12647.

Thresh, J. M., & Owusu, G. K. (1986). The control of cocoa swollen shoot disease in Ghana: An evaluation of eradication procedures. *Crop Protection, 5*, 41–52.

Toojinda, T., Tragoonrung, S., Vanavichit, A., Siangliw, J. L., Pa-In, N., Jantaboon, J., et al. (2005). Molecular breeding for rainfed lowland rice in the Mekong region. *Plant Production Science, 8*, 330–333.

Ullstrup, A. J. (1972). The impacts of the southern corn leaf blight epidemics of 1970-1971. *Annual Review of Phytopathology, 10*, 37–50.

Van Bueren, E. L., Jones, S. S., Tamm, L., Murphy, K. M., Myers, J. R., Leifert, C., & Messmer, M. (2011). The need to breed crop varieties suitable for organic farming, using wheat, tomato and broccoli as examples: A review. *NJAS-Wageningen Journal of Life Sciences, 58*, 193–205.

Walters, D. R., Ratsep, J., & Havis, N. D. (2013). Controlling crop diseases using induced resistance: Challenges for the future. *Journal of Experimental Botany, 64*, 1263–1280.

Welz, H. G., & Geiger, H. H. (2000). Genes for resistance to northern corn leaf blight in diverse maize populations. *Plant Breeding, 119*, 1–4.

Withers, L. A. (1989). In vitro conservation and germplasm multiplication. In A. H. D. Brown, O. H. Frankel, D. R. Marshall, & J. T. Williams (Eds.), *The use of plant genetic resources* (pp. 309–334). Cambridge: Cambridge University Press.

Wolfe, M. S., & Barrett, J. A. (1979). Disease in crops: Controlling the evolution of plant pathogens. *Journal of the Royal Society of Arts, 127*, 321–333.

Wolfe, M. S., & Gessler, C. (1992). The use of resistance genes in breeding epidemiological considerations. In T. Boller & F. Meins (Eds.), *Genes involved in plant defense* (pp. 3–23). Wien: Springer.

Wyss, U. (1981). Ectoparasitic root nematodes: Feeding behavior and plant cell responses. In B. M. Zuckerman & R. A. Rhode (Eds.), *Plant parasitic nematodes* (Vol. III, pp. 325–351). New York: Academic.

Zadoks, J. C. (2001). Plant disease epidemiology in the twentieth century. A picture by means of selected controversies. *Plant Disease, 85*, 808–816.

Chapter 4
Synthetic Chemicals: Major Component of Plant Disease Management

Imran Ul Haq, Muhammad Kaleem Sarwar, Anjum Faraz, and Muhammad Zunair Latif

Abstract Direct protection using synthetic chemicals is one of the basic principles of plant disease management. Historical perspectives of using chemicals for plant diseases control Include application of effective methods for controlling plant diseases. Fungicides, Bactericides and Nematicides are applied through different methods such foliar, slurry, drench, paste etc.). Fungicides or Fungistatics, can be classified based on mode of action, usage and composition. Limitations of pesticide usage occur in plant disease management, due to health hazards and pesticide impact on the environment. Insurgence of fungicidal resistance in plant pathogens is also a significant threat. Efficacy of chemicals compounds is also affected by climate changes. Recent trends in the development and use of synthetic chemicals (broad spectrum and new chemistry fungicides) in plant disease control Consider a comparison between pesticides and alternative plant disease control methods, fungicide marketing policies and procedures.

Keywords Foliar fungicides · Synthetic fungicides · Viricides · Fungicidal resistance

4.1 Historical Prospective of Chemicals for Plant Disease Control

A number of chemicals that we are using today are being applied by farmers since 100 years ago, against pests causing serious losses in food productions (Rhoades 1963). It is hypothesized that organized agriculture started with wheat and barley cultivation about 5000 years BC, in the Middle East (Behrens 1957). It is likely and

I. Ul Haq (✉) · M. K. Sarwar · A. Faraz · M. Z. Latif
Department of Plant Pathology, University of Agriculture Faisalabad, Faisalabad, Pakistan

© Springer Nature Switzerland AG 2020

I. Ul Haq, S. Ijaz (eds.), *Plant Disease Management Strategies for Sustainable Agriculture through Traditional and Modern Approaches*, Sustainability in Plant and Crop Protection 13, https://doi.org/10.1007/978-3-030-35955-3_4

possible that these crops were infected with plant pathogens already in the same era. However, the cause of plant diseases remained a mystery for several millennia. Damage to plants was (and still is) closely associated with natural phenomena. It is believed that the first management attempts focused on weather conditions around the year, and that they continued in this way for centuries. This practice resulted in naming the disorders and diseases as Blight, Blast and Mildews by the majority of people. Rapid destruction of plants was termed as Blast/Blight while a slow and more visible growth appearance of the pathogen on the plant surface were termed as "mildews". Local farmers are still using these terminology (Mildew and Blight) for any disease that they experience in their fields or orchards. In the early ages of organized agriculture, people began to focus only to those diseases and disorders that were directly related to their interest as they directly affected them. Let's have a look at the commercial development of chemicals for crop protection from these humble beginnings until the modern era.

About 500 AD people in India used fumigants, obtained by animal (dog, cow) bones and cat excretions, on cucumbers. In the seventeenth century, microorganisms (fungi, bacteria) were already detected on diseased plants. However, their role in causing plant diseases remained unknown. In 1761, the role of copper sulphate in the reduction of bunt disease was observed by Schulthess. In 1807, it was assured that wheat bunt was caused by a fungus and copper sulphate was used for its control (Prevost 1807). This attempt resulted in suppression of fungal structures but was not accepted by the majority of people. When Irish famine (1845–1849) occurred, people named bunt as a cause but later on the epidemic was identified as due to attacks of a fungus, which at that time was termed *Botrytis infestans* (actually *Phytophthora* infestans). Attempts were made to control this disease by soil application of copper sulphate, lime and salt mixtures. However, these attempts failed. Unfortunately, foliar applications of these chemicals were not adopted. Meanwhile the concept of seed treatment against smut and bunt diseases was introduced. Later on, arsenic and copper sulphate were tried. In the eighteenth century adding lime addition was considered to reduce the phytotoxicity of copper sulphate on cereal seeds. Various products were developed and tested as fungicides. In nineteenth century, several attempts were made to exactly identify what may be considered as a true fungicide. First fungicides started with the application of elemental sulphur for the management of fruit crops diseases such as powdery mildews and few other diseases of grapevines.

In mid nineteenth century lime-sulphur solutions were used against powdery mildew of vines by Kenrick. The development of synthetic fungicides progressed rapidly in 1930s. In 1942, thiram, zineb and nabam (introduced in 1943) were followed by maneb (in 1955). In 1961 mancozeb (manganese and zinc) was introduced (Agrios 2005). Major chemical classes i.e. (phthalimides, guanidines, methyl benzimidazole carbamates, SBI morpholines and 2-aminopyrimidines) were introduced from 1945 to 1970, in an era dominated by the crop protection industry. During this period ornamental, vegetable and fruit diseases were focused. Experiments were conducted by foliar applications of newly introduced fungicides (maneb, ethirimol, benomyl and tridemorph) to control powdery mildew of barley throughout Europe (Russell 2005).

4.2 Basic Principles of Plant Disease Management with Synthetic Chemicals

The excessive use of synthetic chemicals in agriculture has been initiated in the last few decades and the major groups of chemicals were developed in the last 60 years (Martínez 2012). The role of synthetic chemicals in food security and safety has become more prominent as they promised economic and social benefits for global economy. Hence, these chemicals have become an important component of the global agrochemical business.

Losses caused by plant pathogens are the main issue in management of field crops and during postharvest storage. The use of chemicals is one of the most effective methods of controlling plant diseases, either in field or during storage, transportation and marketing. This includes the application of chemicals that directly inhibit the pathogens' growth, either upon plant surfaces or within host plant tissues. These chemicals either have adverse effect on the pathogens' life cycle or are completely injurious to them. This approach has been widely adopted due to its effectiveness in controlling the plant diseases.

Synthetic fungicides played a major role in crop protection by controlling fungal diseases across the world. This has been the most appropriate and quick method to control fungal diseases, particularly post-harvest diseases. In ancient times, most farmers were unable to assess the losses caused by fungi but in the changing scenarios of growing world population the losses have become unacceptable. That's why producers are still dependent on synthetic chemicals for managing fungal diseases. However, the need for selectivity, systemic and therapeutic properties in new fungicides has been increased. Fungicides penetration into host plant is necessary, either to eradicate established infections and to be redistributed throughout the plant. Therefore, they show selectivity towards pathogen.

Another approach considers the use of chemicals that do not target a given pathogen but the disease itself.

Depending on the kind of pathogen they affect, the chemicals are called fungicides, bactericides, antibiotics, nematicides, viricides.

Fungicides: used for killing fungi.
Bactericides: used for killing of bacteria.
Antibiotics: substances that are produced by one microorganism and used to kill or inhibit other microorganisms particularly bacteria, at low concentrations (Madigan et al. 2008).
Nematicides: used for killing nematodes.
Viricides: used for an antagonistic action towards viruses.

4.2.1 Fungicides

Although chemicals are applied to control a wide range of plant pathogens or managing stress sources, their role in controlling fungal diseases has been more prominent. Fungicides are broadly classified based on their chemical nature, their mode of action and usage. Since their discovery in 1807, fungicides are successfully used for disease management in agricultural crops (Leadbeater 2015), to control foliage diseases, for disinfection of bulbs, seeds, tubers, and soil or to eliminate the established infection within the vegetative plant material. Some chemicals are used for wound treatment, protection of fruits and vegetables while others are applied to control insect vectors, that spread diseases. The majority of chemicals are only effective before infection starts. They are effective only in those parts of the plant where they have been applied, and are unable to translocate or to be absorbed by the plants. In their majority they are used as foliar sprays or as dusting. Their effectiveness largely depends upon the ability to be absorbed by and act on the pathogen. However, they must have ability to be insoluble to avoid rain depletion, for a longer protection of the treated plant tissues. Fungicides must be capable to cover and protect the whole infected area and have good adhering ability, so their effect will lasts for a longer time. Moreover, they must be toxic to the pathogen but not to the plants and consumers. As they are used as protectant and not as curative, their application must precede the pathogen arrival or at least be performed before the establishment and germination of the pathogen into the host tissues. Fungicides are also infused or injected into high economic value tree species, such as elm trees, to reduce severity. These chemicals, however, need periodic applications for effective and long term prevention.

Fungicides inhibiting demethylation and sterol biosynthesis are of much importance, but must diffuse into the plant tissues to eliminate recently established infections. Earliest fungicides were inorganic compounds, which were based on simple elements or metallic molecules. Organic products (i.e. thiram and captan), developed and introduced into the market in early-mid-1900s, had a broad spectrum and were used either as protectant or contact fungicides. Protectant fungicides are effective against a wide range of fungal pathogens and protect the plant parts to which they are applied. In the early 1960s, systemic fungicides began to be developed. They have the ability to be absorbed by the plant surfaces without causing any damage to it, and to be transported to the infection site where they eradicate or control the infection. However, most of them are not fully systemic, as they have only one way of action. At present, only one fungicide, fosetyl aluminium, is truly systemic, being characterized by an upward and downward distribution.

4.2.2 Antibiotics

The role of antibiotics in plant disease management is very prominent. About 40 antibiotics of fungal or bacterial origin were screened during the 1950s for the control of plant pathogens (Goodman 1959). Since 1950, antibiotic have been in practice to control bacterial diseases of important ornamental plants, fruits and vegetables (McManus et al. 2002). However, they have limited applications in the management of bacterial and fungal diseases, as fungi and bacteria have ability to develop resistance against these chemicals. Historically, antibiotics have been widely used for the control of apple and peach fire blight disease (McManus and Stockwell 2001).

Actually, only a few antibiotics are commonly in use, or permitted, to control plant diseases. Since its introduction as a plant protectant in 1955, streptomycin is the most commonly used antibiotic, effectively applied against bacterial canker, bacterial spot of stone fruit, bacterial speck of tomato, and fire blight of apple and pear (McManus et al. 2002). It is also effective as seed dressing on cotton, tomatoes and beans. However, due to phytotoxicity its application, particularly on beans, has made its use limited. This antibiotic has been successfully used in the treatment of stem rot of *Dieffenbachia* cuttings, and in the control of other diseases (lethal yellowing of palms and pear decline) caused by phytoplasmas. Also oxytetracycline is considered as one of the most important antibiotics, as development of resistance by plant pathogens to this chemical is rare. It is used against bacterial pathogens (*Pseudomonas* spp., *E. amylovora* and *Xanthomonas* spp.) of apple, pear and various vegetables. Additionally, it is also effective in the suppression of lethal yellowing of elm trees and palms (McCoy 1982).

Gentamycin is used, in some Latin American countries, to control bacterial diseases of fruits and vegetables caused by *Ralstonia*, *Pectobacterium*, *Erwinia*, *Xanthomonas* and *Pseudomonas* spp. (McManus et al. 2002). In Israel, oxolinic acid is commercially used to control fire blights of fruits and other related plants (Shtienberg et al. 2001). Most of the commercially available antibiotics are fungitoxic. In Japan several fungi toxic antibiotics against bacterial (bacterial leaf blight) and fungal (rice blast) diseases of rice, fruits and vegetable crops, have been developed. Some antifungal compounds (cyloheximide and blatomycin-s) have been classified as antibiotics and are used against rice blast and some fungal diseases. Cycloheximide is used as a protectant or eradicant fungicide against powdery mildew and rust diseases, on various crops. However it is more suitable for woody plants as compared to herbaceous ones, due to its phytotoxic effects. In Europe, a less toxic antibiotic, griseofulvin, has been widely used to control powdery mildews and *Botrytis* sp. on greenhouse vegetables. Antibiotics used in agriculture are usually formulated as a powder, having 17–20% active ingredient, by dissolved or suspended in water to an adjusted final concentration of 50–300 ppm.

4.2.3 Nematicides

Nematicides are highly toxic and expensive and are used mostly on high return crops. They are fumigant or non-volatile compounds. Various nematicides are available for effective control of nematode pests of annual crops, but their use is justifiable only on high-value economic crops (Gowen 1997). Since their discovery about 50–60 years ago, several formulations and products have been available worldwide. However, the nematicide industry progressed slowly perhaps due to a lack of knowledge and awareness about nematode pests among farmers' communities, and the few reports available in the past on the economic losses caused by nematodes (Hague and Gowen 1987).

Development and use of nematicides can be divided into 3 periods, basically an ancient (1854–tile World War-I), a median (1919–1942) and a modern (1943 to onward) eras (Taylor 2003). In the second half of the nineteenth century, carbon disulphide was discovered as first synthetic chemical having nematicidal activity. The sugarbeet nematode *Heterodera schachtii* was reported in 1859 and attempts were made for its control by carbon bisulphide applications (Schacht 1859). This ancient period was dominated by carbon bisulphide (an insecticide acting against soil pests). Chloropicrin was also used as a nematicide after World War I. In England, chloropicrin was successfully used against soil pathogens including nematodes (Mathews 1919). In Hawaii, it minimized RKN densities in pineapple fields (Johnson and Godfrey 1932; Godfrey 1935). D-D (a mixture of dichloropropane-dichloropropene) was discovered in 1940s as effective in controlling soil populations of most plant parasitic nematodes (Sikora and Hartwig 1991). This discovery laid the foundation of the development of other nematicides. In 1943, the D-D mixture was used as a soil nematicide n pineapple crops (Carter 1943). In 1944, ethylene dibromide was evaluated as soil nematicide (Thorne and Jensen 1946).

In 1946, nematicides got a quick popularity on a commercial scale and their filed applications increased rapidly. McBeth and Bergeson (1955), reported the nematicidal activity of 1, 2-dibromo-3-chloropropane (DBCP). Furthermore, this chemical was less toxic and effective against nematodes of living trees, grape vines and citrus which can be applied before and after planting. In 1956, Vapam (sodium N-methyl dithiocarbomate dihydrate) was introduced in the market as a herbicide, fungicide and nematicide. The next nematicidal chemical in this series introduced in 1957 was V-C 13 (0–2, 4-dichlorophenyl 0, 0-diethyl phosphoro-thioate). In 1960s, a new generation nematicides such as organophosphates and carbamates was discovered, acting as contact nematicides. Many of them are systemic within plants. Oxamyl is the only systemic nematicide available as a commercial product. Other nematicidal soil fumigants developed were halogenated hydrocarbons and volatile compounds. Since 1960, nematicides got much popularity as their application was easy, and most of the formulations were granular. A high concentration of nematicides is essential for effective control of nematodes on the plant roots. However, it becomes more difficult to control nematodes through chemicals once they penetrate the host roots (Taylor 2003).

4.2.4 Viricides

Some viruses are important plant pathogens causing major economic losses in agricultural and horticultural crops. Chemotherapy against plant viruses has been developed about 50 years ago as a result of a series of incidental observations (Dawson 1984). In 1925, it was found that application of various plant extracts was useful in the inhibition of TMV. Several other chemicals, dyes, plant extracts, analogs azaguanuine and thiouracil were subsequently found effective in virus biosynthesis inhibition (Duggar and Armstrong 1925). In 1961, Quak work laid the foundation of applied chemotherapy against plant viruses. He applied thiouracil on virus infected leaves as a foliar application, and found that these chemicals were not only effective in the reduction of the virus infection, but could also prevent their replication without inhibiting cell metabolism (Quak 1961). In the mid 1960s the first antiviral drug was identified (Galasso 1984). It was proved that plant virus chemotherapy could only be possible if the system became non-infective through the process of aging. However, the role of chemotherapy is mostly curative rather than preventive, and these chemicals can be applied to systemically infected plants. As most crop viruses are seed borne, only seed chemical treatments can be suitable for protection and management. The chemicals used vs plant viruses are toxic and may act either as fumigants or disinfectants (Hansen and Stace-Smith 1989).

4.3 Application of Chemicals

Plant diseases have been of much importance from ancient times as they badly affected the global economy of the countries and have changed the fate of many nations. In this regard Irish famine due to late blight of potato, Bengal famine due to brown leaf spot of rice, chestnut blight and southern leaf blight of corn are the most prominent, as they resulted in severe economic and social losses. They have been responsible of losses not only in field crops and storage commodities, but also on landscape.

The core objective of chemicals application is to reduce losses caused by plant pathogens below a given threshold level, traditionally known as plant disease control. However, this term has been replaced by "plant disease management" and "integrated plant disease management". Correct and true identification of the disease and pathogen is necessary for the development of any management strategic plan. Application of chemicals is called chemotherapy, and is an integral part of crop disease management, worldwide. Some toxic chemicals are used as fumigants, sterilants and disinfectants for eradication of pathogenic organisms, as well as control of diseases that have an economic importance.

Over the years various chemicals have been in practice to control pathogens or diseases, formulated to target one or more organisms. These chemicals must be applied to such sites where they can come into contact with the pathogen, or from where they can be absorbed and translocate to the whole plant tissues, such as seeds, foliage or growing parts. Seeds can be treated with liquid or solid chemicals such as dusts to suppress or kill the mycoflora and nematodes, or limit the eggs hatching, either on the seed surface or in the surrounding soil. They are also used for seed dressings or pelleting. Liquid sprays are most commonly used for the foliage of growing plants. Application of chemical dusts is most common in dry areas.

The exposed plant surfaces (after mechanical operations) can be painted with suitable chemicals to restrict pathogen entry into the tissues. Dipping of propagative materials in the chemical solution is also a common practice. Internal tree infections can be treated by injecting chemicals into internal host tissue, by a hole in the stem.

4.4 Human Civilization and Fungicides

Human civilization and crop cultivation are linked together. Both have been threatened many times in the course of centuries, by severe plant disease epidemics as recorded in Biblical and early Greek and Roman times. Most recently, plant disease epidemic outbreaks such as the Irish (1846–1850) and Bengal famine (1943), left long-lasting and significant impacts on society and economy, with loss of about 5.5 million lives (Kislev 1982; Deising et al. 2002). With an ever increasing world population and a finite, decreasing per capita agricultural land area, the increasing pace in crop yields should continue exponentially. However, this trend has been constantly challenged by many biotic and abiotic stressors (Cleland 2013; Yildiz 2017). Plant pathogens induced crop decrease was estimated to reach 16–20% of yield. This food is lost and could be attained in case of pathogens exclusion. Losses due to fungi have the highest share, that may reach up to 80% (Savary et al. 2006; Moore et al. 2011). Plant diseases have severe detrimental impact on yields in almost any crop, and this situation could be worsen under favorable conditions if no control is applied. This is especially true in developing countries, having limited resources to manage crops. Human intervention is indispensable to get the desired yields by controlling disease and keeping their effect on yield potential at a minimum level (Shurtleff et al. 2018).

Human civilization has faced disease issues and acknowledged its importance since primitive times and had used more spiritual than practical, maybe not knowingly, methods of disease control. But starting from nineteenth to twentieth century more practical methods and chemicals were developed and used extensively. Simple plant protective agents, copper and sulfur, laid the foundation of today's billion dollar vast growing pesticide industry, based on complex chemistries with various mode of actions against different groups of fungi (Hewitt 2000; Deising et al. 2002). Fungicides may control a disease during the developmental crop stages of and increase its productivity. They may also increases its market value by saving the produce from spots and blemishes, in field and storage conditions (McGrath 2004). Fungicides as pesticidal agents may be chemical or biological, and are marketed to kill or inhibit the fungal growth or spore germination. Modern fungicides instead of killing, inhibit fungal development for a specific period of time by interfering with routine metabolic processes.

4.5 Classification of Fungicides

Fungicides can be classified into different classes/groups on the basis of their chemical structure, mode of action, type of crop, mode of application, method of absorption/mobility etc.

4.5.1 *Mode of Action*

To delay resistance against fungicides and to avoid detrimental impacts on plant growth, it is effective to use specific fungicide or mixture of fungicides, with diverse mode of actions. These may target fungal growth by inhibiting spore germination, colonization, reproduction, at any or all stages of disease development. Various biochemical processes or structures of target fungi are suppressed or disturbed by fungicides. Various unknown and known modes affect cell membranes, cell division, synthesis of proteins, respiration, signaling etc. (Hewitt 2000; Mueller et al. 2008; Yang et al. 2011).

(a) **Cell membrane components**

Organization and function of porous cell membrane in fungi is similar to that of higher eukaryotes, including: selectivity to ionic conductivity and organic molecules, signaling, adhesion, shape maintenance, protection of cell contents from noxious substances etc. Ergosterol (ERG), a special sterol membrane lipid and an important component of the cell membrane biogenesis, derived its name from ergot, the common name of *Claviceps* spp., with a critical role in the fluidity and permeability of the fungal cell membrane (Alberts et al. 2002; Iwaki et al. 2008). Being a special fungal membrane component, alternative to cholesterol in animals, ergosterol serves as an effective and important target for fungicides, disintegrating plasma membrane or disturbing ERG biosynthesis. Sterol inhibitors are the most effective and broadest used fungicides, that can act as protectant as well as suppressers of fungal growth by affecting various developmental stages. Even after decades from their discovery, sterol biosynthesis inhibitors account for a 1/fifth share in global market, due to their broad spectrum and mobility within plant. Among them, triazole acts by dismantling the integrity of the cell membrane with an ergosterol-specific site of action. Due to its reserve in spore, ERG specific fungicides have no effect on germination and germ tube development of spores. Aromatic hydrocarbon (AH) fungicides, supposedly, disturb mycelial growth by interfering with the biogenesis of lipids in the cell membrane, whereas De Methylation Inhibitors (DMI) disturb different reactions in the ERG production pathway, by inhibiting the enzyme demethylase, necessary for ergosterol biosynthesis, eventually killing the fungus (Hewitt 2000. Fishel 2005; Mueller et al. 2008; Vagi et al. 2013; Yang et al. 2015).

When treated with AH fungicides such as dicloran and etridiazole, lysis of fungal cell membrane occurs by destruction of the membrane linoleic acid and by phospholipids hydrolysis, respectively (Radzuhn and Lyr 1984). Apart from impact on plasma membrane, certain fungicides, such as acriflavine, affect intracellular membranes such as the mitochondrial membrane, by reducing its permeability. This causes an imbalance of the proton gradient across the membrane, which results in the reduction of ATP synthesis and ultimately in the fungal cell death (Kawai and Yamagishi 2009).

(b) **Signal transduction**

In fungi, cellular signaling is mainly mediated by the Mitogen Activated Protein (MAP) kinase pathway, necessary for responding to environmental stimuli and intracellular signal transmission. Fungicides affecting osmotic signaling retard growth and differentiation in fungi by interfering with spore germination and mycelial growth (Yoshimi et al. 2005). Fungicides affecting cell membrane and its components also disturb signal transduction pathways, which take place on the cytoplasm and plasma membrane interface (Yang et al. 2011). Two group of fungicides, phenylpyrroles (PP) and dicarboximides, although characterized by a different structure, target the same osmotic signal transduction pathway in fungi. Due to their effectiveness, they have been used worldwide against a range of phytopathogenic fungi (Tanaka and Izumitsu 2010). PP fungicide fludioxonil interferes with the osmoregulatory signaling in *Botrytis* spp. by inhibiting spore germination and germ tube extension (Kim et al. 2007; Rosslenbroich and Stuebler 2000). Iprodione, a dicarboximide group fungicide, causes malfunctions during the sterol biosynthesis, lowers growth rate, mycelium elongation and disturbs hexoses and chitin productions (Ochiai et al. 2002). Both groups primarily target the histidine kinase pathway, in which improper signaling finally causes abnormal phosphorylation of MAP kinase. This is involved in the expression of genes responsible for hyperosmotic environmental adaptations (Hagiwara et al. 2007; Vargas-Pérez et al. 2007).

(c) **Respiration**

Fungal respiration is inhibited by various groups of fungicides having different modes of actions. In fungi, the mitochondrial respiratory complex I transports electrons to ubiquinone from the reduced form of NADH (Joseph-Horne et al. 2001). Complex I inhibitors disturb respiration by retarding NADH oxidation-reduction activity. Diflumetorim, an effective C1 fungicide, control powdery mildew and rust disease pathogens in ornamental plants by targeting an oxido-reductase (Fujii and Takamura 1998; Tomlin 2006). Oxidation of succinate and its further coupling with ubiquinone is carried out by respiratory complex II system, which is translated by a complex of four SDH genes (Bullis and Lemire 1994; Daignan-Fornier et al. 1994; Joseph-Horne et al. 2001). When targeted by commonly used complex II inhibitors such as boscalid, carboxin and flutolanil, SDH function in electron transport and tricarboxylic (TC) cycle are disabled, which causes reduced respiration and fungal growth (Motoba et al. 1988; Matsson and Hederstedt 2001; Spiegel and Stammler 2006). SDH inhibitors effectiveness against diseases is testified by tremendous yield increase reported (Smith et al. 2008; Bittencourt et al. 2007).

Compex III of mitochondrial respiration is inhibited by fungicides by inactivating cytochrome bc1 at Quonine outside and inside (Qo and Qi) sites (Gisi et al. 2002). Qo and Qi inhibitor fungicides share the same target enzyme, but have distinct binding sites as name indicates. These fungicides hinder the energy producing potential of the fungus by limiting respiration, which results in death (Hewitt 1998; Mueller et al. 2008). Newest, widely used and important strobilurin fungicides belong to QoI family (Vincelli 2002).

While most fungicides under the umbrella of respiration mode of action target enzyme complexes, some hinder respiration process through other targets of energy conversion as uncoupling and ATP synthase, by upsetting oxidative phosphorylation. Fluazinam has an unusual uncoupling activity by joining with glutathione, interrupting ATP synthesis and disturbing a number of metabolic pathways (Guo et al. 1991; Brandt et al. 1992).

(d) **Amino acid and protein synthesis**

Proteins are synthesized by long or short chains of amino acids which perform many vital roles in cell functioning. Functions of fungal proteins in fungi include cell wall strengthening and structure, sensing changes in local environment, signal transduction, biochemical reactions etc. Cell wall synthesis enzymes of fungi are specific and are used as target for fungicides (Lodish et al. 2000; Hall et al. 2013). Fungicides affect protein synthesis on various stages i.e. initiation, elongation, and termination steps of protein synthesis or by disturbing genes of methionine biosynthesis, as in anilino-pyrimidines (AP) fungicides. Streptomycin, having both fungicide and bactericide properties, affect synthesis of amino acids, translation process and also disturbs the 70S ribosome in *E. coli*, by adding isoleucine in polypeptide chains. Oxytetracycline disturbs amino-acyl complexes of tRNAs on the ribosome and ultimately retards bacterial community by affecting ectoenzyme activity (Old and Gorini 1965; Halling-Sørensen et al. 2002; Carr et al. 2005).

(e) **Mitosis and cell division**

Cell division is imperative for continuation of life, necessary for growth and reproduction in multicellular and for reproduction in unicellular organisms. Different groups of fungicides cause death of fungal pathogens by interrupting cell division and mitosis, at various steps (Seiler 1975; McCarroll et al. 2002). Methyl benzimidazole carbamates inhibit processes in which tubulin monomers react together to form microtubule polymers (Gupta et al. 2004; Davidse 1986; Koo et al. 2009). Three compounds of benzimidazoles, namely carbendazim, benomyl and thiabendazole, inhibit polymerization and proliferation by targeting recombinant β_2 tubulin up to 93.5%, 92.6% and 81.6% respectively. They are known for inhibiting mitosis in *Fusarium* spp. Benzimidazole targets the microtubules forming process and is ineffective against polymerized cytoskeleton or spindle microtubules (Zhou et al. 2016). Cytoskeleton microtubules perform vital functions in the cell, whereas spindle microtubules arrange chromosomes at the metaphase plate in a linear fashion. Applications also cause chromatid loss due to instability between spindles and kinetochore connections (Rathinasamy and Panda 2006).

(f) **Nucleic acid synthesis**

Among fungicides affecting nucleic acid metabolism and synthesis, phenylamides (PA) group include several effective fungicides with RNA polymerase I as action site. Metalaxyl affects the RNA chain, inhibiting the uridine incorporation. RNA synthesis is more affected than DNA (Fisher and Hayes 1982; Sukul and Spiteller 2000). Interference in systemic activity of RNA transferase occurs during the nucleic acid synthesis, by affecting the uridine transcription process. Activity of the enzyme adenosine deaminase is inhibited by ethirimol, belonging to hdroxypyrimidine group. It inhibits the activity of nucleotides as adenine and metabolic agents as inosine (Hollomon and Chamberlain 1981). Ethrimol also disturbs the nucleotide pool balance by increasing the adenine salvage pathway enzyme phospho-ribosyl-transferase. It also inhibits the activity of adenosine deaminase, which results in ceased inosine synthesis and impaired nucleic acids production (Brown and Simpson 1994). DNA/RNA synthesis is inhibited by fungicide of the heteroaromatic group such as hymexazol. When applied it reduces the thymidine incorporation ratio as compared to uridine in RNA, reducing colony growth by targeting DNA synthesis (Kamimura et al. 1976).

(g) **Multisite activity**

Multi-target fungicides are widely used in agriculture but may have detrimental impact on other non-target organisms. Chlorothalonil, due to its multi biochemical action sites, is considered as a successful fungicide, reducing enzymatic activity and blocking the transformation of glutathione into its various structures (Chen et al. 2001). Another broadly used fungicide, mancozeb, shows a multisite activity inhibiting metabolism of targeted cells (Cycoń et al. 2010). Other examples of multisite fungicides are thiram, captan, propineb, maneb and many copper-based compounds (Milenkovski et al. 2010).

4.5.2 *Classification Based on General Uses*

(a) **Seed and planting material**

Discovery of carboxin, first systemic fungicide, was a milestone in the control of seed borne diseases. It is effective against surface and deeply penetrating pathogens, especially in controlling loose smut, a major disease of wheat and barley. Due to its active redistribution ability within the plant, carboxin replaced less effective products previously applied in seed treatments such as the organomercurial fungicides (Kulka and Von Schmeling 1987; Maude 1996; Klittich 2008). Systemic fungicides protect plants in its early, tender age by reducing the spread of fungal compounds toxic to young plant parts such as cotyledons, leaves and seedlings. Systemic fungicides are widely used in the seed treatment market and are more effective in controlling seed borne diseases in comparison to non-systemic fungicides. The latter are still used as successful protectants. Seed treatments with captan and diathane M-45 increased the germination and reduced the seed associated mycoflora effectively, as compared to control (Mahal 2014). Protectant fungicides i.e. iprodione and vinclozin, can also effectively eradicate seed borne pathogens such as *Sclerotinia* sp. from infected sunflower seeds, equally to systemic fungicide such as benomyl (Herd and Phillips 1988). The effectiveness of protectants, equivalent to that of systemic fungicides, is due to the location of the seed borne pathogens spores or mycelium, that may be mostly found on the seed surface, in superficial tissues and/or in the pericarp. Only few pathogens reside in the inner seed parts as endospore or in the embryo.

Pathogens which develop close to the seed surface, such as *Sclerotinia*, *Fusarium*, *Alternaria* etc., are controlled by both protectants and systemic fungicides. For protectants to be effective it is needed an ability to reach the internal seed tissues (Maude 1996). The chemotherapeutic effects on treated seeds is maximum as the seed and inoculum volumes are minimal, as compared to other plant parts. Due to these concentration factors the active ingredient remains at high levels for a long period of time, and ultimately increases the chances to eradicate the seed borne pathogens. Planting materials such as cuttings, bulbs, tubers etc., that also are infected by severe pathogens, may remain symptomless for a longer period of time. This makes it difficult to formulate a disease-free crop. Processes and factors in the control of planting material pathogens are similar to those applied for seed borne pathogens, except the increased volume as compared to seed, the higher inoculum amounts, and the higher concentration and exposure time required bu. the fungicide for pathogen control (Ivic 2010). Treatments of *Phytophthora*-infected potato tubers with the systemic fungicide thiophanate methyl plus mancozeb as a preventive measure before infection, increased plants emergence (Inglis et al. 1999). Similarly, a reduced effect of the fungal pathogen *Phaeomoniella* sp. was observed in benomyl-treated grapevine cuttings, when compared with control. Fungicide treatment is also effective for planting material. Myclobutanil-treated chrysanthemum cuttings showed no symptoms of white rust as compared to the 90% prevalence found in control (Bonde et al. 1995). Strawberry seedlings infected with *Colletotrichum* when treated with prochloraz, propiconazole and difenconazole treatment, showed lower mortalities in comparison to the 80% rate found in the control (Freeman et al. 1997).

(b) **Soil fungicides**

Soil borne pathogens show their presence through aboveground symptoms, that become visible long after the infection occurs. This causes considerable losses to root or crown parts. Symptoms of soil borne pathogens are complex and nonspecific, difficult to diagnose accurately and are challenging for control through fungicides. Few systemic fungicides show the ability of downward translocation, such as fosetyl-Al, that effectively controls the stem canker of avocado caused by *Phytophthora* spp., when soil drenched. Exact doses for soil treatments are difficult to calculate, as plants uptake small amounts of the fungicide, as compared to the foliar application.

Fungicides can be used before and after cultivation begins. Most of the time they are preventive but may also be curative. The amounts of a.i. applied with soil treatments is relatively high, because of the crop biomass that usually already develops, prior the aboveground symptoms are visible (El-Hamalawi et al. 1995; Ivic 2010). For example, Phytophthora root and crown rot of apples in plants early developmental stages were controlled by matalaxyl soil drenching. Soil drenched apple trees remained alive as compared to control trees which died due to the rot (Utkhede 1987). Carbendazim, a benzimidazole fungicide, when used as soil drench recovered 15–20 years old apples trees from *Rosellinia* root rot (Gupta 1977). Pre- and post-sowing soil drenching with carbendazim, carboxin and thiram was investigated against wilt and rot diseases of cotton. Pre-sowing drenching was most effective as compared to post-planting (Chauhan et al. 1988). Tridemorph when applied as a soil drench gave promising results against *Rosellinia* rot of rubber, cocoa and mulberry (Mappes and Hiepko 1984). *Phytophthora* root rot has been studied extensively as compared to other similar diseases. However, high efficacy of fluazinam was also observed on *Rosellinia* white root rot of grapevine, after soil drench at different concentrations, even when the soil was highly infested (Kanadani et al. 1998; Hoopen and Krauss 2006). Three fungicides, azoxystrobin, trifloxystrobin and kresoxim methyl, were tested for their root uptake ability, in pearl millet against downy mildew. Azoxystrobin showed a systemic activity to some extents, while the other two lacked root uptake, being not systemic (Sudisha et al. 2005).

Methyl bromide (MeBr) has been used extensively, until its ban, in strawberry nurseries for soil borne pathogens such as *Phytophthora*, *Pythium*, *Rhizoctonia*, *Fusarium* and *Verticillium* spp. in Mediterranean countries, such as Italy and Spain. Chloropicrin, dazomet and metam sodium are newer soil fumigants being used and found to be effective (De Cal et al. 2005). Di-nitrogen tetraoxide, as compared to methyl bromide, is not phytotoxic and is less dispersed in the atmosphere. These properties and its effectiveness in killing all fungi present in infested soil within 10–20 min in pre-planting soil fumigation make it a better alternative to MeBr (Tadmor et al. 2005). Paraformaldehyde when also used as a substitute for MeBr, reducing soil pathogens from 4000 to 40 colony forming units/g increasing seed germination (Al-Khatib et al. 2017). Treatments of infested soil and potting media with metam sodium at the high concentration of 1.0 ml/L eliminated all pathogens including *Phytophthora*, *Pythium*, *Thielaviopsis* and *Cylindrocladium* spp. effectively (Linderman and Davis 2008).

(c) **Foliar fungicides**

Fungicides are efficient in controlling diseases if applied prior to pathogen establishment in host or before symptoms appear. Systemic fungicides having curative effect can eradicate the pathogen even after its establishment in the host tissue. However, as for seed treatments, the time of application after pathogenesis is very critical, in order to control foliar diseases, especially when the mycelium grows deeper in tissues and reaches high densities (Ivic 2010). Effectiveness of strobilurin and trifloxystrobin, decreased from 89 to 60% when applied 24 and 96 h after tomato inoculation with *Cladosporium fulvum*, respectively. The decrease of the tomato leaf mold disease was also significant at various concentration of trifloxystrobin (Veloukas et al. 2007). Effectiveness of myclobutanil as foliar treatment against white rust of chrysanthemum was evaluated at different periods of post infection. Susceptible plants resulted in few pustules as compared to the untreated control, when exposed to the inoculum and sprayed with myclobutanil after 10, 15 and 20 days. Sprays applied 5 days before inoculation reduced the infection, but did not prevent it (Bonde et al. 1995). High level of pathogen inoculum reduce the ability of the fungicide to control a disease, especially in post infection, when the pathogen development and sporulation are relatively high, due to favorable environmental conditions.

In an experiment to control blossom blight caused by *Monilinia* fructicola in sour cherry, nine different fungicides were tested. Four of them namely: tebuconazole, propiconazole, iprodione and vinclozoline were effective when sprayed at inoculum concentration of 5×10^3 conidia/ml, 48 hours after infection. When the inoculum concentration was increased to 5×10^4, the disease reduction was less, as compared to control (Wilcox 1990). An inoculum dependence was also observed in black rot of grapes, when inoculum concentration increased to 1×10^6 from 2×10^4 conidia/ml, showing a much reduced effectiveness of azoxystrobin (Hoffman and Wilcox 2003). Foliar efficacy of three strobilurins were evaluated against downy mildew of pearl millet. Although all three a.i. were highly significant, only azoxystrobin appeared the most effective, with a 91% disease reduction, as compared to control (Sudisha et al. 2005). Azoxystrobin and pyraclostrobin were evaluated against strawberry leather rot disease. Protectant and curative effects of both fungicides were tested and found to be at same level, when plants were inoculated with 10^5 zoospores/ml of *Ph. cactorum* (Rebollar-Alviter et al. 2007).

(d) **Injection**

Vascular wilts, wood rot and cankers are difficult to control, due to the same reasons given for root and crown rot diseases. In case of vascular wilt, the pathogens can be present in various plant parts through the xylem vessels. Wood rotting fungi could cause serious damages to lignin and cellulose before the trees start to show symptoms. Due to the various ways of entry of vascular wilt and wood rotting fungi, such as through soil, wounds or insect vectors etc., their penetration period is hard to determine. These disease are difficult and expensive to control as can develop throughout the year, in particular cankers which are, however, easier to detect. With development of systemic fungicides and new method of use these disease can be managed, as shown by trunk injections (Ivic 2010). Cocoa stem canker caused by *Ph. palmivora* was effectively controlled by trunk injection with potassium phosphonate, as compared to ridomil spraying application and control (Guest et al. 1994). When metalaxyl and fosetyl Al were used as trunk paints, they effectively reduced *Phytophthora* trunk rot in peach (Taylor and Washington 1984). Thiabendazole injected in trunk was effective in managing Dutch elm disease. Thiabendazole and propiconazole are also effective in controlling the disease, but only when it is not too severe (Lanier 1987; Scheffer et al. 2008). Propiconazole injected in oak trees showed reduction in crown losses, as compared to control, Disease prevalence was reduced from 100 to 41%, and the results were more effective when plants were treated prior to symptoms appearance (Appel and Kurdyla 1992). Vines injected with cyproconazole against trunk disease esca showed a significant control of the disease, and increased production (Calzarano et al. 2004). Symptoms of esca were reduced when the trunk was injected with thiabendazole, propiconazole and difenoconazole (Dula et al. 2007). A similar control was observed when tetraconazole, penconazole and flusilazole were used (Di Marco et al. 2000). Apple trees trunk injected with prohexadione carboxylic acid against fire blight disease, caused by *Erwinia amylovora*, showed a reduced primary infection at blossom (Düker and Kubiak 2011). Ash dieback, a serious forest problem, was effectively controlled by thiabendazole and allicin trunk injections, which reduced necrosis (Dal Maso et al. 2014). Single injection of phosphites against apple scab showed significant and effective result as compared to propiconazole and penthiopyrad (VanWoerkom et al. 2014).

4.6 Recent Trends in Development of Synthetic Chemicals

Synthetic fungicide with novel site of action with low risk of resistance development have a key role in controlling well established plant pathogens. Important fungicides are those which can reduce disease spread in modern agriculture, when the same variety is grown on a large area with a high risk of epidemics development. Fungicides with systemic, curative, known mode of action and long term control are preferred. Chemicals which have a major share in the global fungicide market are at medium to high risk as concerns resistance development. Focus is those which are environment-friendly and effective at low doses. Trend shifted from multisite and site-specific chemicals to novel fungicides of various classes which are ecofriendly, due to low doses as compared to earlier products (Kumar and Gupta 2012). Fungicides with various chemistries and modes of action against a broad range of diseases have been introduced. These new generation of chemicals are safer and selective, mostly having a specific, single action site, with high potency of disease control (Leadbeater 2012). Focus is on the development of efficient chemicals with improved formulations that are environmentally safe, proceed from a natural source, with a low dosage and a reduced number of treatments (Nabi et al. 2017). As pathogens can resist to different chemicals having the same mode of action, they can be effectively controlled by compounds targeting a different action site. Focus is on finding different mode of actions. For control of ascomycetes nine mode of actions have been made available, which ensures plant disease control with a minimum risk of resistance development (Hollomon 2015a, b).

4.7 Broad Spectrum and New Chemistry Fungicides in Plant Disease Control

In spite of the broad range of fungicides available on the market, innovative chemicals having novel and robust modes of actions are needed. New chemistry fungicides discovered with available or new mode of actions are necessary especially for *Pythium* and *Fusarium* soil borne diseases, and bacterial and, possibly, viral diseases as these are a continuous challenge for crops. As discussed earlier, resistance management and control of adapted plant pathogens is effectively performed by fungicides having novel mode of actions, which are important because of their systemic and curative capability, and longevity (Leadbeater 2015). Among the 57 mode of action groups known thus far, the major market share, almost 70%, belongs to few groups. Among them, some fungicides with a high to medium induced resistance risk have more share, as compared to low resistance risk fungicides (McDougall 2014). This shows that there is great need for a continued availability of diverse and effective mode of actions in the market for resistance management and effective plant disease control.

Contact or systemic fungicides are simultaneously effective against a range of pathogens belonging to different classes of ascomycetes, basidiomycetes, oomycetes etc. When they also act on different hosts they are called "broad spectrum fungicides". These can manage disease when the causal organism is uncertain or when it derives from a complex of pathogens, involved in its development. Mancozeb, a dithio-carbamate introduced in 1962, is an important broad spectrum fungicide registered and used in more than 120 countries worldwide. Introduced in late 1970s and still playing major role in fungicide market, in 2004 it was the primary active ingredient with a maximum sale. Even in 2007 mancozeb was second with 500 million US dollar sale after tebuconazole, another important fungicide. Mancozeb development, commercialization and use on different crops and diseases resulted in various formulations. If co-formulations are included, used to further broaden its spectrum, then this sale figure in 2007 reached 740 million US dollars (AgroSciences 2008). Mancozeb is under review in the EU due to reproductive and developmental toxicity (Runkle et al. 2017).

In agriculture, azoxystrobin, a model broad spectrum member of the strobilurin group, played a significant role with registration on more than 80 crops, in various countries. Key for such appreciation and popularity is its effective action against four major fungal classes of ascomycetes, basidiomycetes, deuteromycetes and oomycetes. Almost all strobilurin members, i.e. trifloxystrobin, kresoxim methyl and metominostrobin, have a broad spectrum activity with varying levels of control. Trifloxystrobin and kresoxim methyl provide a moderate and a poor to moderate control against oomycete and basidiomycete diseases, respectively. However, metominostrobin is exclusive against oomycete diseases on rice and turf hosts (Bartett et al. 2001, 2002). Benzovindiflupyr, a broad spectrum chemical, is effective against apple, grapes and strawberry diseases, applied alone or in combination with other fungicides (Ishii et al. 2016). When compared to boscalid against *Botrytis* and *Alternaria* spp., it was found to be promising in reducing conidia germination, and was also effective against *Colletotrichum* spp. (Vega and Dewdney 2015). Strobilurins outperform all other fungicides in the market share except DMI. Data indicate that fungicides belonging to new chemistries with an excellent performance in disease control are demanded and accepted by farmers all over the world, at least where it was registered. This is also due to the reduced performance of previously applied fungicides and resistance development (Morton and Staub 2008).

4.8 Fungicide Market, Policies and Procedures

The world population growth is increasing the demand of fungicide and of disease-free crop productions and food security. These are driving force increasing the fungicides demand globally (McDougall 2018a, b). Introduction of novel compounds targeting respiration, cell and cell membrane components, signal transduction etc. have effectively managed plant diseases threatening many crop productions. Despite the effectiveness of old fungicides, there is a need to develop more classes of fungicides, as site-specific and systemic novel compounds, in order to tackle resistance risks (Nabi et al. 2017). Introduction of new chemistries in the market is necessary for a better and continuous management of diseases. This is a lengthy process involving various time-requiring steps, including research, formulation, trials, registration etc. all requiring huge economic efforts.

In 2010, the time required from new product development to first sale was almost 10 years, involving around 260 million USD. This effort, powered by extensive studies, is required for: (i) research and science to find an active ingredient, (ii) formulation, (iii) safety to humans, other organisms and environment, (iv) optimization of production so that a newly introduced chemical can maintain itself in the market for longer period of time, ensuring safety to consumers. In recent times focus has been given to chemicals which are highly effective at low doses and to development of new application methods. Finding a new active ingredient is not the only requirement to develop a new fungicide. In general, one a.i. out of ten is accepted, the others being rejected due to many obstacles on the way to marketing. Over the past few years many products have been introduced in the world market, some of them, such as DMI and stobilurin-type, are confined to local markets (McDougall 2010; Leadbeater 2015). Cost requirement of a new product increased by 55%, from 184 to 286 million USD, since the start of this century. This increase is due to shifts of priorities since 1960s. Initially the focus was on yield increases and disease control, now it is on the fungicide efficacy, prior to marketing. Recently, much focus has been given to human and environment which are at risk, due to unsafe chemicals. Extensive research and development, strict registration policies and other regulatory decisions increased time requirement from 8.3 to 11.3 years (Carson 1962; McDougall 2018a, b).

The sale of crop protection chemicals in 2014 was 56.7 billion USD, corresponding to almost 73% of total agribusiness. Share of fungicides in crop protection sales was 25.9%, with 14.7 billion USD. In 2017, crop protection sale was 57.7 billion USD which is expected to increase with a compound annual sale growth (CAGR) of 5% to 77.3 billion USD within 5 years (McDougall 2014; Research and Market 2018). Fungicide sale showed an increasing market trend and is expected to reach 17.58 billion USD from 2017 to 2022 (List 2018). According to Statista (2016), fungicide market in 2023 will reach 21 billion USD. Almost 90% of the global agribusiness sale is achieved by 10 leading industrial companies. Agribusiness, especially the crop protection market, is facing various challenges, ranging from increasing food demand to resistant risk, which are faced by various existing popular chemicals worldwide. These can be tackled by heavy investment in research and development departments of the pesticide industry. Outcome of this heavy investment in RD will be the invention of new sustainable chemistries, with less environmental impact (Maienfisch and Stevenson 2015).

The first comprehensive law regarding pesticide distribution in USA was passed in 1947 (Federal Insecticide, Fungicide, and Rodenticide Act, FIFRA). It required labelling and product registration with USDA, and was mostly concerned about a chemical effectiveness, rather than its harmful impact on humans or the environment (Rohrman 1968). The present act is still known as FIFRA, although there have been major changes in the law since then, especially with the FEPCA amendments in 1972. For convenience, we will refer to the pre-1972 version as the "Old FIFRA". This law in its original form was just a formality. In 1964 few revisions in the registration process were added to authorize the Secretary to cancel the registration of new or existing chemicals. In the late 1960smany environmental groups were active against over use of harmful agricultural chemicals such as DDT, Aldrin, Mirex etc. (Case et al. 2011; Schierow and Esworthy 2004). Main purpose of FIFRA is the registration of chemicals, from manufacturing to import. A pesticide has to be registered with the U.S. Environmental Protection Agency (EPA), that requires data from a many years study, which needs a significant economic investment, as discussed earlier. Registration was given for a limited period of 5 years. After that period additional data had to be provided for further sale and marketing of the product. This timing was extended in 1996 to 15 years, however products registered earlier had to face new strict standards which, according to the pesticide industry, discourage RD and new chemistries research (Miller 2014).

USA shifted regulatory pesticide powers from USDA to EPA in the 1970s, during the transition towards a socially derived regulatory model. Brazil, in 1990s adopted a new and unique model of pesticide regulation, by forming a troika of agriculture, health and environmental ministries with predominant economic interests rather than checking socio-environmental issues. A pesticide regulation framework was adopted recently in 2011 is by the EU, which has prevalent concern of socio-environment regulations (Pelaez et al. 2013). Different countries impose different pesticide regulations, including limits of residues in food, prerequisites for product registration and restrictions on usage. In USA, there is a specific limit of pesticide residues allowed in a given crop which are properly regulated. This means that the amounts of pesticide to apply may vary from country to country, for the same crop. Food or other organic commodities imported are subject to pesticide regulations of the accepting country. For USA any pesticide or chemical, subject to import or export, must be registered with EPA, and an EPA approved label is required before export. Chemicals manufactured in the USA, but only for export purposes, may not be registered with EPA, but for sale and use within the country it has to go through state laws more stringent than federal regulations (NPIC 2018).

4.9 Fungicide Resistance and Plant Pathogens

As discussed previously, earlier inorganic fungicides such as those based on sulphur and copper, (most popular among them was the Bordeaux mixture), were used extensively until the discovery between 1940s and 1970 of organic fungicides, that were mostly multisites, with a broad spectrum (Bernard and Gordon 2000; Morton and Staub 2008). Site-specific fungicides, benzimidazoles, were discovered in late 1960s. A few years after their discovery resistant plant pathogens were observed, along with a decreased fungicidal activity. *Botrytis cinerea* was the first plant pathogen for which resistance to fungicides was described, raising increasing awareness on this major issue.

Chemistry and MOA of fungicides, along with biology and mode of fungus reproduction, are the main factors involved in the insurgence of resistance (Brent and Hollomon 1998). Short life-cycle and abundant sporulation of *B. cinerea* make it a high risk resistant pathogen. Risk increases when the conditions are favorable for grey mold and fungicide applications are frequently repeated (Brent and Hollomon 1995). *Botrytis cinerea* as model organism developed both target site-specific and efflux transport resistance. As a result, strains having multi-resistance have been observed. This erosion of efficacy is a major threat to *B. cinerea* management. The same phenomenon of resistance development can be also observed in other fungal pathogens (Hahn 2014). If a favourable mutation occurs in the encoding gene of a fungicide target protein, it will help the fungal cell to survive. Target site resistance is most important, as it is observed in all site specific fungicides. This leads mostly to cross resistance, towards fungicides having the same MOA (Cools and Fraaije 2008).

Product having MOA of the same type may interact differently from each other if particular changes occur at specific sites of action. The outcome may be a range of resistance levels with a further evolution. Different mutations at the target sites, alone or in combinations, will result in different patterns of resistance towards the same group of fungicides (Parker et al. 2011). Overexpression of target gene, as in the citrus and apple scab pathogen, and of fungicide efflux transporters are other major resistance types found in fungi (Ma and Michailides 2005). Each fungicide class has a specific resistance risk behavior. Phthalimides, copper-based fungicides, dithiocarbamates, mancozeb, thiram etc. have rarely faced the risk of resistance insurgence. On the contrary strobilurin, benzimidazoles such as benomyl, metalaxyl, pyraclostrobin etc., face high resistance risks after 2–10 years of market introduction. Cross-resistance correlates with all QoI fungicides, targeting Qo center, which supports the idea that resistance occurs when changes occur at the target site or the metabolic pathway related to it. A site which has resistance against a specific MOA will then show resistance against all fungicides having that same mode of action, either in use fungicide or new The risk of resistance development is in effect higher for site-specific fungicides, as compared to multi-site fungicides (Brent and Hollomon 2007a, b).

Contrary to the site-specific type, fungi can develop efflux transport resistance against different groups of fungicides having different MOA. Efflux transporters have a key role in fungicide cross resistance of (Reimann and Deising 2005). The behavior of efflux mechanism resistance towards several fungicide groups is also known as MDR (multi-drug resistance) and has been mostly observed in laboratory assays (Kretschmer 2012). Pathogen exposure to the fungicide significantly enhances the resistance risk, at a dose of chemical applied which should be sufficient to kill or inhibit the pathogen wild type. When the pathogen is exposed to higher doses, especially when these are applied for eradication, it starts a survival competition favoring the fittest within the population and increasing the overall resistance risk. To manage this threat to fungicide efficacy, low dose rates with a steady use for long period of time appear effective (Ishii and Holloman 2015).

Despite the significant understanding of various resistance development processes, and of their implication of management strategies, problem may always occur. Previously established anti-resistance strategies, with the exploration of new mode of actions within the legislation umbrella, may allow a success if new and old strategies are combined successfully as in IDM systems. Activation of host resistance through systemic acquired resistance (SAR) is a new trend in IDM, as was observed for probenazole in the rice blast case. IDM systems always imply a judicious use of all available options, including synthetic chemicals, bio-fungicides, SAR, conventional breeding and GM technology (Hollomon 2015a, b).

References

Agrios, G. N. (2005). *Plant pathology* (5th ed., pp. 79–103). Burlington: Elsevier Academic Press.

AgroSciences. (2008). *Internal market research based on panel data supplied by Agrobase-Logigram*. Archamps, France.

Alberts, B., Bray, D., Lewis, J., Raff, M., Roberts, K., & Watson, J. D. (2002). *Molecular biology of the cell*. London: Garland Publishing, Inc.

Al-Khatib, M. T., Shequarah, M. N., & Alsmadi, S. A. (2017). Control of soil-borne pathogens by soil fumigation with paraformaldehyde (fogidesfarm) as alternative to methyl bromide. *Asian Journal of Plant Pathology, 11*, 81–88.

Appel, D. N., & Kurdyla, T. (1992). Intravascular injection with propiconazole in live oak for oak wilt control. *Plant Disease, 76*, 1120–1124.

Bartett, D. W., Clough, J. M., Godfrey, C. R., Godwin, J. R., Hall, A. A., Heaney, S. P., & Maund, S. J. (2001). Understanding the strobilurin fungicides. *Pesticide Outlook, 12*, 143–148.

Bartlett, D. W., Clough, J. M., Godwin, J. R., Hall, A. A., Hamer, M., & Parr-Dobrzanski, B. (2002). The strobilurin fungicides. *Pest Management Science, 58*, 649–662.

Behrens, H. (1957). Archäoligische überlegungen zur frage nach dem entstehungsgebiet der landwirtschaft. In *Beträge zur frühgeschichte der landwirtschaft S11*. Berlin: Akademie-Verlag.

Bernard, B. K., & Gordon, E. B. (2000). An evaluation of the common mechanism approach to the Food Quality Protection Act: Captan and four related fungicides, a practical example. *International Journal of Toxicology, 19*, 43–61.

Bittencourt, S. R. M. D., Menten, J. O. M., Araki, C. A. D. S., Moraes, M. H. D. D., Rugai, A. D. R., Dieguez, M. J., & Vieira, R. D. (2007). Efficiency of the fungicide carboxin+ thiram in peanut seed treatment. *Revista Brasileira de Sementes, 29*, 214–222.

Bonde, M. R., Peterson, G. L., Rizvi, S. A., & Smilanick, J. L. (1995). Myclobutanil as a curative agent for chrysanthemum white rust. *Plant Disease, 79*, 500–505.

Brandt, U., Schubert, J., Geck, P., & von Jagow, G. (1992). Uncoupling activity and physicochemical properties of derivatives of fluazinam. *Biochimica et Biophysica Acta (BBA) – Bioenergetics, 1101*, 41–47.

Brent, K. J., & Hollomon, D. W. (1995). *Fungicide resistance in crop pathogens: How can it be managed* (FRAC Monograph No. 1, 2nd Rev. Ed., pp. 1–48). Brussels: Fungicide Resistance Action Committee. Monograph 1 GCPF, FRAC.

Brent, K. J., & Hollomon, D. W. (1998). *Fungicide resistance: The assessment of risk* (pp. 1–48). Brussels: Global Crop Protection Federation.

Brent, K. J., & Hollomon, D. W. (2007a). *Fungicide resistance in crop pathogens: How can it be managed?* (2nd Rev. Ed.). Online. Brussels: Fungicide Resistance Action Committee (FRAC). Crop Life Int.

Brent, K. J., & Hollomon, D. W. (2007b). *Fungicide resistance: The assessment of risk* (Fungicide Resistance Action Committee FRAC monograph no. 2, pp. 7–8). Aimprint UK.

Brown, J. K., & Simpson, C. G. (1994). Genetic analysis of DNA fingerprints and virulences in *Erysiphe graminis* f.sp. *hordei*. *Current Genetics, 26*, 172–178.

Bullis, B. L., & Lemire, B. D. (1994). Isolation and characterization of the *Saccharomyces cerevisiae* SDH4 gene encoding a membrane anchor subunit of succinate dehydrogenase. *Journal of Biological Chemistry, 269*, 6543–6549.

Calzarano, F., Di Marco, S., & Cesari, A. (2004). Benefit of fungicide treatment after trunk renewal of vines with different types of esca necrosis. *Phytopathologia Mediterranea, 43*, 116–124.

Carr, J. F., Gregory, S. T., & Dahlberg, A. E. (2005). Severity of the streptomycin resistance and streptomycin dependence phenotypes of ribosomal protein S12 of *Thermus thermophilus* depends on the identity of highly conserved amino acid residues. *Journal of Bacteriology, 187*, 3548–3550.

Carson, R. (1962). *Silent spring*.

Carter, W. W. (1943). A promising new soil amendment and disinfectant. *Science, 97*, 383–384.

Case, D. R., Bell, C. L., & Steinway, D. M. (2011). *Environmental law handbook*. Rockville: Government Institutes.

Chauhan, M. S., Yadav, J. P. S., & Gangopadhyay, S. (1988). Chemical control of soilborne fungal pathogen complex of seedling cotton. *International Journal of Pest Management, 34*, 159–161.

Chen, S. K., Edwards, C. A., & Subler, S. (2001). Effects of the fungicides benomyl, captan and chlorothalonil on soil microbial activity and nitrogen dynamics in laboratory incubations. *Soil Biology and Biochemistry, 33*, 1971–1980.

Cleland, J. (2013). World population growth; past, present and future. *Environmental and Resource Economics, 55*(4), 543–554.

Cools, H. J., & Fraaije, B. A. (2008). Are azole fungicides losing ground against Septoria wheat disease? Resistance mechanisms in Mycosphaerella graminicola. *Pest Management Science: Formerly Pesticide Science, 64*(7), 681–684.

Cycoń, M., Piotrowska-Seget, Z., & Kozdrój, J. (2010). Responses of indigenous microorganisms to a fungicidal mixture of mancozeb and dimethomorph added to sandy soils. *International Biodeterioration & Biodegradation, 64*(4), 316–323.

Daignan-Fornier, B., Valens, M., Lemire, B. D., & Bolotin-Fukuhara, M. (1994). Structure and regulation of SDH3, the yeast gene encoding the cytochrome b560 subunit of respiratory complex II. *Journal of Biological Chemistry, 269*(22), 15469–15472.

Dal Maso, E., Cocking, J., & Montecchio, L. (2014). Efficacy tests on commercial fungicides against ash dieback in vitro and by trunk injection. *Urban Forestry & Urban Greening, 13*(4), 697–703.

Davidse, L. C. (1986). Benzimidazole fungicides: Mechanism of action and biological impact. *Annual Review of Phytopathology, 24*, 43–65.

Dawson, W. O. (1984). Effects of animal antiviral chemicals on plant viruses [tobacco mosaic virus, cowpea chlorotic mottle virus]. *Phytopathology, 74*, 211–213.

De Cal, A., Martinez-Treceno, A., Salto, T., López-Aranda, J. M., & Melgarejo, P. (2005). Effect of chemical fumigation on soil fungal communities in Spanish strawberry nurseries. *Applied Soil Ecology, 28*, 47–56.

Deising, H. B., Reimann, S., Peil, A., & Weber, W. E. (2002). Disease management of rusts and powdery mildews. In F. Kempken (Ed.), *The Mycota XI. Application in agriculture* (pp. 243–269). Berlin: Springer.

Di Marco, S., Mazzullo, A., Cesari, C. F., & A. (2000). Control of esca: Status and perspectives. *Phytopathologia Mediterranea, 39*, 232–240.

Duggar, B. M., & Armstrong, J. K. (1925). The effect of treating the virus of tobacco mosaic with the juices of various plants. *Annals of the Missouri Botanical Garden, 12*, 359–366.

Düker, A., & Kubiak, R. (2011). Stem injection of prohexadione carboxylic acid to protect blossoms of apple trees from fire blight infection (*Erwinia amylovora*). *Journal of Plant Diseases and Protection, 118*, 156–160.

Dula, T., Kappes, E. M., Horvath, A., & Rabai, A. (2007). Preliminary trials on treatment of esca-infected grapevines with trunk injection of fungicides. *Phytopathologia Mediterranea, 46*, 91–95.

El-Hamalawi, Z. A., Menge, J. A., & Adams, C. J. (1995). Methods of fosetyl-Al application and phosphonate levels in avocado tissue needed to control stem canker caused by *Phytophthora citricola*. *Plant Disease, 79*, 770–778.

Fishel, F. M. (2005). *Pesticide toxicity profile: Neonicotinoid pesticides*. IFAS, University of Florida.

Fisher, D. J., & Hayes, A. L. (1982). Mode of action of the systemic fungicides furalaxyl, metalaxyl and ofurace. *Pesticide Science, 13*, 330–339.

Freeman, S., Nizani, Y., Dotan, S., Even, S., & Sando, T. (1997). Control of *Colletotrichum acutatum* in strawberry under laboratory, greenhouse, and field conditions. *Plant Disease, 81*, 749–752.

Fujii, K., & Takamura, S. (1998). Pyricut® (difulmetorim, UBF-002EC) – A new fungicide for ornamental use. *Agrochemicals Japan, 72*, 14–15.

Galasso, G. J. (1984). Future prospects for antiviral agents and new approaches. *Journal of Antimicrobial Chemotherapy, 14*(Suppl_A), 127–136.

Gisi, U., Sierotzki, H., Cook, A., & McCaffery, A. (2002). Mechanisms influencing the evolution of resistance to Qo inhibitor fungicides. *Pest Management Science, 58*, 859–867.

Godfrey, G. H. (1935). Experiments on the control of the root-knot nematode in the field with chloropicrin and other chemicals. *Phytopathology, 25*, 67–90.

Goodman, R. N. (1959). The influence of antibiotics on plants and plant disease control. In H. S. Goldberg (Ed.), *Antibiotics: Their chemistry and non-medical uses* (pp. 322–421). Princeton: Van Nostrand.

Gowen, S. R. (1997). *Chemical control of nematodes: Efficiency and side-effects*. FAO Plant Production and Protection Paper (FAO). http://www.fao.org/3/v9978e/v9978e08.htm

Guest, D. I., Anderson, R. D., Foard, H. J., Phillips, D., Worboys, S., & Middleton, R. M. (1994). Long-term control of *Phytophthora* diseases of cocoa using trunk-injected phosphonate. *Plant Pathology, 43*, 479–492.

Guo, Z. J., Miyoshi, H., Komyoji, T., Haga, T., & Fujita, T. (1991). Uncoupling activity of a newly developed fungicide, fluazinam [3-chloro-N-(3-chloro-2, 6-dinitro-4-trifluoromethylphenyl) -5-trifluoromethyl-2-pyridinamine]. *Biochimica et Biophysica Acta (BBA)-Bioenergetics, 1056*, 89–92.

Gupta, V. K. (1977). Possible use of carbendazim in the control of *Dematophora* root rot of apple. *Indian Phytopathology, 30*, 527–531.

Gupta, K., Bishop, J., Peck, A., Brown, J., Wilson, L., & Panda, D. (2004). Antimitotic antifungal compound benomyl inhibits brain microtubule polymerization and dynamics and cancer cell proliferation at mitosis, by binding to a novel site in tubulin. *The Biochemist, 43*, 6645–6655.

Hagiwara, D., Matsubayashi, Y., Marui, J., Furukawa, K., Yamashino, T., Kanamaru, K., et al. (2007). Characterization of the NikA histidine kinase implicated in the phosphorelay signal transduction of *Aspergillus nidulans*, with special reference to fungicide responses. *Bioscience, Biotechnology, and Biochemistry, 71*, 844–847.

Hague, N. G. M., & Gowen, S. R. (1987). Chemical control of nematodes. In R. H. Brown & B. R. Kerry (Eds.), *Principles and practice of nematode control in crops* (pp. 131–178). Sydney: Academic.

Hahn, M. (2014). The rising threat of fungicide resistance in plant pathogenic fungi: *Botrytis* as a case study. *Journal of Chemical Biology, 7*, 133–141.

Hall, R. A., Lenardon, M. D., Alvarez, F. J., Nogueira, F. M., Mukaremera, L., & Gow, N. A. R. (2013). The *Candida albicans* cell wall: Structure and role in morphogenesis and immune recognition. *The Fungal Cell Wall*, 1–26.

Halling-Sørensen, B., Sengeløv, G., & Tjørnelund, J. (2002). Toxicity of tetracyclines and tetracycline degradation products to environmentally relevant bacteria, including selected tetracycline-resistant bacteria. *Archives of Environmental Contamination and Toxicology, 42*, 263–271.

Hansen, A. J., & Stace-Smith, R. (1989). Antiviral chemicals for plant disease control. *Critical Reviews in Plant Sciences, 8*, 45–88.

Herd, G. W., & Phillips, A. J. L. (1988). Control of seed-borne *Sclerotinia sclerotiorum* by fungicidal treatment of sunflower seed. *Plant Pathology, 37*, 202–205.

Hewitt, H. G. (1998). *Fungicides in crop protection*. New York: CAB International.

Hewitt, G. (2000). New modes of action of fungicides. *Pesticide Outlook, 11*, 28–32.

Hoffman, L. E., & Wilcox, W. F. (2003). Factors influencing the efficacy of myclobutanil and azoxystrobin for control of grape black rot. *Plant Disease, 87*, 273–281.

Hollomon, D. W. (2015a). Fungicide resistance: 40 years on and still a major problem. In *Fungicide resistance in plant pathogens* (pp. 3–11). Tokyo: Springer. https://doi.org/10.1007/978-4-431-55642-8.

Hollomon, D. W. (2015b). Fungicide resistance: Facing the challenge – A review. *Plant Protection Science, 51*, 170–176.

Hollomon, D. W., & Chamberlain, K. (1981). Hydroxypyrimidine fungicides inhibit adenosine deaminase in barley powdery mildew. *Pesticide Biochemistry and Physiology, 16*, 158–169.

Hoopen, G. M., & Krauss, U. (2006). Biology and control of *Rosellinia bunodes*, *Rosellinia necatrix* and *Rosellinia pepo*: A review. *Crop Protection, 25*, 89–107.

Inglis, D. A., Powelson, M. L., & Dorrance, A. E. (1999). Effect of registered potato seed piece fungicides on tuber-borne *Phytophthora infestans*. *Plant Disease, 83*, 229–234.

Ishii, H., & Holloman, D. W. (Eds.). (2015). *Fungicide resistance in plant pathogens*. Tokyo: Springer. https://doi.org/10.1007/978-4-431-55642-8.

Ishii, H., Zhen, F., Hu, M., Li, X., & Schnabel, G. (2016). Efficacy of SDHI fungicides, including benzovindiflupyr, against *Colletotrichum* species. *Pest Management Science, 72*, 1844–1853.

Ivic, D. (2010). Curative and eradicative effects of fungicides. In O. Clarisse (Ed.), *Fungicides*. IntechOpen. https://doi.org/10.5772/13766.

Iwaki, T., Iefuji, H., Hiraga, Y., Hosomi, A., Morita, T., Giga-Hama, Y., & Takegawa, K. (2008). Multiple functions of ergosterol in the fission yeast Schizosaccharomyces pombe. *Microbiology, 154*, 830–841.

Johnson, M. O., & Godfrey, G. H. (1932). Chloropicrin for nematode control. *Industrial & Engineering Chemistry, 24*, 311–313.

Joseph-Horne, T. I. M., Hollomon, D. W., & Wood, P. M. (2001). Fungal respiration: A fusion of standard and alternative components. *Biochimica et Biophysica Acta (BBA)-Bioenergetics, 1504*, 179–195.

Kamimura, S., Nishikawa, M., & Takahi, Y. (1976). Mode of action of soil fungicide hymexazol, 3-hydroxy-5-methylisoxazole, on *Fusarium oxysporum* f. *cucumerinum*. *Japanese Journal of Phytopathology, 42*, 242–252.

Kanadani, G., Date, H., & Nasu, H. (1998). Effect of fluazinam soil-drench on white root rot of grapevine. *Japanese Journal of Phytopathology, 64*, 139–141.

Kawai, M., & Yamagishi, J. I. (2009). Mechanisms of action of acriflavine: Electron microscopic study of cell wall changes induced in *Staphylococcus aureus* by acriflavine. *Microbiology and Immunology, 53*, 481–486.

Kim, J. H., Campbell, B. C., Mahoney, N., Chan, K. L., Molyneux, R. J., & May, G. S. (2007). Enhancement of fludioxonil fungicidal activity by disrupting cellular glutathione homeostasis with 2, 5-dihydroxybenzoic acid. *FEMS Microbiology Letters, 270*, 284–290.

Kislev, M. E. (1982). Stem rust of wheat 3300 years old found in Israel. *Science, 216*, 993–994.

Klittich, C. J. (2008). Milestones in fungicide discovery: Chemistry that changed agriculture. *Plant Health Progress, 9*, 31.

Koo, B. S., Park, H., Kalme, S., Park, H. Y., Han, J. W., Yeo, Y. S., et al. (2009). α-and β-tubulin from *Phytophthora capsici* KACC 40483: Molecular cloning, biochemical characterization, and antimicrotubule screening. *Applied Microbiology and Biotechnology, 82*, 513–524.

Kretschmer, M. (2012). Emergence of multi-drug resistance in fungal pathogens: A potential threat to fungicide performance in agriculture. In T. S. Thind (Ed.), *Fungicide resistance in crop protection: Risk and management* (pp. 251–267). Oxfordshire: CABI.

Kulka, M., & Von Schmeling, B. (1987). Carboxin fungicides and related compounds. In H. Lyr (Ed.), *Modern selective fungicides: Properties, applications, mechanisms of action* (pp. 119–131). London: VEB Gustav Fisher Verlag, Jena and Longman Group.

Kumar, S., & Gupta, O. M. (2012). Expanding dimensions of plant pathology. *JNKVV Research Journal, 46*, 286–293.

Lanier, G. N. (1987). Fungicides for Dutch elm disease: Comparative evaluation of commercial products. *Journal of Arboriculture, 13*, 189–195.

Leadbeater, A. (2012). The role of FRAC in resistant management. *Journal of Mycology Plant Pathology, 42*, 25.

Leadbeater, A. (2015). Recent developments and challenges in chemical disease control – A review. *Plant Protection Science, 51*(4), 163–169.

Linderman, R. G., & Davis, E. A. (2008). Eradication of *Phytophthora ramorum* and other pathogens from potting medium or soil by treatment with aerated steam or fumigation with metam sodium. *HortTechnology, 18*, 106–110.

List, F. C. (2018). *Fungicides sorted by mode of action (including FRAC Code numbering)*. Fungicides market. (Chloronitriles, Dithiocarbamates, Triazoles, Strobilurins, Benzimidazoles, Mancozeb and others) for fruit vegetables, oilseeds pulses, cereals grains and other crops: Global industry perspective, comprehensive analysis and forecast, 2016–2022. https://www.zionmarketresearch.com/report/fungicides

Lodish, H., Berk, A., Zipursky, S. L., Matsudaira, P., Baltimore, D., & Darnell, J. (Eds.). (2000). *Molecular cell biology* (4th ed.) New York: W. H. Freeman. ISBN:10: 0-7167-3136-3.

Ma, Z., & Michailides, T. J. (2005). Advances in understanding molecular mechanisms of fungicide resistance and molecular detection of resistant genotypes in phytopathogenic fungi. *Crop Protection, 24*, 853–863.

Madigan, M. T., Martinko, J. M., Dunlap, P. V., & Clark, D. P. (2008). Brock biology of microorganisms, 12th Ed. *International Microbiology, 11*, 65–73.

Mahal, M. F. (2014). Effects of fungicides and plant extracts on seed germination and seed associated mycoflora of *Lens arietinum* L. and *Lathyrus sativus*. *Journal of Bio-Science, 22*, 101–110.

Maienfisch, P., & Stevenson, T. M. (2015). Modern agribusiness–markets, companies, benefits and challenges. In *Discovery and synthesis of crop protection products* (pp. 1–13). Washington, DC: American Chemical Society.

Mappes, D., & Hiepko, G. (1984). New possibilities for controlling root diseases of plantation crops. Mededelingen van de Faculteit Landbouwwetenschappen. *Rijksuniversiteit Gent, 49*, 283–292.

Martínez J. A. (2012). Natural fungicides obtained from plants. In: *Fungicides for plant and animal diseases*. InTech. www.intechopen.com. https://doi.org/10.5772/26336.

Mathews, D. J. (1919). Report of the work of the W. B. Randall research assistant. Nursery and Market Garden Industry Development Society, Ltd. Experiment and Research Station, Cheshunt, Herts, U.K. *Annual Report, 5*, 18–21.

Matsson, M., & Hederstedt, L. (2001). The carboxin-binding site on *Paracoccus denitrificans* succinate: Quinone reductase identified by mutations. *Journal of Bioenergetics and Biomembranes, 33*, 99–105.

Maude, R. B. (1996). *Seedborne diseases and their control: Principles and practice*. Wallingford: CAB International.

McBeth, C. W., & Bergeson, G. B. (1955). 1, 2-dibromo-3-chloropropane: A new nematocide. *Plant Disease Reporter, 39*, 223–225.

McCarroll, N. E., Protzel, A., Ioannou, Y., Stack, H. F., Jackson, M. A., Waters, M. D., & Dearfield, K. L. (2002). A survey of EPA/OPP and open literature on selected pesticide chemicals: III. Mutagenicity and carcinogenicity of benomyl and carbendazim. *Mutation Research/Reviews in Mutation Research, 512*, 1–35.

McCoy, R. E. (1982). Use of tetracycline antibiotics to control yellows diseases. *Plant Disease, 66*, 539–542.

McDougall, P. (2010). *The cost of new agrochemical product discovery, development and registration in 1995, 2000 and 2005–8*. Available at: http://www.croplife.org/view_document.aspx?docId=2478

McDougall, P. (2014). *Trends in industry research and development* (AgriFutura No 174, April 2014). Available at: www.phillipsmcdougall.com

McDougall, P. (2018a). *Evolution of the crop protection industry since 1960*. Available at: www.Agribusinessintelligence.informa.com

McDougall, P. (2018b, June). *The global crop protection industry in 2017*. AgrAspire. Available at: www.Agribusinessintelligence.informa.com

McGrath, M. T. (2004). What are fungicides. *The Plant Health Instructor*. https://doi.org/10.1094/PHI-I-2004-0825-01. Updated 2016.

McManus, P. S., & Stockwell, V. O. (2001). Antibiotic use for plant disease management in the United States. *Plant Health Progress, 2*, 14.

McManus, P. S., Stockwell, V. O., Sundin, G. W., & Jones, A. L. (2002). Antibiotic use in plant agriculture. *Annual Review of Phytopathology, 40*, 443–465.

Milenkovski, S., Baath, E., Lindgren, P. E., & Berglund, O. (2010). Toxicity of fungicides to natural bacterial communities in wetland water and sediment measured using leucine incorporation and potential denitrification. *Ecotoxicology, 19*, 285–294.

Miller, M. L. (2014). Pesticides. In Case et al. (Eds.), *Environmental law handbook* (22nd ed., pp. 870–875). Lanham: Bernan Press.

Moore, D., Robson, G. D., & Trinci, A. P. (2011). *21st century guidebook to fungi with CD*. Cambridge: Cambridge University Press.

Morton, V., & Staub, T. (2008). *A short history of fungicides*. St. Paul: APSnet Features.

Motoba, K., Uchida, M., & Tada, E. (1988). Mode of antifungal action and selectivity of flutolanil. *Agricultural and Biological Chemistry, 52*, 1445–1449.

Mueller, D., Bradley, C. A., & Nielsen, J. (2008). *Field crop fungicides for the North Central United States*. Ames: Agricultural Experiment Station, Iowa State University.

Nabi, S. U., Raja, W. H., Dar, M. S., Kirmani, S. N., & Magray, M. M. (2017). New generation fungicides in disease management of horticultural crops. *Indian Horticulture Journal, 7*, 1–07.

NPIC (National Pesticide Information Center). (2018). *International pesticide regulations*. http://npic.orst.edu/reg/intreg.html

Ochiai, N., Fujimura, M., Oshima, M., Motoyama, T., Ichiishi, A., Yamada-Okabe, H., & Yamaguchi, I. (2002). Effects of iprodione and fludioxonil on glycerol synthesis and hyphal development in *Candida albicans*. *Bioscience, Biotechnology, and Biochemistry, 66*, 2209–2215.

Old, D., & Gorini, L. (1965). Amino acid changes provoked by streptomycin in a polypeptide synthesized in vitro. *Science, 150*, 1290–1292.

Parker, J. E., Warrilow, A. G., Cools, H. J., Martel, C. M., Nes, W. D., Fraaije, B. A., et al. (2011). Mechanism of binding of prothioconazole to Mycosphaerella graminicola CYP51 differs from that of other azole antifungals. *Applied and Environmental Microbiology, 77*(4), 1460–1465.

Pelaez, V., da Silva, L. R., & Araújo, E. B. (2013). Regulation of pesticides: A comparative analysis. *Science and Public Policy, 40*, 644–656.

Prevost, B. (1807). *Memoir on the immediate cause of caries or coal of the bles, and several. Other diseases of plants, and on the condoms of caries*. Paris: Bernard Quai des Augustins. [Translation by Keitt G.W. (1939). *Memoir on the immediate cause of bunt or smut of wheat, and of several other diseases of plants, and on preventives of bunt*. Menasha: American Phytopathological Society, Phytopathological Classics No. 6].

Quak, F. (1961). Heat treatment and substances inhibiting virus multiplication in meristem culture to obtain virus-free plants. *Advances in Horticultural Science Application, 1*, 144–148.

Radzuhn, B., & Lyr, H. (1984). On the mode of action of the fungicide etridiazole. *Pesticide Biochemistry and Physiology, 22*, 14–23.

Rathinasamy, K., & Panda, D. (2005). Suppression of microtubule dynamics by benomyl decreases tension across kinetochore pairs and induces apoptosis in cancer cells. *The FEBS Journal, 273*, 4114–4128.

Rebollar-Alviter, A., Madden, L. V., & Ellis, M. A. (2007). Pre-and post-infection activity of azoxystrobin, pyraclostrobin, mefenoxam, and phosphite against leather rot of strawberry, caused by *Phytophthora cactorum*. *Plant Disease, 91*, 559–564.

Reimann, S., & Deising, H. B. (2005). Inhibition of efflux transporter-mediated fungicide resistance in *Pyrenophora tritici-repentis* by a derivative of 4′-hydroxyflavone and enhancement of fungicide activity. *Applied and Environmental Microbiology, 71*, 3269–3275.

Research and Market. (2018). *Crop protection chemicals market: Global industry trends, share, size, growth, opportunity and forecast 2018–2023*. https://www.prnewswire.com/news-releases/global-crop-protection-chemicals-herbicides-fungicides-insecticides-and-others-market-to-2023-300745519.html

Rhoades, W. C. (1963). The history and use of agricultural chemicals. *Florida Entomologist, 46*, 275–277.

Rohrman, D. F. (1968). The law of pesticides: Present and future. *Journal of Public Law, 17*, 351.

Rosslenbroich, H. J., & Stuebler, D. (2000). *Botrytis cinerea* – History of chemical control and novel fungicides for its management. *Crop Protection, 19*, 557–561.

Runkle, J., Flocks, J., Economos, J., & Dunlop, A. L. (2017). A systematic review of Mancozeb as a reproductive and developmental hazard. *Environment International, 99*, 29–42.

Russell, P. E. (2005). A century of fungicide evolution. *Journal of Agricultural Science, 143*, 11–25.

Savary, S., Teng, P. S., Willocquet, L., & Nutter, F. W., Jr. (2006). Quantification and modeling of crop losses: A review of purposes. *Annual Review of Phytopathology, 44*, 89–112.

Schacht, H. (1859). About some enemies and diseases of the sugar beet. *Journal of the Association for the Beet Sugar Industry in the Zollverein, 9*, 239–250.

Scheffer, R. J., Voeten, J. G. W. F., & Guries, R. P. (2008). Biological control of Dutch elm disease. *Plant Disease, 92*, 192–200.

Schierow, L. J., & Esworthy, R. (2004). *Pesticide law: A summary of the statutes*. Washington, DC: Congressional Research Service, Library of Congress.

Seiler, J. P. (1975). Toxicology and genetic effects of benzimidazole compounds. *Mutation Research/Reviews in Genetic Toxicology, 32*, 151–167.

Shtienberg, D., Zilberstaine, M., Oppenheim, D., Herzog, Z., Manulis, S., Shwartz, H., & Kritzman, G. (2001). Efficacy of oxolinic acid and other bactericides in suppression of *Erwinia amylovora* in pear orchards in Israel. *Phytoparasitica, 29*, 143–154.
Shurtleff, M. C., Pelczar, M. J., Kelman, A., & Pelczar, R. M. (2018). *Plant disease*. Encyclopædia Britannica, Inc. https://www.britannica.com/science/plant-disease
Sikora, R. A., & Hartwig, J. (1991). Mode-of-action of the carbamate nematicides cloethocarb, aldicarb and carbofuran on Heterodera schachtii: 2. Systemic activity. *Rev Nématol, 14*, 531–536.
Smith, D. L., Garrison, M. C., Hollowell, J. E., Isleib, T. G., & Shew, B. B. (2008). Evaluation of application timing and efficacy of the fungicides fluazinam and boscalid for control of Sclerotinia blight of peanut. *Crop Protection, 27*, 823–833.
Spiegel, J., & Stammler, G. (2006). Baseline sensitivity of *Monilinia laxa* and *M. fructigena* to pyraclostrobin and boscalid. *Journal of Plant Diseases and Protection, 113*(5), 199–206.
Statista. (2016). *Fungicide market value worldwide from 2014 to 2023* (in billion U.S. dollars). https://www.statista.com/statistics/586532/fungicide-market-value-worldwide
Sudisha, J., Amruthesh, K. N., Deepak, S. A., Shetty, N. P., Sarosh, B. R., & Shetty, H. S. (2005). Comparative efficacy of strobilurin fungicides against downy mildew disease of pearl millet. *Pesticide Biochemistry and Physiology, 81*, 188–197.
Sukul, P., & Spiteller, M. (2000). Metalaxyl: Persistence, degradation, metabolism, and analytical methods. *Reviews of Environmental Contamination and Toxicology, 164*, 1–26.
Tadmor, Z., Sachs, K., Chet, I., Spiegel, Y., Ravina, I., Mano, G., & Yannai, S. (2005). An innovative approach: The use of di-nitrogen tetroxide for soil fumigation. *Biosystems Engineering, 91*, 413–419.
Tanaka, C., & Izumitsu, K. (2010). Two-component signaling system in filamentous fungi and the mode of action of dicarboximide and phenylpyrrole fungicides. In *Fungicides*. IntechOpen.
Taylor, A. L. (2003). Nematocides and nematicides – A history. *Nematropica, 33*, 225–232.
Taylor, P. A., & Washington, W. S. (1984). Curative treatments for *Phytophthora cactorum* in peach trees using metalaxyl and phosethyl Al. *Australasian Plant Pathology, 13*, 31–33.
Thorne, G., & Jensen, V. (1946). A preliminary report on the control of sugar beet nematode with two chemicals, DD and Dowfume W15. In *Proceedings of the American Society of Sugar Beet Technologist*, pp. 322–326.
Tomlin, C., & British Crop Protection Council. (2006). *The e-pesticide manual: A world compendium*. Alton: BCPC.
Utkhede, R. S. (1987). Chemical and biological control of crown and root rot of apple caused by *Phytophthora cactorum*. *Canadian Journal of Plant Pathology, 9*, 295–300.
Vagi, P., Preininger, E., Kovacs, G. M., Kristof, Z., Boka, K., & Böddi, B. (2013). *Structure of plants and fungi* (pp. 1–109). Eötvös Loránd University, Budapest.
Van Woerkom, A. H., Aćimović, S. G., Sundin, G. W., Cregg, B. M., Mota-Sanchez, D., Vandervoort, C., & Wise, J. C. (2014). Trunk injection: An alternative technique for pesticide delivery in apples. *Crop Protection, 65*, 173–185.
Vargas-Pérez, I., Sánchez, O., Kawasaki, L., Georgellis, D., & Aguirre, J. (2007). Response regulators SrrA and SskA are central components of a phosphorelay system involved in stress signal transduction and asexual sporulation in *Aspergillus nidulans*. *Eukaryotic Cell, 6*, 1570–1583.
Vega, B., & Dewdney, M. M. (2015). Sensitivity of *Alternaria alternata* from citrus to boscalid and polymorphism in iron-sulfur and in anchored membrane subunits of succinate dehydrogenase. *Plant Disease, 99*, 231–239.
Veloukas, T., Bardas, G. A., Karaoglanidis, G. S., & Tzavella-Klonari, K. (2007). Management of tomato leaf mould caused by *Cladosporium fulvum* with trifloxystrobin. *Crop Protection, 26*, 845–851.
Vincelli, P. (2002). Q o I (strobilurin) fungicides: Benefits and risks. *Plant Health Instructor*.
Wilcox, W. F. (1990). Postinfection and antisporulant activities of selected fungicides in control of blossom blight of sour cherry caused by *Monilinia fructicola*. *Plant Disease, 74*, 808–811.

Yang, C., Hamel, C., Vujanovic, V., & Gan, Y. (2011). Fungicide: Modes of action and possible impact on non-target microorganisms. *ISRN Ecology, 2011*, 130289.

Yang, H., Tong, J., Lee, C. W., Ha, S., Eom, S. H., & Im, Y. J. (2015). Structural mechanism of ergosterol regulation by fungal sterol transcription factor Upc2. *Nature Communications, 6*, 6129.

Yildiz, M. (2017). Crop production for increasing population. *Journal of Plant Genetics and Breeding, 1*, e102.

Yoshimi, A., Kojima, K., Takano, Y., & Tanaka, C. (2005). Group III histidine kinase is a positive regulator of Hog1-type mitogen-activated protein kinase in filamentous fungi. *Eukaryotic Cell, 4*, 1820–1828.

Zhou, Y., Xu, J., Zhu, Y., Duan, Y., & Zhou, M. (2016). Mechanism of action of the benzimidazole fungicide on *Fusarium graminearum*: Interfering with polymerization of monomeric tubulin but not polymerized microtubule. *Phytopathology, 106*, 807–813.

Chapter 5
Biological Antagonism: A Safe and Sustainable Way to Manage Plant Diseases

Yasir Iftikhar, Ashara Sajid, Qaiser Shakeel, Zohaib Ahmad, and Zia Ul Haq

Abstract Biological control is a viable alternatives to the use of synthetic chemicals for plant pathogens management, based on application of microbial antagonists as biological control agents (BCA). Plant health is significantly affected in many ways by a wide variety of pathogens. Cross protection, predation, hyperparasitism, induced resistance, antibiosis and competition are different mechanisms used by BCA. Knowledge is required for successful application of biocontrol in intensive management approaches. BCA can be applied at the site of infection directly or in each crop year, at sites in which they will multiply and spread to other field parts. To keep pathogen populations below economic threshold levels, occasional or one time applications can be adopted. However, due to different environmental conditions, biological control has not always produced encouraging results. To improve the BCA performance in the field, work on formulations is needed. For marketing, strains with better adaptability and field survival should be prospected. Most of biological control work has been centered on management of soil borne or seed borne pathogens. Most of the products containing BCA are applied as seed treatments for protecting major crops such as wheat, rice, sugar beet, corn and cotton. BCA are also used in foliar sprays to manage downy and powdery mildew, leaf spot and blight. Antagonistic microorganisms have also been used against few post-harvest pathogens. In spite of all significant improvements, this area still needs due

Y. Iftikhar · A. Sajid
Department of Plant Pathology, College of Agriculture, University of Sargodha, Sargodha, Pakistan

Q. Shakeel (✉) · Z. Ahmad
Discipline of Plant Pathology, University College of Agriculture and Environmental Sciences, The Islamia University of Bahawalpur, Bahawalpur, Pakistan
e-mail: qaiser.shakeel@iub.edu.pk

Z. Ul Haq
Department of Plant Pathology, Faculty of Agriculture Sciences and Technology, Bahauddin Zakariya University, Multan, Pakistan

© Springer Nature Switzerland AG 2020
I. Ul Haq, S. Ijaz (eds.), *Plant Disease Management Strategies for Sustainable Agriculture through Traditional and Modern Approaches*, Sustainability in Plant and Crop Protection 13, https://doi.org/10.1007/978-3-030-35955-3_5

consideration for developing more reliable and stable formulations, especially when for field applications. In this view, more research is required on innovative formulations by exploring novel microorganisms, using nano- and biotechnologies for their improvement, studying the impact of environmental conditions and the mass production of BCA. With a growing of biocontrol demand by growers, the future outlook of biocontrol is bright. By improving biocontrol research it is possible to completely replace chemical pesticides by BCA, improving yields, protection technologies and the environment, leading to a more sustainable agriculture.

Keywords Hyperparasitism · Biological antagonism · Entomopathogenic · Mycoparasitism · Obligate parasite

5.1 Introduction

Biotic agents such as pathogens, harmful pests and weeds are major yield-limiting factors in agriculture. To improve the agricultural products quantitatively and qualitatively, these pest constraints need to be managed. Different agricultural practices are being used to manage pests (Benhamou 2004; Heydari 2007; Cook 1993; Agrios 1988; Islam et al. 2005; Baker 1987; Chisholm et al. 2006; Kloepper et al. 2004; Bargabus et al. 2002, 2004). Most commonly, plant diseases are controlled by different cultural practices and pesticides (Baker 1987; Agrios 1988). However, pesticide pollution is a major, actual concern which led to the development and application of strict regulations towards the use of pesticides. There is also pressure for removal from the market of chemicals which are not eco-friendly. The carcinogenicity and effects on non-target hosts due to their extensive use also raised public pesticide concerns. A further issue about pesticides use regards the insurgence of resistant pathogens, and the difficulty in registration of new chemicals (Cook 1993). All human health and environmental protection concerns forced researchers to rethink and develop alternative strategies for disease management which are eco-friendly (Cook 1993; Baker 1987).

Cook (1993) stated that best alternative to pesticides is biological control. Biological controls means management of one organism's development by the exploitation of other living organisms (Cook 1993; Baker 1987). Advantages related to adaptation of biological control are: environment protection which is eco-friendly, self-sustaining and long lasting; support to existing communities of beneficial species (Cook 1993). Mechanism of action of biological control might include the supplemented release of exotic species or creating favourable conditions for the multiplication of naturally existing microbes or the use of non-pathogenic strains (Schouten et al. 2004; Cook 1993).

According to Agrios (1988), more than 70% of diseases are caused by fungal pathogens. Worldwide, fungal pathogens are the main reason of major annual yield losses in agriculture. Many fungal diseases such as flower and leaf diseases (powdery mildew), vegetables and fruit diseases (*Botrytis* spp.), pathogens of cut and pruning wounds and soil borne diseases, have been successfully managed by BCA (Heydari and Misaghi 1998, 1999, 2003; Cook 1993; Baker 1987; Agrios 1988; Heydari et al. 2004). The interaction among plant, BCA, environment and people still need extensive consideration for better and effective development of biological control strategies. This chapter will (i) illustrate different definitions as well as basic biocontrol mechanisms, (ii) explain the biological control and microbial diversity interaction, (iii) present the recent position of study and implementation of biological controls, and (iv) concisely list forthcoming guidelines that may results in progression of additional and efficient biological control against various fungal diseases (Kessel et al. 2005).

5.2 Biological Control

Biological control (syn. biocontrol) has been used especially in plant pathology and entomology. The use of antagonists or host-specific weed pathogens are also included in biological control. The organisms used to manage other species are termed BCAs. The fermented or extracted natural products (bio-fertilizers or biopesticides) also may be considered as biological control (Cook 1993). The definitions of biological control depend on type, number and source of BCA, targets and human timing and involvement. The biocontrol of plant diseases varies from biocontrol of insect in the following ways (Table 5.1).

5.2.1 Terminology

Predator: organism that normally preys on other or free-living or pest species that are directly devoured. A huge number of preys are killed amongst its entire lifetime.

Table 5.1 Comparison of bio-control for plant diseases and insect pests

Disease	Insects
Control of disease is mainly accomplished by predators and parasites, hyper parasites, competition and antibiosis	Mostly by predators and parasites
Antagonists are not mobile and are broadly passive. Contact of the pathogen is unintentional.	Parasites are dynamic, versatile and look for their prey.
This technique depends largely on local organisms.	Predators/ parasites are normally introduced from different countries
Pathogen free planting materials and seeds are widely used.	Healthy seeds (having no pest) are not utilized.
Mass effect (single types of pathogen many competitors/ antagonists are available.	Single parasite or predator for single prey.

Entomopathogenic: microorganisms that can act as parasites of different insects inducing a disease that seriously deactivates or kills them. **Antagonism**: the activity of any microorganism that overcomes the action of a plant pathogen. ***Parasitoid:*** highly specialized insects that lay their eggs in or on the body of an insect host, which is then utilized as food for the larva development. The host is eventually killed. **Pathogen**: bacteria, fungi, and viruses that disabled or kill their host and are relatively host-specific.

5.3 Biological Management of Fungal Plant Pathogens

5.3.1 Beneficial Microbes and Plant Interaction

Plant health has been significantly affected in various ways by the interactions between plants and their pathogens, in many cases throughout their entire life cycle (Fitter and Garbaye 1994; Katska 1994; Agrios 1988; Chisholm et al. 2006; Bull et al. 2002; McSpadden-Gardener and Weller 2001). To study biocontrol mechanisms of action, different interactions must be studied, starting from the direct contact between interacting organisms.

The different types of interactions that can exist between two organisms are: parasitism, mutualism, competition, protocorporation, predation, commensalism, amensalism and neutralism (Hoitink and Boehm 1999; Chisholm et al. 2006; Bull et al. 2002; Fitter and Garbaye 1994; Katska 1994; Bankhead et al. 2004). Both at micro- and macroscopic level, any of these interactions can be observed. For diseases development both plants and pathogens, interact at multiple scales, being the disease the starting point for the development of biological control (Chisholm et al. 2006; Bull et al. 2002; Fitter and Garbaye 1994). Specific or non-specific interactions resulting in a positive, crop beneficial way is considered as a successful biological control from the plants and farmers' view points (Cook 1993; Weller et al. 2002).

Different mechanisms characterizing ecosystem processes that contribute to biological control can be classified and functionally outlined. The beneficial association of two or several species in which every contributing species receives a benefit is called mutualism (Fitter and Garbaye 1994; Kerry 2000; Biermann and Linderman 1983; Chisholm et al. 2006; Garcia-Garrido and Ocampo 1989; Duchesne 1994; Katska 1994). An example is the link of arbuscular mycorrhizal fungi with plants, yielding associations which represent an obligate, biochemical and physical interaction (Fitter and Garbaye 1994; Chisholm et al. 2006; Katska 1994). The association can also be facultative, as in the case of *Rhizobium* bacteria. Rhizobia can survive both in soil and in mutualistic interactions within root nodules of leguminous crops. These associations help plants by improving their nutrition through nitrogen fixation and manipulation/activation of defense mechanisms (Fitter and Garbaye 1994; Chisholm et al. 2006). The prevailing environmental conditions have marked effects on disease suppression and host development, and many BCA are classified as facultative, mutualistic microorganisms (Cook 1993).

The interaction in which one species is neither benefitted nor harmed, while the other one gets a benefit is called commensalism (Fitter and Garbaye 1994). Most of soil inhibiting microbes are commensals, because they are get benefits while the plants rarely show any positive or negative impact (Katska 1994; Chisholm et al. 2006). Commensals raise great challenges to plant pathogens and are responsible for lowering their populations, and ultimately the disease severity (Cook 1993).

Neutralism is a mechanism in which a member of one species has no effect on another one (Berg et al. 2005; Chisholm et al. 2006), while the negative interaction between organisms is defined as antagonism (Cook 1993). the inability of one organism to regulate the population dynamics of another organism (pathogen) is a kind of neutralism, while the antagonistic competition for food, shelter or space may decrease the activity, growth and fecundity of a target species.

The prolonged symbiotic interaction in which a species coexists for a specific time period during which one gets a benefit and the other is harmed, is called parasitism (Lo et al. 1997; Cook 1993; Chisholm et al. 2006). The activity of hyperparasites (parasitizing i.e. a plant parasite) results in biocontrol, as they feed and may regulate densities of plant pathogens (Lo et al. 1997). The stimulation of defense systems in the host after inoculation with a relatively avirulent strain may result in biocontrol and regulation of a virulent strain, an interesting aspect of biological control (Cook 1993). The last (but not least), mechanism of biological control is predation, in which one species consumes and obtains nourishment from another organism, usually by hunting and killing it. This term is applied to animals in higher trophic levels, mesofauna and protists (Cook 1993), but may be also observed at the microcale, including nematodes and bacteria.

As discussed earlier, the type of biological control interaction depends on prevailing environmental conditions. Generally mutualism and antagonism are manipulated for biocontrol of plant pathogenic microorganisms (Chisholm et al. 2006; Bull et al. 2002; Fitter and Garbaye 1994; Katska 1994).

5.3.2 Mechanism Involved in Biological Control

Different experimental approaches were studied to characterize most basic mechanisms of biological control (Howell et al. 1988; Audenaert et al. 2002; Homma et al. 1989; Van Dijk and Nelson 2000; Heydari et al. 1997; De Meyer and Hofte 1997; Ryu et al. 2004; Meziane et al. 2005; Elad and Baker 1985; Islam et al. 2005). In all studies it was revealed that each pathogen was antagonised by other microbes. Selectivity of the antagonist for a given target pathogen and physical contact result in direct antagonism. The most direct antagonism is hyperparasitism, because in this case a suppressive effect is exerted by a single microorganism and no other species is required (Linderman 1994; Harman et al. 2004). In indirect antagonism rather than directly targeting a pathogen, an activation of the host defense pathways

Table 5.2 Antagonistic mechanisms exploited in biocontrol activity

Type	Mechanism	Examples
Direct competition	Predation /Hyperparasitism	*Trichoderma virens*
		Ampelomyces quisqualis
		Pasteuria penetrans
		Lysobacter enzymogenes
		Use of mycoviruses
Mixed-paths	Production of antibiotics	Cyclic lipopeptides
		2,4-diacetylphloroglucinol
		Phenazines
	Lytic enzymes	Proteases, glucanases, chitinases
	Unregulated waste products	Hydrogen cyanide
		CO_2
		Ammonia
	Physical/chemical interference	Soil pores blockage
		Blockage of communication signals
		Remote sensing
Creating harmful environment	Food and space struggle	Consumption of soil resources
		Siderophore foraging
		Destruction of niche
	Activation of resistance of the host	Cell wall degradation
		Destruction of molecular signaling pathway

is produced. An elicited and improved host defense is achieved by non-pathogenic microorganisms (Silva et al. 2004; Leeman et al. 1995; Kloepper et al. 1980; Maurhofer et al. 1994; Lafontaine and Benhamou 1996). There is an increasing attention on studying and establishing the mechanism of biocontrol in a particular host-pathogen interaction. There are a few examples of different mechanisms of biological control that may be operating in the same specific interaction (Table 5.2).

5.4 Mycoparasitism

Specific BCA directly attack the pathogen cells or propagules. Hyperparasites may be categorized into four major groups, including Predators, Facultative parasite, Hypoviruses and Obligate bacterial pathogens. Viruses infecting *Cryphonectria parasitica* (the causal agent of the chestnut blight) provide these examples of hypoparasites. These hypoparasites cause the reduction of pathogens virulence and pathogenicity. Hyperparasites have been reported as very successful in managing chestnut blight in different areas (Milgroom and Cortesi 2004).

The interaction of all factors involved in specific biological control, such as hyperparasites, host, pathogen and environment, will determine the success or failure of any specific BCA. Sclerotia-infecting hypoparasites i.e. *Coniothyrium minitans* and those that infect fungal hyphae i.e. *Pythium oligandrum*, have been identified. Multiple hyperparasites can attack a single fungal species. Pathogens causing powdery mildew are been parasitized by a few hyperparasites such as *Gliocladium virens, Ampelomyces quisqualis, Acremonium alternatum, Cladosporium oxysporum* and *Acredontium crateriforme* (Milgroom and Cortesi 2004). In other cases, in which available nutrition is limited, BCA may exhibit a predatory behaviour. *Trichoderma* spp. release a range of cell wall degrading enzymes affecting plant pathogens. If *Trichoderma* develops in a nutrient rich environment it will not directly attack a pathogen such as *Rhizoctonia solani*. In contrast, if available nutrients are limited such as during bark decomposition, or in a condition of limited cellulose availability, *Trichoderma* spp. may release chitinases and will directly parasitize *R. solani* (Benhamou and Chet 1997).

5.5 Antibiosis

Antibiotic are toxins released by different microorganism that can kill other microbes even at low concentrations. The production of compounds with antibiotic activity has been reported for many microorganisms (Shanahan et al. 1992; Homma et al. 1989; Thomashow and Weller 1988; Islam et al. 2005; Thomashow et al. 1990; Howell and Stipanovic 1980; Leclère et al. 2005; Shahraki et al. 2009). The effectiveness of antibiotics in management of plant pathogens and diseases has been studied by different researchers (Howell and Stipanovic 1980; Thomashow and Weller 1988; Shanahan et al. 1992; Homma et al. 1989; Islam et al. 2005; Thomashow et al. 1990, 2002). In all conditions, especially *in-vitro* and *in-situ*, all the studied antibiotic were effective in controlling pathogens and their diseases. It was also demonstrated that any antibiotic which is effective must be produced in required doses to kill pathogens.

The production of antibiotics by biocontrol bacteria has been studied. Comparison of mutant strains showed that those without capacity to produce phloroglucinols and phenazines can colonize the rhizosphere but are incapable of inhibiting the growth of soil borne pathogens. While the corresponding wild strain can colonize and manage these pathogen and the root disease with a significant impact. Also, there are biocontrol strains that produce more than one antibiotic for their biocontrol activity. These types of BCA are supposed to control a wide range of pathogens and allow management of the induced disease (Keel et al. 1989; Thomashow and Weller 1988; Homma et al. 1989; Islam et al. 2005; Howell and Stipanovic 1980; Thomashow et al. 1990; Shanahan et al. 1992).

5.6 Production of Metabolites

BCA produce several metabolites other than antibiotics, that can harm growth and reproduction of target pathogens. Many polymeric compounds, i.e. cellulose, chitin, hemicellulose, proteins and DNA, can be broken by these metabolites such as lytic enzyms (Wilhite et al. 2001; Anderson et al. 2004; Press et al. 2001; Loper and Buyer 1991; Howell et al. 1988; Ordentlich et al. 1988). Many studies showed that these metabolites can directly suppress the development and growth of plant pathogens. The breakdown of complex polymers is important to get carbon which is necessary in the antagonistic activity. For example, chitinase expression by *Serratia marcescens* can suppress growth of *Sclerotium rolfsii* (Ordentlich et al. 1988). In other cases, the production of lytic enzymes by Myxobacteria and Lysobacteria leads to the development of effective biocontrol of many pathogenic fungi (Bull et al. 2002). Howell et al. (1988) showed that the defenses system of the host can be induced by oligosaccharides, fungal cell wall derivatives. The composition and availability of carbon and nitrogen based metabolites in rhizosphere soil are important factors for effectiveness. For example, *Fusarium oxysporum* f.sp. *radices-lycopersici*, the causal organism of root rot, can be managed by application of chitosan (the polymer of β1,4 glucosamine produced from chitin by alkaline deacylation, which is also biodegradable and non-toxic) (Benhamou 2004; Lafontaine and Benhamou 1996).

Another secondary microbe metabolite such as hydrogen cyanide is very effective in controlling plant diseases. The picomolar concentration of HCN is toxic for aerobic microbes and can cause a blockage in the cytochrome-oxidase-pathway (Ramette et al. 2003; Phillips et al. 2004). This is the case of *Pseudomonas fluorescens* which produces both siderophores and HCN. However, it is believed that the biocontrol activity of *P. fluorescens* against *Thielaviopsis basicola* is mainly due to the production of HCN.

Other secondary metabolites are also effective in plant disease control. Howell et al. (1988) reported that suppression of *Pythium ultimum* (the causal agent of cotton damping off) by *Enterobacter cloacae* is due to volatile compounds containing ammonia.

5.7 Competition

Competition for food and shelter is a common phenomenon of wildlife. Similarly, microbes also compete for available food resources in the root rhizosphere. The competition of one organism vs another is an important aspect of biocontrol. For a successful biological control activity, a microorganism must compete, after colonization, for obtaining nutrition from senescent tissues, exudates, waste products of insects and leachates (Loper and Buyer 1991; Elad and Baker 1985; Keel et al. 1989). The competition for food of soil (i.e. *Fusarium, Pythium* spp.) and foliar microbes that germinate by producing appressoria and infection, is a more critical factor (Loper and Buyer 1991; Elad and Baker 1985; Keel et al. 1989). In a study on suppression of Fusarium wilt, *Pseudomonas putida* was capable of producing agglutinin. Compared to *P. putida* mutant strain deficient in agglutinin production revealed that *P. putida* producing agglutinin was able to better colonize the rhizosphere, yielding ultimately a better protection as compared to the mutant strain, also highlighting the role of this lectin (Anderson et al. 1988; Tari and Anderson 1988). Similarly, soil microbial communities can protect plants with higher efficacy because of their ability for a quick colonization, thus limiting the resources for other pathogenic microbes.

The availability of micronutrients such as iron, is limited depending on soil pH, oxidization state and aeration. Iron may be present in soil in its ferric form, which is extremely insoluble in water, lowering its concentration. Competition of this essential nutrient is very important for a BCA success. Almost all microorganisms produce siderophores which have ability of bind iron for their survival (Shahraki et al. 2009). There is a direct correlation between biocontrol ability of pseudomonads and siderophore production.

5.8 Resistance: Induced or Systemic

Many stimuli such as light, water or physical stress, nutrients availability and temperature are important as they may induce or affect resistance of host plants (Moyne et al. 2001; Van Wees et al. 1997; Vallad and Goodman 2004; Leeman et al. 1995). Depending on the source, amount and type of stimulus, resistance can be systemic or local (Van Loon et al. 1998; De Meyer and Hofte 1997; Zhang et al. 2002; Audenaert et al. 2002; Kloepper et al. 1980; Leeman et al. 1995; Vallad and Goodman 2004; Van Wees et al. 1997; Van Peer and Schippers 1992). The pathway of systemic or local resistance induced by BCA has been characterized recently. The first pathway relies on the production of salicylic acid after infection by pathogens, which ultimately increases the expression of pathogenesis-related portions (PR proteins) yielding a systemic acquired resistance (Vallad and Goodman 2004). The second pathway is characterized by the production of jasmonic acid or ethylene after infection by a mild or localized pathogen or parasite, which results in induced systemic resistance (Van Wees et al. 1997; Van Loon et al. 1998; Leeman et al. 1995; Audenaert et al. 2002; Klcepper et al. 1980; De Meyer and Hofte 1997; Zhang et al. 2002; Van Peer and Schippers 1992). When multiple stimuli are received and processed, plants may activate the various pathways of resistance. The strength and duration of the host resistance very likely changes over time. If we can control the stimulus then we can control the induction of resistance. The microbial community associated with plants is detected and plants respond to the quantitative or qualitative changes and signals. This interaction also indicates that further induction stimuli exerted by any microbes or chemical will not improve the resistance, health or productivity set in place by the plant (Vallad and Goodman 2004).

5.9 Application Methods

Extensive research, management and knowledge are required for successful application of BCA (Shah-Smith and Burns 1997; Baker 1987; Heydari et al. 2004; Cook 1993). The profitability, appreciation and crop depending requirements need to be considered for biocontrol application. Overall application includes biological products such as microbial fungicides. When these are applied, growth and development of natural soil inhabitants may also be indirectly supported, which in turn further reduces the pathogens activity (Cook 1993: Shah-Smith and Burns 1997; Heydari et al. 2004).

Direct application of BCA to the infection site is another strategy, such as for antagonistic bacteria and fungi for seed treatment and coating. *Trichoderma harzianum* and *P. fluorescens* are applied to stored fruits for protection. Direct application is widely used for application of BCA as shown by success in management of some pathogens (El-Ghaouth et al. 2000; De Capdeville et al. 2002; Janisiewicz and Peterson 2004). Another way of application of fungal biocontrol is one place application. This methods depends on the BCA multiplication and spread. Once are applied at one place with a low population density they will develop and spread to other parts of the plant to protect. In general, hypovirulent applications are considered to be applied once and to develop with time, ultimately spread in whole field without requiring yearly applications. However, to maintain the population of BCAs occasional treatments may be needed (Milgroom and Cortesi 2004).

5.10 Future Prospects

Since 1970s, biological control has become a mature science and got support from both public and private sectors. This is shown by the increasing literature data on biological control published on journals of both plant pathology and entomology, and by the funding from national agencies (i.e. USDA) for biocontrol research. Some research grants include USDA IR-4, Section 406 program, Regional IPM grants and integrated organic programs. Over the last year, researcher learnt much more about biological control. There is still a need to develop new and different strategies that can provide answers on emerging issues and invasive pathogen species. With the advancement of science, researcher are able to characterize both BCA and pathogens increasing the general understanding of many pathosystem. Cellular and molecular studies encouraged researcher to develop new techniques. Some of the areas which need to be addressed for developing fruitful biological control method are indicated as follows.

5.10.1 Antagonistic Microbes Ecology

The establishment of biocontrol agents largely depends on a number of ecological factors, also affecting the activity and performance of the microorganisms applied. At this regard some questions need to be answered and clarified:

1. How are the antagonists and their target pathogens distributed in the environment?
2. How important environmental conditions affect the BCA activity?
3. How do different management practices affect the naturally existing and introduced microorganisms?
4. Which factors regulate the suppressive activity of a biocontrol agent?

5.10.2 Application Method

For enhancing the BCA activity work on application methods is still needed. The investigation must focus on the development of application methods which can increase the BCA effectiveness. The works needed to be done is as follow:

1. Searching for new strains and their variants.
2. Using advanced techniques i.e. genetic engineering of microbes for increasing applications.
3. Developing proper formulations.

5.10.3 New Strains and New Mechanisms

As fungal pathogens always pose a great threat to crops, it must be remarked that each pathogen is different and it ability to cause disease differs consequently. Therefore, it is very important to explore the natural diversity of species to find new strains with different mechanism of biocontrol. The following aspects must hence be also investigated.

1. Characterization of new strains and their use.
2. Genetic study of BCAs to explore new genes or combination of genes that can be manipulated.
3. Instead of using single strain, focus should be given to the development of a combination of strains with diverse mechanisms.

5.10.4 Integrated Pest Management

As integration of biological control strategies with other disease management strategies is very important, some aspects to consider are as follow:

1. Cropping patterns should be chose to get maximum benefit from biological control.
2. IPM and biocontrol strategies must fit each other.
3. Compatibility of BCA and plant cultivars must be considered during breeding programs.

5.11 Research and Development

Biological control can fulfil the gap originated by farmers' reluctance to use chemical pesticides and by the search for new strategies of disease management. Actual lines of action such as crop rotation, breeding programs, use of tillage and/or resistant cultivars etc. are not always sufficient for a successful management. Next step is the application of BCAs as amendments, inoculants or of active ingredients derived directly from natural sources. Biological control has no or just a little effect on environment and other non-pathogenic species (Jacobsen et al. 2004; Guetsky et al. 2001). If growers cannot sustain their production then they can still use less harmful and less specific chemical toxins. However, as biocontrol is very successful under lab condition, when it comes to the commercial application there are still constraints including stability, efficacy, cost and mode of application. As more research and a better understanding of biological control are needed, the research on adoption of BCA as part of IPM is expected to increase in the years to come.

5.12 Biological Control of Nematodes

In the last several nematicides have been withdrawn from the market due to health and environmental hazards associated with production and use. Due to increasing public concern, there has been interest in the development of alternative methods of control, including use of BCA. A number of studies showed that nematophagous fungi and bacteria increase under perennial crops and monocultures. As such they may control some nematode pests, including cyst and root-knot nematodes (Stirling 1985; Stirling 2011).

Nematode-suppressive soils have been reported from around the world and include documented cases of effective biological control (Yang et al. 2012; Giné et al. 2016). Finally, a number of commercial products based on nematophagous fungi and bacteria have been developed.

5.13 Biological Control of Plant Parasitic Nematodes

In soil plant parasitic nematodes are attacked by natural enemies which can be exploited for practical use in field conditions. Many predators such as fungi, nematodes or other predacious organisms such as insects and mites have been identified. Parasitic fungi and bacteria have been investigated as promising BCA. Their development in nematode biocontrol has been reviewed (Stirling 2011; Timper 2014; Devi and George 2018).

5.14 Fungi as Biocontrol Agents of Nematodes

Among various microorganisms which parasitize or prey on plant parasitic nematodes, fungi have vital position and possess great biocontrol potential. In various soil types plant parasitic nematodes are destroyed by fungi on continuous basis. The biocontrol potential of fungi against females of cyst nematodes was observed firstly by Kuhn in 1877. There are more than 70 genera and 160 fungal species which are associated with nematodes.

5.14.1 Predacious Fungi

There are more than 50 species of this group of fungi which kill nematodes. The nematode trapping efficiency decreases with the life span of the fungi. Their efficacy can be increased by soil amendments with organic matter. According to predacious activities, they can be classified as endoparasitic or trapping fungi.

5.14.1.1 Endoparasitic Fungi

These are mostly specific to single species or group of nematodes. Being obligate parasites, therefore, they are difficult to culture in absence of the host. *Hirsutella, Meria, Nematophthora* and *Nematoctonus* are ideal BCAs against nematodes. These fungi in general attack the nematode host by adhesive spores, from which a germ tube develops which later penetrates the nematode cuticle. The fungal hyphae divide and multiply throughout the nematode body and absorb its contents. The hyphae then emerge from the dead carcass. *Catenaria vermicola* often attacks *Heterodera schachtii*, while *Nematophthora* and Pchonia chlamydosporia have been reported as parasites of *H. avenae* (Kerry 2000). These fungi have a key role in natural regulation of the population dynamics of plant nematodes, in some soils.

5.14.1.2 Trapping Fungi

The nematode trapping fungi develop adhesive networks, sticky knobs or constricting rings formed by the mycelium. All are specialized hyphal structures capable to capture nematodes. The fungi then digest the nematode internal tissues. The nematode trappers may be grouped as follows.

(I) *Sticky branches*: the mycelium bears small lateral branches which join to form loops (anastomose). The plant nematodes are trapped in this loop, as those produced, for example, by *Dactylella lobata.*

(II) *Sticky networks*: the hyphae curl around and anastomose forming similar branches. These loops form three dimensional structures. Nematodes are trapped in the network due to the hyphae sticky surface as those, for example, of *Arthrobotrys oligospora*.

(III) *Sticky knobs*: small spherical or sub-spherical lobes are present on short lateral hyphae, the terminal lobe being sticky to trap nematodes. *Dactylella ellipsospora* illustrates this mechanism of trapping.

(IV) *Constricting rings*: the short branch of a fungal hypha anastomoses with its base to form a ring. Whenever a nematode enters the ring, a swelling of the inner ring wall occurs, dramatically reducing the ring cavity and eventually strangling and/or immobilizing the nematode. Subsequently, hyphae penetrate and kill the prey, as for example in *Dactylaria bembicodes*.

(V) *Non-constricting ring:* This trap is similar to the previous one but the ring develops as an infective structure and kills the nematode. *Dactylaria candida* forms such kind of ring.

Trapping fungi are easy to produce *in vitro* and have wide host ranges. Although nematode trapping fungi did not attain much popularity A commercial product named Royale 300® formulated from one isolate of *Arthrobotrys* sp. has been commercialized for some time for nematode management of *Ditylenchus myceliophagous* in mushroom production. Another product based on *Arthrobotrys* spp. (Royale 350®) has been commercialized for control of root knot nematodes (Cayrol 1983).

5.14.2 Parasitic Fungi

A number of fungi from this group has elicited more interest in management of plant parasitic nematodes, as compared to other fungal groups discussed previously. They can survive even in absence of their hosts and can be cultured axenically. These parasitic fungi can be isolated from eggs, juveniles or adult nematodes. Many of them have preferential hosts, and have a certain degree of nematode density regulation. They are also called as opportunistic fungi because they parasitize some nematode stages whenever in contact. Sedentary stages of cyst and root knot nematodes are susceptible to these fungi, present in soil or as endophytes in roots. Whenever egg masses or nematode cysts come in contact with such fungi, these develop on them and eventually parasitize eggs. *Cylindrocarpon, Exophila, Fusarium, Gliocladium, Paecilomyces, Phoma* and *Pochonia chlamydosporia* are most common examples of such fungi. Their damaging action occurs through the enzymatic disruption of nematode structural elements such as the egg shells or cuticle. Physiological effects on nematodes, or when endophytic, on roots, may also occur including, but not limited to, the biosynthesis of toxic metabolites.

5.15 Bacteria as Biocontrol Agents of Nematodes

Nematode parasitic bacteria may be grouped into the following categories:

- Obligate parasites
- Rhizosphere bacteria
- Antagonistic bacteria

5.15.1 Obligate Parasite

Pasteuria spp. are Gram+ obligate bacterial parasites, forming durable endospores. They are parasitic to a number of nematodes, including plant parasitic and free living species. Species of *Pasteuria* include *P. penetrans*, *P. thornei, P. usgae* and *P. nishizawe* parasitizing different nematode hosts. *Pasteuria* spp. are worldwide in distribution and have been reported from 323 nematode species belonging to several different genera of free-living, predatory and plant-parasitic nematodes (Stirling 2011). *Pasteuria penetrans* is one of the most important natural antagonist of root-knot nematodes (*Meloidogyne* spp.) and is highly specific, even at the host population level. The host specificity biocontrol potential of these bacteria has been revealed on many crops. *Pasteuria nishizawe* attacks the soybean cyst nematode *Heterodera glycines*, whereas *P. usgae* is associated to sting nematodes (*Belonolaimus* spp.). These bacteria have a significant role in some suppressive soils. An immediate control root-knot nematode can be achieved by applying up to 10^5 endospores /g of soil, while at 1000–5000 endospores/ g of soil it may take 3 years for establishment in soil (Chen and Dickson 1998; Stirling 2011; Kokalis-Burelle 2015). Field studies, however, showed that *P. penetrans* and other similar species may persist in soil for a long time (Ciancio and Quénéhervé 2000).

The endospores of *Pasteuria* spp. are non-motile and remain in soil, and get attached to the cuticle of passing nematode. Several hundreds of such spores may attach to the cuticle of a single a nematode. However the host may become infected by only one such propagule. The endospore germinates and the germ tube penetrates the cuticle, producing micro colonies in the nematode body. Parasitized nematodes become sterile because as the reproductive system does not develop. Moreover, spore-encumbered juveniles may also fail to reach the root These bacteria are also compatible with certain nematicides i.e. no impact was found for some nematicides on endospore survival and infectivity.

Prior to the development of industrial artificial methods for mass culturing, *P. penetrans* was produced for experimental purposes using the host *Meloidogyne* spp., on a suitable host plant. The nematode-infected roots containing females filled with endospores are powdered, sieved through fine mesh and used as a powder (Stirling 2011). Formulates based on *P. penetrans* endospores for seed coating are now commercially available.

5.15.2 Rhizosphere Bacteria

Another strategy used for nematode biocontrol is based on the introduction of bacteria colonizing the host plant rhizosphere, called rhizobacteria. They grow in the rhizosphere providing a certain defense from pathogens attacks and are considered ideal as biocontrol agent. Some rhizobacteria also have positive effects on plant growth. They are known as plant growth promoting rhizobacteria (PGPR) or plant health promoting rhizobacteria (PHPR). Applications to sugar beet and potato seed significantly lowered early root infestation by the sugar beet cyst nematode *H. schachtii* and the potato cyst nematode *Globodera pallida*.

Many bacteria antagonistic to nematodes are from genus *Pseudomonas*. Others belong to *Agrobacterium, Anthrobacter*, *Bacillus* (i.e. *B. subtilis*, *B. cereus, B. sphaericus*), and *Serratia*. The role of *Ps. fluorescence* as biocontrol agent has been investigated as it appeared effective against both root-knot and cyst nematodes. Bare root dip treatments of tomato seedlings in a suspension of *P. fluorescence* proved to be effective against root-knot nematodes. *Agrobacterium radiobacter* and *B. sphaericus* produce toxic metabolites which affect penetration of *G. pallida*, consequently increasing crop production. *Azotobacter*, including aerobic, Gram- and nitrogen fixing species, is also gaining importance in management of plant parasitic nematodes.

5.15.3 Bacterial Antagonists

Many soil bacteria produce butyric acid, cyanide, exotoxins and hydrogen sulphides. These compounds are antagonistic to nematodes. Compounds such as ammonia and hydrogen sulphide have poisonous effects on root- knot nematodes of rice. *Bacillus thuringiensis* var. *thuringiensis*, *Chromobacterium* spp., *Clostridium butyricum* and *Desulfovibrio desulfuricans* are important antagonistic bacteria against plant parasitic nematodes. *Bacillus thuringiensis* possesses a biocontrol potential. It is an aerobic, Gram+ and produces endospores. There are more than 200 isolates of *B. thuringiensis.* Although it is well known for its pathogenicity to insects, some strains have been reported to be effective against the eggs and juveniles of root-knot nematodes, or against other nematodes such as seed gall, leaf and bud, and lesion nematodes, free living and animal parasitic species (Zuckerman et al. 1993; Sharma 1994; Leyns et al. 1995; Wei et al. 2003).

5.16 Nematodes as BCA of Nematodes

Predatory nematodes may contribute to biocontrol of plant parasitic nematodes. In 1917 Cobb reported about the effectiveness of *Mononchus* sp. against plant parasitic nematodes. Predatory nematodes may offer some advantage over fungi and bacteria as BCA because of their active movement and host searching ability. These nematodes are provided with specialized teeth to catch and swallow the prey. Addition of organic amendments helps to increase their multiplication, given the increase in free living species. Predatory nematodes belong to four orders i.e. Aphelenchida, Diplogasterida, Dorylaimida and Mononchida, differing for their feeding parts, searching behavior and feeding mechanisms (Aatif et al. 2015).

5.16.1 Aphelenchida

The members of this order have piercing and sucking sort of stylet. Their prey nematodes are *Acrobeloides* spp., *Bursilla labiate* and other *Aphelenchoides* spp., such as *A. richardsoni*.

5.16.2 Dorylaimida

They occur in many soil types and habitats and are characterized by piercing and sucking stylets. They actively search their preys. *Eudorylaimus obtusicaudatus* was reported as feeding on eggs inside *H. schachtii* cysts. The host range includes nematodes such as *Aphelenchus avenae*, *Panagrellus redivivus*, *Anguina tritici* and *Tylenchulus semipenetrans* (Bilgrami 2008).

5.16.3 Diplogasterida

Members of this group have cutting-sucking mouths, and may feed also on bacteria. *Diplogaster* or *Koerneria* spp. may be multiplied on both prey nematode and different bacterial species, either *in vitro* and *in vivo*, because of their facultative feeding habit. *Mononchoides fortidens* was cultured on *Rhabditis* spp. on agar with skim milk powder (Devi and George 2018).

5.16.4 *Mononchida*

These predators have cutting and sucking mouths, and feed on nematodes such as *Meloidogyne* spp., *Pratylenchus* spp., *Paratylenchus* spp., *Meloidodera* spp. and *Tylenchorhynchus* spp. which may even be swallowed entire. A single *Mononchus papillatus* was found to destroy more than one thousand juveniles of *H. radicicola* during its life (Bilgrami 2008; Devi and George 2018).

5.16.5 *Symbionts of Entomopathogenic Nematodes*

Entomopathogenic nematodes (EPN) are also used in biocontrol programs of plant nematodes. EPN of genera *Heterorhabditis* and *Steinernema* are BCA of many insects, and are mutually associated with endosymbiotic bacteria *Photorhabdus* and *Xenorhabdus*, respectively. A potential antagonistic effect of this symbiosis has been reported for plant parasitic nematodes, that were suppressed by the production of secondary metabolites with a nematicidal effect Such as ammonia, indole and stilbene derivatives. These products are toxic to J_2 stages and eggs of root-knot nematode and adults and J_4 of pine wood nematodes (Hu et al. 1999). Some other features such as competition among nematode groups for root exudates and space in the rhizophere support the EPN suppressive effect on plant parasitic nematodes. Some nematodes are also available commercially as NemAttack™, NemaSeek™ etc.

5.17 Environmental Concerns

The establishment and activity of nematode natural enemies depend on various factors such as species, density and rate of development of the natural enemy, soil conditions, and host plant. Temperature and relative humidity are important factors affecting the biocontrol effectiveness. The understanding of these interactions is essential for effective biocontrol program. Temperature directly affects sporulation and growth of most BCA as well as of targeted species. Soil humidity also influences the survival and growth of bacteria as well as nematodes. In most cases, however, it does not limit growth of fungi. Soil structure and texture also affect the activity of nematode, as well as the growth and spread of microorganisms.

The incorporation of BCA in soil is difficult, and broadcasting methods are not worth due to costs. The use of agents in cereal crops depends on the organisms present in the root zone, and BCA are applied either on seed or in rows. Residual soil microflora also has a static effect, opposing introduced species and competing for energy sources. Consequently, this process may affect the BCA performance. As plant parasitic nematodes are mostly mobile, at least during their larval stages, to infect these pests the BCA evolved adhesive spores or traps. Sedentary nematodes may be parasitized by fungal hyphae in the root zone without forming special infective structures. Cyst nematode females may be destroyed by fungi when they are exposed on the root surface, or the fungi can reduce their fertility rate or parasitize the eggs. In the case of root knot nematodes the eggs are exposed in the root zone, whereas the females remain inside roots. Therefore, the egg parasitic fungi must rapidly kill their eggs to avoid competition by other soil inhabitants. The degradation of soil amendments by non-parasitic microbes releases nematicidal compounds. For example chitin degrading bacteria releases NH_3, which is lethal to many nematodes.

5.18 Future Prospects

The BCA of plant parasitic nematodes have an important role in regulating their population densities. *Pasteuria penetrans, Ps. fluorescens* and nematicidal strains of *B. thuringiensis* have potential to act as effective BCA. The obligate parasite *P. penetrans* possess essential capacities for biocontrol, except its high host specificity. Its obligate behavior confers a certain degree of independence from other soil bacteria, as concerns competition for food sources. Also a number of opportunistic fungi such as *Trichoderma* and *Pochonia* are suitable as BCA of plant parasitic nematodes. They are ubiquitous with a rapid dispersal, and are abundant in the rhizosphere. They can be easily produced in axenic culture for introduction into soil.

The development of integrated pest management program encouraged the integration of multiple control practices, including use of BCA. Researchers must focus on the following features for the management of plant parasitic nematodes, with the help of potential BCA, specific for any concerned ecosystem.

(i) Identification and selection of effective strains of natural enemies.
(ii) Development of a standardized and effective rearing, culturing, storage, handling, release and evaluation procedures.
(iii) Understanding the biology, ecology, physiology, genetic behavior of BCA.
(iv) Identifying most efficient host genotypes-symbionts.
(v) Developing mass culture techniques for field applications.
(vi) Full demonstration and assessment of BCA benefits under field conditions.
(vii) Education of public for BCA effective utilization of.

5.19 Major Developments in Biomanagement of Plant Viruses

Plant viruses are devastating pathogens of the plants. They are unique and distinguished pathogens because of their intrinsic properties. Unlike other plant pathogens, once infected by a virus, the plant will support the virus multiplication throughout its whole life-cycle. The viral infection may have a systemic nature. So far, control and eradication of plant viruses through chemicals is not successful due to the virus biology. Although, biological control of plant diseases caused by fungi, bacteria and nematodes is gaining importance, plant viruses still need new technologies to get any breakthrough.

There is a lot of literature available for the suppression and management of bacteria, fungi and nematodes, but few data are available regarding biocontrol of virus diseases.

Plant viruses, unlike other pathogens, are difficult to control but only can be managed by different methods under an integrated disease management strategy. Selective and non-selective approaches to their control are referred as i) cultural practices and legislations, ii) host resistance and biological control, respectively. There is also a vital interaction between the insect vectors, their BCA use and the application of chemicals or physical (i.e. insect traps, nets etc.) methods.

The progress of viral infection depends on the initial source of inoculum. When biocontrol and host resistance are applied, they will only work for the specific virus targeted (Bos 1992; Buddenhagen 1977; Thresh 1980, 1981, 1982; Jones 2006).

For virus management, the virus-vector relationship need to be understood. Among BCA, PGPR, including a mixture of microorganisms which are beneficial for the plant health (Pal and McSpadden Gardener 2006), play an important role in strengthening the host by improving its nutritional status. Success biocontrol stories against viruses are under the way and need to be investigated for effective management of virus epidemics.

Tomato leaf curl virus (ToLCV) is an important and destructive virus in India and major tomato growing areas in Asia. It is one leading limiting factor in tomato production and is responsible of substantial losses. The virus is transmitted through whitefly (*Bemisia tabaci*) (Muniyappa and Veeresh 1984). The major symptoms are upward curling, yellowing of leaves and abortion of flowers (Gafni 2003). ToLCV belongs to the geminivirus group, family *Geminiviridae* (subgroup – III) (Saikia and Muniyappa 1989; Harrison et al. 1991). Management was attempted through different approaches (Mishra et al. 2014), including the use of agrochemicals vs the insect vectors, that may also imbalance the natural microbial community of beneficial species. However, the virus biocontrol is difficult as there is a complex interaction with the vectors. Therefore, although chemicals may control the vectors in the short term, they may induce the evolution of new resistant insects, in absence of beneficial organisms in the field. To avoid the use of chemicals, resistant plant varieties represent a more favourable option that can be referred to as biological control. However, lack of dominant resistant genes and emerging of new virus races are important constraints in using new varieties (Mishra et al. 2014).

PGPR commonly used for growth promotion may sustain effective biocontrol for virus management. Elicitors were considered to sustain the plant physiology, as shown by comparison between healthy and virus infected samples. The elicitors usually trigger defense mechanisms against fungi, bacteria and viruses. Among them, chitosan was the most important helping in regulating resistant genes. Chitin and chitosan performed well in controlling some plant viruses (Abdelbasset et al. 2010).

Streptomyces, *Pseudomonas*, *Gliocladium*, *Bacillus*, and *Trichoderma* spp. with different PGPR showed a potential against viruses in tomato and other crops (Kandan et al. 2007; Srinivasan et al. 2005; Kavino et al. 2008; Kirankumar 2008). ToLCV was managed through different rhizobacterial isolates with or without PGPR and chitosan treatments (Mishra et al. 2014). Three different PGPRs viz., *Pseudomonas* 206(4), *Pseudomonas* B-15 and *Pseudomonas* JK-16 were used against ToLCV. Chitosan in combination with *Pseudomonas* spp. not only suppressed the virus but also greatly improved plant growth, biomass, chlorophyll content, and yield. High phenolic compounds were observed in chitosan and rhizobacterial treatments against ToLCV. Additionally, peroxidase (PO) and phenylalanine ammonia lyase (PAL) activities were also increased. Polyphenol oxidase (PPO) and chitinase activities were also high in plants treated with chitosan and PGPRs (Mishra et al. 2014).

Kandan et al. (2007) observed an increase in phenolic compounds that contributed in protecting cowpea against **Tomato spotted wilt virus** when treated with *P. fluorescens*. PAL activity plays an important role in defense reactions such as the phenyl propanoid metabolism (Harish 2005). Harish (2005), managed **Banana bunchy top virus** (BBTV) through a biocontrol approach based on the increase in peroxidase activity. Production of hydrogen peroxides and lignification is linked with PO and PPO activities which in turn directly inhibit the pathogen or indirectly restrict its development (Silva et al. 2004). Chitosan was used in potato plants and reported a response involving callose, ribonuclease and β-1,3 glucanase against **Potato virus X** (PVX).

Several *Streptomyces* spp. were used for management of **Cucumber mosaic virus** (CMV) in cucumber plants, in relation to a hypersensitive response. Isolates of *Streptomyces* could inhibit the production of local lesions in treated cucumber plants, as compared to control. Induced systemic resistance was detected through different biological assays (El-Dougdoug et al. 2012).

Mixture of bacterial isolates has been used for management of **Cotton leaf curl virus** disease (CLCuD) (Ramzan et al. 2016). *Bacillus* spp. and *Pseudomonas* spp. antimicrobial activities were reported for the phosphate solubilization and production of indole-acetic acid.

Some rhizosphere species may result beneficial for the host plant growth. Sometimes they are used to reduce the impact of a disease (Murphy et al. 2000; Kandan et al. 2005). These microorganisms may provide protection to the plants from viruses through different mechanisms, such as by improving growth or indirectly through antibiosis, production of lytic enzymes, and induced systemic resistance (Haas and Keel 2003; Jeger et al. 2009). Management of CLCuD through BCAs is not a common practice as only one attempt has been reported (Ramzan et al. 2016).

References

Aatif, H. M., Javed, N., Khan, S. A., Ahmed, S., & Raheel. (2015). Virulence of entomopathogenic nematodes against *Meloidogyne incognita* for invasion, development and reproduction at different application times in brinjal roots. *International Journal of Agriculture and Biology, 17*, 995–1000.

Abdelbasset, E. H., Adam, L. R., Hadrami, I. E., & Daayf. (2010). Chitosan in plant protection. *Marine Drugs, 8*, 968–987.

Agrios, N. A. (1988). *Plant pathology* (3rd ed., pp. 220–222). Cambridge, MA: Academic Press.

Anderson, A. J., Habibzadegah-Tari, P., & Tepper, C. S. (1988). Genetic studies on the role of an agglutinin-in root colonization by *Pseudomonas putida*. *Applied and Environmental Microbiology, 54*, 375–380.

Anderson, L. M., Stockwell, V. O., & Loper, J. E. (2004). An extracellular protease of *Pseudomonas fluorescens* inactivates antibiotics of *Pantoea agglomerans*. *Phytopathology, 94*, 1228–1234.

Audenaert, K., Pattery, T., Cornelis, P., & Hofte, M. (2002). Induction of systemic resistance to *Botrytis cinerea* in tomato by *Pseudomonas aeruginosa* 7NSK2: Role of salicylic acid, pyochelin and pyocyanin. *MPMI, 15*, 1147–1156.

Baker, K. F. (1987). Evolving concepts of biological control of plant pathogens. *Annual Review of Phytopathology, 25*, 67–85.

Bankhead, S. B., Landa, B. B., Lutton, E., Weller, D. M., & Gardener, B. B. (2004). Minimal changes in rhizosphere population structure following root colonization by wild type and transgenic biocontrol strains. *FEMS Microbiology Ecology, 49*, 307–318.

Bargabus, R. L., Zidack, N. K., Sherwood, J. E., & Jacobsen, B. J. (2002). Characterization of systemic resistance in sugar beet elicited by a non-pathogenic, phylosphere-colonizing *Bacillus mycoides*, biological control agent. *Physiological and Molecular Plant Pathology, 61*, 289–298.

Bargabus, R. L., Zidack, N. K., Sherwood, J. E., & Jacobsen, B. J. (2004). Screening for the identification of potential biological control agents that induce systemic acquired resistance in sugar beet. *Biological Control, 30*, 342–350.

Benhamou, N. (2004). Potential of the mycoparasite, *Verticillium lecanii*, to protect citrus fruit-against *Penicillium digitatum*, the causal agent of green mold: A comparison with-the effect of chitosan. *Phytopathology, 94*, 693–705.

Benhamou, N., & Chet, I. (1997). Cellular and molecular mechanisms involved in the interaction between *Trichoderma harzianum* and *Pythium ultimum*. *Applied and Environmental Microbiology, 63*, 2095–2099.

Berg, G., Krechel, A., Ditz, M., Sikora, R. A., Ulrich, A., & Hallmann, J. (2005). Endophytic and ectophytic potato-associated bacterial communities differ in structure and antagonistic function against plant pathogenic fungi. *FEMS Microbiology Ecology, 51*, 215–229.

Biermann, B., & Linderman, R. G. (1983). Use of vesicular-arbuscular mycorrhizal roots, intraradical vesicles and extraradical vesicles as inoculum. *The New Phytologist, 95*, 97–105.

Bilgrami, A. L. (2008). Biological control potentials of predatory nematodes. In A. Ciancio & K. G. Mukerji (Eds.), *Integrated management and biocontrol of vegetable and grain crops nematodes* (pp. 3–28). Dordrecht: Springer.

Bos, L. (1992). New plant virus problems in developing countries: A corollary of agricultural modernisation. *Advances in Virus Research, 38*, 349–407.

Buddenhagen, I. W. (1977). Resistance the vulnerability of tropical crops in relation to their evolution and breeding. *Annals of the New York Academy of Sciences, 287*, 309–326.

Bull, C. T., Shetty, K. G., & Subbarao, K. V. (2002). Interactions between Myxobacteria, plant pathogenic fungi and biocontrol agents. *Plant Disease, 86*, 889–896.

Cayrol, J. C. (1983). Biological control of *Meloidogyne* by *Anthrobotrys irregularis*. *Revue de Nematologie, 6*, 265–273.

Chen, Z. X., & Dickson, D. W. (1998). Review of *Pasteuria penetrans*: Biology, ecology and biological control potential. *Journal of Nematology, 30*, 313–340.

Chisholm, S. T., Coaker, G., Day, B., & Staskawicz, B. J. (2006). Host-microbe interactions: Shaping the evolution of the plant immune response. *Cell, 124*, 803–814.
Ciancio, A., & Quénéhervé, P. (2000). Population dynamics of *Meloidogyne incognita* and infestation levels by *Pasteuria penetrans* in a naturally infested field in Martinique. *Nematropica, 30*, 77–86.
Cook, R. J. (1993). Making greater use of introduced microorganisms for biological control of plant pathogens. *Annual Review of Phytopathology, 31*, 53–80.
De Capdeville, G., Wilson, C. L., Aist, B. S. V., & J.R. (2002). Alternative disease control agents induce resistance to blue mold in harvested Red Delicious apple fruit. *Phytopathology, 92*, 900–908.
De Meyer, G., & Hofte, M. (1997). Salicylic acid produced by the rhizobacterium *Pseudomonas aeruginosa* 7NSK2 induces resistance to leaf infection by *Botrytis cinera* on bean. *Phytopathology, 87*, 588–593.
Devi, G., & George, J. (2018). Predatory nematodes as bio-control agents against plant parasitic nematodes: A review. *Agricultural Reviews, 39*, 55–61.
Duchesne, L. C. (1994). Role of ectomycorrhizal fungi in biocontrol. In F. L. Pfleger & R. G. Linderman (Eds.), *Mycorrhizae and plant health* (pp. 27–45). Paul: APS Press.
Elad, Y., & Baker, R. (1985). Influence of trace amounts of cations and siderophore-producing pseudomonads on chlamydospore germination of *Fusarium oxysporum*. *Phytopathology, 75*, 1047–1052.
El-Dougdoug, K. A., Ghaly, M. F., & Taha, M. A. (2012). Biological control of cucumber mosaic virus by certain local Streptomyces isolates: inhibitory effects of selected five Egyptian isolates. *International Journal of Virology, 8*, 151–164. https://doi.org/10.3923/ijv.2012.151.164.
El-Ghaouth, A., Smilanick, J. L., Brown, G. E., Ippolito, A., Wisniewski, M., & Wilson, C. L. (2000). Application of *Candida saitoana* and glycolchitosan for the control of post harvest diseases of apple and citrus fruit under semi-commercial conditions. *Plant disease, 84*, 243–248.
Fitter, A. H., & Garbaye, J. (1994). Interactions between mycorrhizal fungi and other soil organisms. *Plant and Soil, 159*, 123–132.
Gafni, Y. (2003). Tomato yellow leaf curl virus, the intracellular dynamics of a plant DNA virus. *Molecular Plant Pathology, 4*(1), 9–15.
Garcia-Garrido, J. M., & Ocampo, J. A. (1989). Effect of VA mycorrhizal infection of tomato on damage caused by *Pseudomonas syringae*. *Soil Biology and Biochemistry, 21*, 165–167.
Giné, A., Carrasquilla, M., Martínez-Alonso, M., Gaju, N., & Sorribas, F. J. (2016). Characterization of soil suppressiveness to root-knot nematodes in organic horticulture in plastic greenhouse. *Frontiers in Plant Science, 7*, 164.
Guetsky, R., Shtienberg, D., Elad, Y., & Dinoor, A. (2001). Combining biocontrol agents to reduce the variability of biological control. *Phytopathology, 91*, 621–627.
Haas, D., & Keel, C. (2003). Regulation of antibiotic production in rootcolonizing *Pseudomonas* spp. and relevance for biological control of plant disease. *Annual Review of Phytopathology, 41*, 117–153.
Harish, S. (2005). *Molecular biology and diagnosis of Banana bunchy top virus and its management through induced systemic resistance*. Ph.D. Thesis. Coimbatore: Tamil Nadu agricultural University.
Harrison, B. D., Swanson, M. M., McGrath, P. F., & Fargette, D. (1991). *Patterns of antigenic variation in whitefly-transmitted geminiviruses*. Report of the Scottish Crop Research Institute for 1990, pp. 88–90.
Harman, G. E., Howell, C. R., Viterbo, A., Chet, I., & Lorito, M. (2004). *Trichoderma* species-opportunistic, avirulent plant symbionts. *Nature Reviews. Microbiology, 2*, 43–56.
Heydari, A. (2007). Biological control of turfgrass fungal diseases. In M. Pessaraki (Ed.), *Turfgrass management and physiology*. Boca Raton: CRC press.
Heydari, A., & Misaghi, I. J. (1998). Biocontrol activity of *Burkholderia cepacia* against *Rhizoctonia solani* in herbicide-treated soils. *Plant and Soil, 202*, 109–116.

Heydari, A., & Misaghi, I. J. (1999). Herbicide-mediated changes in the populations and activity of root associated microorganisms: A potential cause of plant stress. In M. Pessarakli (Ed.), *Handbook of plant and crop stress* (2nd ed.). New York: Marcel Dekker Press.

Heydari, A., & Misaghi, I. J. (2003). The role of rhizosphere bacteria in herbicide-mediated increase in *Rhizoctonia solani*-induced cotton seedling damping-off. *Plant and Soil, 257*, 391–396.

Heydari, A., Misaghi, I. J., & McCloskey, W. B. (1997). Effects of three soil-applied herbicides on populations of plant disease suppressing bacteria in the cotton rhizosphere. *Plant and Soil, 195*, 75–81.

Heydari, A., Fattahi, H., Zamanizadeh, H. R., Zadeh, N. H., & Naraghi, L. (2004). Investigation on the possibility of using bacterial antagonists for biological control of cotton seedling damping-off in greenhouse. *Applied Entomology and Phytopathology, 72*, 51–68.

Hoitink, H. A. J., & Boehm, M. J. (1999). Biocontrol within the context of soil microbial communities: A substrate-dependent phenomenon. *Annual Review of Phytopathology, 37*, 427–446.

Homma, Y., Sato, Z., Hirayama, F., Konno, K., Shirahama, H., & Suzui, T. (1989). Production of antibiotics by Pseudomonas cepacia as an agent for biological control of soilborne plant pathogens. *Soil Biology and Biochemistry, 21*(5), 723–728.

Howell, C. R., & Stipanovic, R. D. (1980). Suppression of *Pythium ultimum* induced damping-off of cotton seedlings by *Pseudomonas fluorescens* and its antibiotic pyoluteorin. *Phytopathology, 70*, 712–715.

Howell, C. R., Beier, R. C., & Stipanovic, R. D. (1988). Production of ammonia by *Enterobacter cloacae* and its possible role in the biological control of Pythium pre-emergence damping-off by the bacterium. *Phytopathology, 78*, 1075–1078.

Hu, K., Li, J., & Webster, J. M. (1999). Nematicidal metabolites produced by *Photorhabdus luminescens* (Enterobacteriaceae), bacterial symbiont of entomopathohenic nematodes. *Nematology, 1*, 457–469. https://doi.org/10.1094/PHI-A-2006-1117-02.

Islam, M. T., Hashidoko, Y., Deora, A., Ito, T., & Tahara, S. (2005). Suppression of damping-off-disease in host plants by the rhizoplane bacterium *Lysobacter* sp. strain SB-K88 is-linked to plant colonization and antibiosis against soilborne peronosporomycetes. *Applied and Environmental Microbiology, 71*, 3786–3796.

Jacobsen, B. J., Zidack, N. K., & Larson, B. J. (2004). The role of *Bacillus*-based biological control agents in integrated pest management systems: Plant diseases. *Phytopathology, 94*, 1272–1275.

Janisiewicz, W. J., & Peterson, D. L. (2004). Susceptibility of the stem pull area of mechanically harvested apples to blue mold decay and its control with a biocontrol agent. *Plant and Disease, 88*, 662–664.

Jeger, M. J., Jeffries, P., Elad, Y., & Xu, X. (2009). A generic theoretical model for biological control of foliar plant diseases. *Journal of Theoretical Biology, 256*, 201–214.

Jones, R. A. C. (2006). Control of plant virus diseases. *Advances in Virus Research, 67*, 205–244.

Kandan, A., Ramiah, M., Vasanthi, V., Radjacommare, R., Nandakumar, R., Ramanathan, A., & Samiyappan, R. (2005). Use of *Pseudomonas fluorescens*-based formulations for management of tomato spotted wilt virus (TSWV) and enhanced yield in tomato. *Biocontrol Science and Technology, 15*, 553–569.

Kandan, A., Ramiah, M., Vasanthi, V. J., Radjacommare, R., Nandakumar, R., Ramanathan, A., & Samiyappan, R. (2007). Use of Pseudomonas fluorescens-based formulations for management of tomato spotted wilt virus (TSWV) and enhanced yield in tomato. *Biocontrol Science and Technology, 15*(6), 553–569.

Kavino, M., Harish, S., Kumar, N., Saravanakumar, D., & Samiyappan, R. (2008). Induction of systemic resistance in banana (Musa spp.) against Banana bunchy top virus (BBTV) by combining chitin with root-colonizing Pseudomonas fluorescens strain CHA0. *European Journal of Plant Pathology, 120*(4), 353–362.

Katska, V. (1994). Interrelationship between vesicular-arbuscular mycorrhiza and rhizosphere microflora in apple replant disease. *Biologia Plantarum, 36*, 99–104.

Keel, C., Voisard, C., Berling, C. H., Kahir, G., & Defago, G. (1989). Iron sufficiency is a prerequisite for suppression of tobacco black root rot by *Pseudomonas fluorescens* strain CHA0 under gnotobiotic conditions. *Phytopathology, 79*, 584–589.
Kerry, B. R. (2000). Rhizosphere interactions and the exploitation of microbial agents for the biological control of plant-parasitic nematodes. *Annual Review of Phytopathology, 38*, 423–441.
Kessel, G. J. T., Kohl, J., Powell, J. A., Rabbinge, R., & van der Werf, W. (2005). Modeling spatial characteristics in the biological control of fungi at the leaf scale: Competitive substrate colonization by *Botrytis cinerea* and the saprophytic antagonist *Ulocladium atrum*. *Phytopathology, 95*, 439–448.
Kirankumar, R., Jagadeesh, K. S., Krishnaraj, P. U., & Patil, M. S. (2008). Enhanced growth promotion of tomato and nutrient uptake by plant growth promoting rhizobacterial isolates in presence of tobacco mosaic virus pathogen. *Karnataka Journal of Agricultural Sciences, 21*, 309–311.
Kloepper, J. W., Leong, J., Teintze, M., & Schroth, M. N. (1980). *Pseudomonas siderophores*: A mechanism explaining disease suppression in soils. *Current Microbiology, 4*, 317–320.
Kloepper, J. W., Ryu, C. M., & Zhang, S. (2004). Induced systemic resistance and promotion of plant growth by *Bacillus* spp. *Phytopathology, 94*, 1259–1266.
Kokalis-Burelle, N. (2015). *Pasteuria penetrans* for control of *Meloidogyne incognita* on tomato and cucumber, and *M. arenaria* on Snapdragon. *Journal of Nematology, 47*, 207–213.
Lafontaine, P. J., & Benhamou, N. (1996). Chitosan treatment: An emerging strategy for enhancing resistance of greenhouse tomato plants to infection by *Fusarium oxysporum* f.sp. *radicis-lycopersici*. *Biocontrol Science and Technology, 6*, 111–124.
Leclère, V., Bechet, M., Adam, A., Guez, J. S., Wathelet, B., Ongena, M., Thonart, P., et al. (2005) Mycosubtilin overproduction by *Bacillus subtilis* BBG100 enhances the organism's antagonistic and biocontrol activities. *Applied and Environmental Microbiology, 71*, 4577–4584.
Leeman, M., van Pelt, J. A., den Ouden, F. M., Heinsbroek, M., Schippers, B. P. A. H. M., & B. (1995). Induction of systemic resistance by *Pseudomonas fluorescens* in radish cultivars differing in susceptibility to Fusarium wilt, using novel bioassay. *European Journal of Plant Pathology, 101*, 655–664.
Leyns, F., Borgoni, G., Arnaut, G., & De Waele, D. (1995). Nematicidal activity of *Bacillus thuringiensis* isolates. *Fundamental & Applied Nematology, 18*, 211–218.
Linderman, R. G. (1994). Role of VAM fungi in biocontrol. In F. L. Pfleger & R. G. Linderman (Eds.), *Mycorrhizae and plant health*. St. Paul: APS Press. ISBN-10: 0890541582.
Lo, C. T., Nelson, E. B., & Harman, G. E. (1997). Biological control of *Pythium*, *Rhizoctonia* and *Sclerotinia* infected diseases of turfgrass with *Trichoderma harzianum*. *Phytopathology, 84*, 1372–1379.
Loper, J. E., & Buyer, J. S. (1991). Siderophores in microbial interactions of plant surfaces. *MPMI, 4*, 5–13.
Maurhofer, M., Hase, C., Matraux, J. P., & Defago, G. (1994). Induction of systemic resistance of tobacco to tobacco necrosis virus by the root-colonizing *Pseudomonas fluorescens* strains CHAO: Influence of the *gacA* gene and pyoverdine production. *Phytopathology, 84*, 139–146.
McSpadden-Gardener, B. B., & Weller, D. M. (2001). Changes in populations of rhizosphere bacteria associated with take-all disease of wheat. *Applied and Environmental Microbiology, 67*, 4414–4425.
Meziane, H., van der Sluis, I., van Loon, L. C., Bakker, H. M., & P.A.H.M. (2005). Determinants of *Pseudomonas putida* WCS358 involved in inducing systemic resistance in plants. *Molecular Plant Pathology, 6*, 177–185.
Milgroom, M. G., & Cortesi, P. (2004). Biological control of chestnut blight with hypovirulence: A critical analysis. *Annual Review of Phytopathology, 42*, 311–338.
Mishra, S., Jagadeesh, K. S., Krishnaraj, P. U., & Prem, S. (2014). Biocontrol of tomato leaf curl virus (ToLCV) in tomato with chitosan supplemented formulations of Pseudomonas sp. under field conditions. *Australian Journal of Crop Science (AJCS), 8*, 347–355.

Moyne, A. L., Shelby, R., Cleveland, T. E., & Tuzun, S. (2001). Bacillomycin D: An iturin with antifungal activity against *Aspergillus flavus*. *Journal of Applied Microbiology, 90*, 622–629.

Muniyappa, V., & Veeresh, G. K. (1984). Plant virus diseases transmitted by whiteflies in Karnataka. *Proceedings of Indian Academy of Sciences, 93*, 397–406.

Murphy, J. F., Zehnder, G. W., Schuster, D. J., Sikora, E. J., Polston, J. E., & Kloepper, J. W. (2000). Plant growth promoting rhizobacterial mediated protection in tomato against tomato mottle virus. *Plant and Disease, 84*, 779–784.

Ordentlich, A., Elad, Y., & Chet, I. (1988). The role of chitinase of *Serratia marcescens* in biocontrol of *Sclerotium rolfsii*. *Phytopathology, 78*, 84–87.

Pal, K. K., & McSpadden Gardener, B. (2006). Biological control of plant pathogens. *The Plant Health Instructor*. https://doi.org/10.1094/PHI-A-2006-1117-02.

Phillips, A. D., Fox, T. C., King, M. D., Bhuvaneswari, T. V., & Teuber, L. R. (2004). Microbial products trigger amino acid exudation from plant roots. *Plant Physiology, 136*, 2887–2894.

Press, C. M., Loper, J. E., & Kloepper, J. W. (2001). Role of iron in rhizobacteria mediated induced systemic resistance of cucumber. *Phytopathology, 91*, 593–598.

Ramette, A., Moenne-Loccoz, Y., & Defago, G. (2003). Prevalence of fluorescent-pseudomonads producing antifungal phloroglucinols and/or hydrogen cyanide in soils naturally suppressive or conducive to tobacco root rot. *FEMS Microbial Ecology, 44*, 35–43.

Ramzan, M., Tabassum, B., Nasir, I. A., Khan, A., Tariq, M., Awan, M. F., et al. (2016). Identification and application of biocontrol agents against cotton leaf curl virus disease in *Gossypium hirsutum* under greenhouse conditions. *Biotechnology & Biotechnological Equipment, 30*, 469–478.

Ryu, C. M., Farag, M. A., Hu, C. H., Reddy, M. S., Kloepper, J. W., & Pare, P. W. (2004). Bacterial volatiles induce systemic resistance in *Arabidopsis*. *Plant Physiology, 134*, 1017–1026.

Saikia, A. K., & Muniyappa, V. (1989). Epidemiology and control of tomato leaf curl virus in Southern India. *Tropical Agriculture (Trinidad), 66*, 350–354.

Schouten, A., van den Berg, G., Edel-Hermann, V., Steinberg, C., Gautheron, N., Alabouvette, C., et al. (2004). Defense responses of *Fusarium oxysporum* to 2,4-DAPG, a broad spectrum antibiotic produced by *Pseudomonas fluorescens*. *MPMI, 17*, 1201–1211.

Shahraki, M., Heydari, A., & Hassanzadeh, N. (2009). Investigation of antibiotic, siderophore and volatile metabolites production by *Bacillus* and *Pseudomonas* bacteria. *Iranian Journal of Biology, 22*, 71–85.

Shah-Smith, D. A., & Burns, R. G. (1997). Shelf-life of a biocontrol *Pseudomonas putida* applied to the sugar beet seeds using commercial coatings. *Biocontrol Science and Technology, 7*, 65–74.

Shanahan, P., O'Sullivan, D. J., Simpson, P., Glennon, J. D., & O'Gara, F. (1992). Isolation of 2,4-diacetylphloroglucinol from a fluorescent pseudomonad and investigation of physiological parameters influencing its production. *Applied and Environmental Microbiology, 58*, 353–358.

Sharma, R. D. (1994). *Bacillus thuringiensis* a biocontrol agent of *Meloidogyne incognita* on barley. *Nematologia Brasileira, 18*, 79–84.

Silva, H. S. A., Romeiro, R. D. S., Macagnan, D., Halfeld-Vieira, B. D. A., Pereira, M. C. B., & Mounteer, A. (2004). Rhizobacterial induction of systemic resistance in tomato plants: Non-specific protection and increase in enzyme activities. *Biological Control, 29*, 288–295.

Srinivasan, K., Surendiran, G., & Maathivanan, N. (2005). Pathological and molecular biological investigations on sunflower necrosis virus (SNV) and ISR mediated biological control of SNV by PGPR strains. *Asian Conference on Emerging Trends in Plant- Microbe Interaction*, 8–10 December Chennai India.

Stirling, G. R. (1985). Host specificity of *Pasteuria penetrans* within the genus *Meloidogyne*. *Nematologica, 31*, 203–209.

Stirling, G. R. (2011). Biological control of plant-parasitic nematodes: An ecological perspective, a review of progress and opportunities for further research. In K. Davies & Y. Spiegel (Eds.), *Biological control of plant-parasitic nematodes: Progress in biological control* (Vol. 11, pp. 1–38). Dordrecht: Springer.

Tari, P. H., & Anderson, A. J. (1988). Fusarium wilt suppression and agglutinability of *Pseudomonas putida*. *Applied and Environmental Microbiology, 54*, 2037–2041.

Thomashow, L. S., & Weller, D. M. (1988). Role of a phenazine antibiotic from *Pseudomonas fluorescens* in biological control of *Gaeumannomyces graminis* var. *Tritici*. *Journal of Bacteriology, 170*, 3499–3508.

Thomashow, L. S., Weller, D. M., Bonsall, R., & Pierson, L. S. (1990). Production of de antibiotic phenazine 1-carboxylic acid by fluorescent pseudomonad species in the rhizosphere of wheat. *Applied and Environmental Microbiology, 56*, 908–912.

Thomashow, L. S., Bonsall, R. F., & Weller, D. M. (2002). *Antibiotic production by soil and rhizosphere microbes in situ* (2nd ed., pp. 638–647). Washington DC: ASM Press.

Thresh, J. M. (1980). The origins and epidemiology of some important plant virus diseases. *Applied Biology, 5*, 1–65.

Thresh, J. M. (Ed.). (1981). *Pests pathogens and vegetation*. London: Pitman.

Thresh, J. M. (1982). Cropping practices and virus spread. *Annual Review of Phytopathology, 20*, 193–218.

Timper, P. (2014). Conserving and enhancing biological control of nematodes. *Journal of Nematology, 46*, 75–89.

Vallad, G. E., & Goodman, R. M. (2004). Systemic acquired resistance and induced systemic resistance in conventional agriculture. *Crop Science, 44*, 1920–1934.

Van Dijk, K., & Nelson, E. B. (2000). Fatty acid competition as a mechanism by which *Enterobacter cloacae* suppresses *Pythium ultimum* sporangium germination and damping-off. *Applied and Environmental Microbiology, 66*, 5340–5347.

Van Loon, L. C., Pieterse, B. P. A. H. M., & C.M.J. (1998). Systemic resistance induced by rhizosphere bacteria. *Annual Review of Phytopathology, 36*, 453–483.

Van Peer, R., & Schippers, B. (1992). Lipopolysaccharides of plant-growth promoting *Pseudomonas* spp. strain WCS 417r induce resistance in carnation to *Fusarium wilt*. *Netherlands Journal of Plant Pathology, 98*, 129–139.

Van Wees, S. C., Pieterse, C. M., Trijssenaar, A., van't Westende, Y. A., Hartog, F., & van Loon, L. C. (1997). Differential induction of systemic resistance in *Arabidopsis* by biocontrol bacteria. *MPMI, 10*, 716–724.

Wei, J. Z., Hale, K., Carta, L., Platzer, E., Wong, C., Fang, S. C., & Aroian, R. V. (2003). Bacillus thuringiensis crystal proteins that target nematodes. *Proceedings of the National Academy of Sciences of the United States of America, 100*, 2760–2765. https://doi.org/10.1073/pnas.0538072100.

Weller, D. M., Raaijmakers, J. M., Gardener, B. B. M., & Thomashow, L. S. (2002). Microbial populations responsible for specific soil suppressiveness to plant pathogens. *Annual Review of Phytopathology, 40*, 309–348.

Wilhite, S. E., Lumsden, R. D., & Strancy, D. C. (2001). Peptide synthetase gene in *Trichoderma virens*. *Applied and Environmental Microbiology, 67*, 5055–5062.

Yang, J., Loffredo, A., Borneman, J., & Becker, J. O. (2012). Biocontrol efficacy among strains of *Pochonia chlamydosporia* obtained from a root-knot nematode suppressive soil. *Journal of Nematology, 44*, 67–71.

Zuckerman, B. M., Dicklow, M. B., & Acosta, N. (1993). A strain of Bacillus thuringiensis for the control of plant parasitic nematodes. *Biocontrol Science and Technology, 3*, 41–46.

Zhang, S., Moyne, A., Reddy, M. S., & Kloepper, J. W. (2002). The role of salicylic acid in induced systemic resistance elicited by plant growth-promoting rhizobacteria against blue mold of tobacco. *Biological Control, 25*(3), 288–296.

Chapter 6
Soil Microbes and Plant Health

Farheen Nazli, Najm-ul-Seher, Muhammad Yahya Khan, Moazzam Jamil, Sajid Mahmood Nadeem, and Maqshoof Ahmad

Abstract Soil microbial community is crucial for plant health. They all represent a second much larger genome associated to plants. Microbes vary in their number and diversity which is in order of tens of thousands species in fertile agricultural soils. In general, soil microbial communities include bacteria, fungi, algae, protozoa, nematodes, and microarthropods. Most of them are neutral in relation to their effects on plants. They are important players of the food web as they utilize most of the carbon released by roots as rhizodeposits. Less than ten percent of the total rhizosphere microbes exert beneficial or harmful effects on plants. Pathogenic microorganisms in soil include fungi, oomycetes, bacteria, and nematodes while the beneficial microbial community consists of symbiotic, associative symbiotic and free-living plant growth promoting bacteria (PGPB), arbuscular mycorrhizal fungi and algae. Recent research in plant-microbe interactions showed that host specific microbial species are associated with different plant species in the same soil. The number and diversity of beneficial and deleterious microorganisms depend on the quality and quantity of root exudates which, along with soil physico-chemical properties, give shape to the rhizosphere microbial community structure. This chapter highlights the importance of rhizosphere microbial communities in relation to plant growth. Recent advances on soil-plant-microbe interactions in a balanced and optimized manner are discussed.

Keywords Biocontrol · Mycorrhizal inoculation · Plant microbe interaction · Microbial diversity · Soil microbes

F. Nazli
Department of Soil Science, University College of Agriculture and Environmental Sciences, The Islamia University of Bahawalpur, Bahawalpur, Pakistan

Pesticide Quality Control Laboratory Bahawalpur, Agriculture Department, Government of the Punjab, Bahawalpur, Pakistan

Najm-ul-Seher · M. Jamil · M. Ahmad (✉)
Department of Soil Science, University College of Agriculture and Environmental Sciences, The Islamia University of Bahawalpur, Bahawalpur, Pakistan

M. Y. Khan · S. M. Nadeem
Sub Campus Burewala, University of Agriculture Faisalabad, Burewala, Pakistan

© Springer Nature Switzerland AG 2020
I. Ul Haq, S. Ijaz (eds.), *Plant Disease Management Strategies for Sustainable Agriculture through Traditional and Modern Approaches*, Sustainability in Plant and Crop Protection 13, https://doi.org/10.1007/978-3-030-35955-3_6

6.1 Introduction

To meet the food requirement of rapidly growing world population it is very difficult to develop new high yielding crop varieties, having resistance to biotic and abiotic stresses. Our crops still require fertilizers because nutrients may become inaccessible or may be insufficient in soil. Recent work clearly revealed that microorganisms reduce the use of agricultural inputs regarding inorganic fertilizers. Microorganisms, due to their excessive gene pool, are very useful for soil reactions, i.e. by recycling nutrients needed for plant growth (Li et al. 2017).

Microbial communities of diverse groups are important for agricultural productivity (Sharma et al. 2008). Microorganisms from different genera such as *Bacillus, Pseudomonas*, *Rhizobium*, *Azospirillum*, have been reported for their potential in enhancing both above and belowground plant biomass, being useful tools for sustainable agriculture (Igiehon and Babalola 2018).

The rhizosphere is the region of the soil environment that is maintained by border cells and exudates released by roots (Moe 2013; Igiehon and Babalola 2017). Plants produced mucilages, rhizodeposits, nutrients, and exudates which attract and serve as food sources for microorganisms living in the rhizosphere. Plants different developmental stages control and shape the structure of rhizosphere communities (Hou and Babalola 2013; Chaparro et al. 2013). The difference in rhizosphere microbial community structure is induced by the changes in nature and chemistry of metabolites exuded by roots, at different growth stages.

Plants release exudates that have an effect on the diversity of microorganisms and invertebrates living in the rhizosphere. Root exudates are responsible for increasing their populations in the rhizophere by increasing availability of C as a source of food and energy (Aira et al. 2010) These microorganisms in turn can also affect plants by releasing growth regulatory substances. In this view, the rhizosphere organisms are considered as an external environment for plants (Philippot et al. 2013; Spence et al. 2014).

The physical, chemical and biological properties of rhizospheric soil differ with those of the surrounding soil (Kim et al. 2010), as the number of microorganisms and invertebrates found in the rhizosphere are greater compared to bulk soil. The rhizosphere communities are the second set of genome present in plants and perform several roles for growth and development (Nihorimbere et al. 2011). Soil and rhizosphere microorganisms are affected by many factors such as type of soil, climate, plant species and management practices (Jeffrey et al. 2010).

6.2 Rhizosphere and Root Exudates

Soil is sometimes considered just as a source of nutrients. However, it is a complex ecosystem holding bacteria, protists, fungi, and animals (Muller et al. 2016). The rhizosphere area varies with plant species and soil. The term rhizosphere was first used by Lorentz Hiltner (Hartmann et al. 2008) for the immediate area of soil influenced by the plant roots. It generally covers a 2 mm distance from the root surface, known as the rhizoplane. However, its influence can be found up to 10 mm (Hartmann et al. 2008; Niu et al. 2012). The microbial population in the rhizosphere may vary from thousands to million cells (Nihorimbere et al. 2011). The plant nutrients and exudates are the principle component that changes the microclimate of the rhizosphere (Shukla et al. 2013). Soil microorganisms are C dependent, so they grow well in soil having high amounts of amino acids, organic acids, sugars that are exuded by the plants (Bais et al. 2006).

The presence of microbes in the rhizosphere depends on number of factors such as: (1) plant genotype (Bulgarelli et al. 2012), (2) plant developmental stage (Chaparro et al. 2013), (3) plant hormones (Carvalhais et al. 2013), (4) composition of root exudates (Badri et al. 2013), (5) exposure to disease-suppressive soils (Mendes et al. 2011).

Plant roots release in the rhizosphere a wide variety of chemical compounds that attract soil microorganisms (Huang et al. 2014), that are known as exudates. Root exudates are the products of photosynthesis (Hayat et al. 2017), and include amino acids, organic acids, sugars, enzymes, hormones, mucilage, root cells and C (Dennis et al. 2010). Uren described that about 50% of C fixed by plant is devoted to the root, with 15% respired by the plant and 10% released by the roots as debris (exudates) including border cells (Jones et al. 2009). The compounds released by roots are not organic in nature and may be classified according to their functions such as excretions (H^+, CO_2, HCO_3, ethylene) involved in internal metabolism and secretions (siderophores, enzymes, mucilage, H^+, electrons) required for external processes (Uren 2007).

The rhizosphere is an area of high microbial activity. Several studies reported the effects of roots on microbial process (Kaiser et al. 2010). The roots release higher organic compounds in the rhizosphere due to the microbial biomass whose activities are higher in the rhizosphere than in bulk soil (Jones et al. 2009).

Studies on the microbiome of different plant species revealed that specific exudates perform specific functions, shaping the plant-microbe interaction (Hartmann et al. 2009). They include flavonoids involved in the process of symbiosis between rhizobia and legumes (Abdel-Lateif et al. 2012), sugars and amino acids acting as chemoattractants for microorganisms (Somers et al. 2004) and strigolactones, that enhance branching of mycorrhizal hyphae (Akiyama et al. 2005). Plants attract nematodes, which carry rhizobia toward roots thus contributing to the nodulation process (Horiuchi et al. 2005). The exudates have both positive and negative effects on plants, however, depending on which type of microorganisms are attracted (Hayat et al. 2017).

6.3 Microbial Number and Diversity

In the rhizosphere, there is more cycling of nutrients as well as availability. The solubility of toxic metals also makes it a different microenvironment from bulk soil (Neumann et al. 2009). The type and composition of organic compounds released by roots, i.e. sugars, carbohydrates, nucleotides, vitamins, flavonoids, and stimulators, strongly affect the diversity of species proliferating in the rhizosphere (Dotaniya et al. 2013; Ueno et al. 2007).

The rhizosphere higher amount of nutrients as compared to bulk soil, confer biological and chemical properties due to a large number of macro- and microorganisms which develop a variety of interactions among each other and with roots (Kapoor and Sachdeva 2013). The relationship between plant and microorganisms may be classified as mutualistic, commensalistic or parasitic (Aira et al. 2010; Schirawski and Perlin 2018).

The microbial diversity decreases as the distance from the rhizoplane increases (Chowdhury et al. 2009). It also decreases after long term exposure to older roots, as compared to bulk soil. In general, the bacterial diversity follows the following trend: bulk soil < apical region < basal region (Dennis et al. 2008). According to Hartmann et al. (2009), the root secretions exert both stimulatory as well as inhibitory effects on the microbial community structure and composition. Exudates such as amino acids, organic acids, and carbohydrates have positive effects on chemotactic responses of bacteria. On the other hand roots also release secondary metabolites that inhibit the growth of bacterial and fungal pathogens such as chitosans, jasmonic and salicylic acids (Walker et al. 2003). Plants also release different antibacterial compounds such as ellagic acid, chebulagic acid and norwogonin, that decrease the bacterial number (Miyasaki et al. 2013).

6.4 Plant Microbe Interactions

As agricultural production needs to be doubled in coming year to feed the increasing population of the world, there is also need to reduce the use of chemical fertilizers. To achieve this goal, it is necessary to explore the interaction between plants and microbes, so that microorganisms can be used as biofertilizers, biopesticides as well as bioherbicides (Igiehon and Babalola 2017).

6.4.1 Beneficial Interactions

Plant-associated microorganisms use different mechanisms for influencing and modulating the plant health (Huang et al. 2014). The PGPB *Bacillus amyloliquefaciens* can successfully be used to enhance drought resistance of plants (Su et al.

2017). PGPB isolated from plants growing in metal contaminated soil have the ability to survive at high concentration of zinc and cadmium and can be used for phytoremediation of contaminated sites (Montalbán et al. 2017). Some PGPB protect plants from soil borne diseases by producing toxic compounds (Muller et al. 2013).

Under water deficient conditions, arbuscular mycorrhizal (AM) fungi change the plant water relationship and improve their resistance to water stress (Birhane et al. 2012). AM fungi are well known for their potential to facilitate nutrient uptake, particularly phosphorus, by crops growing in phosphorus deficient soils (Mohammadi et al. 2011), i.e. *Phragmites australis* (Liang et al. 2018) or the AM *Rhizophagus irregularis* that improves plant growth enhancing Verticillium wilt resistance in cotton (Zhang et al. 2018).

Algae enhance soil fertility and promote plant growth by supplying growth promoting substances. The application of algal extract *Oscillatoria* sp. and *Spirogyra* sp. increased seed germination, seedling growth, and number of leaves, and height of *Medicago sativa* (Brahmbhatt and Kalasariya 2015). In another study, Shariatmadari et al. (2013) reported that improvement in plant growth parameters by the application of algal extract was due to the presence of growth promoting substances such as phytohormones.

Actinomycetes are also important soil microorganism contributing a higher proportion of soil biomass. They are capable of producing large quantities of antibiotics, extracellular enzymes, organic acids, bioactive compounds, phytohormones and other secondary metabolites. Several isolates of actinomycetes have been found to promote plant growth via direct and indirect mechanisms (Singh et al. 2018). Sreevidya et al. (2016) reported that four isolates of actinomycetes significantly increased the yield of chickpea. The PGPB *Streptomyces* spp. can be used as biofertilizers, as they help to release nutrients from the complex organic compounds (Vurukonda et al. 2018).

Soil protozoa also affect plant growth by influencing the beneficial potential of PGPB (Weidner et al. 2017).

6.4.2 Negative Interactions

Microorganisms regulate host populations everywhere in the soil environment. They may be pathogenic and may cause diseases in plants. In such interactions, pathogens grow and multiply inside the plant by causing various kind of diseases, and can also move from diseased to healthy plants (Abdul-Kareem 2012).

The pathogenic fungi feed on plant tissues and weaken their defense systems. One of the mechanisms exploited to survive in the host plant is the secretion of effector proteins which interact with plant proteins thereby disturbing their normal function. Recently, effector proteins were identified in fungal pathogens important for Indian agriculture, such as *Microbotryum lychnidis-dioicae* (Kuppireddy et al. 2017) and *Fusarium proliferatum* (Gao et al. 2018). These effector proteins cause destructive tomato diseases by producing dark brown necrotic spots leading to death of the entire plant. The infection on wheat during different growth stages by *Fusarium culmorum* has also been reported by Spanic et al. (2018). Fungal phytopathogens reduce the plant yield although their abundance may decrease in suppressive soils (Lobmann et al. 2016).

Viral infections in plants have also been reported in species of *Chenopodium* and *Nicotiana* grown in common gardens. Viral infections develop necrotic spots and cause reduction in above ground biomass (Kollmann et al. 2007). Changes in the growth and physiological parameters of pepper as affected by *Tobacco mosaic virus* have also been studied. The viral infection causes stunting, necrosis, deformation and defoliation in plants which resulted in reduction of fruit production (Pazarlar et al. 2013). In another study, incidence of Cucumber green mottle mosaic virus, Squash mosaic virus and melon necrotic spot virus was observed which resulted in 10–15% yield reduction (Ling et al. 2014). Cucumber green mottle mosaic virus is a virus disease of cucurbits which can be transmitted via pollen to healthy plants (Liu et al. 2013). Viral diseases in cotton are major cause of yield reduction. According to an estimate, about 30% reduction in cotton is due to cotton leaf curl virus (Hassan et al. 2016).

Yellow vein mosaic disease (YVMD) caused by begomoviruses is the most devastating disease of okra affecting both quality and yield of crops (Sanwal et al. 2016; Venkataravanappa et al. 2017). Another important problem in tomato production caused by a virus is the tomato leaf curl, whose causal agent also belongs to the begomovirus group (Moriones et al. 2017). Microorganisms responsible for food spoilage such are yeasts, moulds and bacteria (Rawat 2015).

6.4.3 Plants as Habitat for Microbial Population

The rhizosphere contains beneficial microorganisms, i.e. rhizobia and AM fungi, which establish associations with the roots, provide fixed nitrogen and nutrients and get in exchange carbon-based compounds. On the other hand, the plants exposed to a wide range of pathogens, such as bacteria, fungi and viruses, evolved several defense mechanisms (Biswas et al. 2016). The roots are covered by layers of microbes in the rhizosphere, and even seed may also contain microorganisms. These complex communities modulate the plant growth, health and development, seed germination, plant ecology and productivity (Chen et al. 2018).

The microorganisms found in association with medicinal plants might be involved in release of secondary metabolite from medicinal plant (Chen et al. 2018). In another study Sanchez-Lopez et al. (2018) investigated whether microorganisms can be transmitted from one generation to another. They isolated *Methylobacterium* sp. from seeds of pioneer *Crotalaria pumila* growing in metal contaminated soil. The seeds associated microbial isolates were capable to colonize the plant up to three generations, including root cortical cells as well as xylem. In another study, the bacterial diversity, plant exudates and physico-chemical properties of rhizosphere soil from young and old tea plants were compared. Although the physico-chemical properties remained similar, changes were observed in catechin and microbial distribution (Arafat et al. 2017).

6.5 Microbes and Plants Growth

Plants affect microbial diversity as the microbiome lives in close association with roots, xylem, stems and leaves. On the other hand, these microorganisms also influence plant health and productivity (Bogino et al. 2013). Microorganisms have been used to enhance crop productivity and soil health as well as for bioremediation of contaminated soils, wastewater treatment and recycling of industrial waste (Ahmad et al. 2011). They play a vital role in nutrients mobility and uptake by plants, promoting growth and protecting plants from diseases (Table 6.1). Fundamental processes include phosphate and sulphate solubilization, nitrogen fixation, denitrification, siderophore production, signal transduction, immune modulation, pathogens control (Prakash et al. 2015). Microorganisms also decompose organic matter and get C and energy for their own growth (Rillig et al. 2007). In anaerobic conditions, microbes immobilize 20–40% of C in the substrate, the remaining being released into the atmosphere (Zak et al. 2000; Rajendiran et al. 2012). Studies recommended the use of microbial-based fertilizers for sustainable agriculture and environment safety, instead of chemical fertilizers alone (Prakash et al. 2015).

Microorganisms-based fertilizers, along with compost, decrease the negative effect of chemical fertilizers and further enhance the quality of crops, as suggested by the enhancement of β-carotene, Brix degrees and vitamin C content reported in tomatoes (El Kiyum 2017). The PGPB colonize roots influencing the plant growth and development (Hayat et al. 2017). The interaction of PGPB with their rhizobial counterparts could result in the enhancement of nodulation efficiency (Guiñazú et al. 2010). Plants release amino acids, carbohydrates, vitamins and lipids, through roots thus enhancing microbial activities in soil. PGPB enhance geochemical cycling of essential nutrients, particularly nitrogen, phosphorus and micronutrients such as copper, zinc, iron, manganese in soil for plant growth and development (Dotaniya and Vasudev 2015). Plant growth promotion by addition of microorganisms in soil i.e., *Pseudomonads* as well as phytohormones productions have been documented (Nassal et al. 2018).

PGPB promote plant growth and crop yields by producing growth hormones, increasing plant nutrient availability, and of microbes that can act as biocontrol agent against pathogens (Dotaniya and Vasudev 2015; Jacoby et al. 2017). Inoculation of plants with AM increased efficiency of water use, plant biomass and chlorophyll content, even under metal stress (Andrade et al. 2015).

Table 6.1 Impact of microbial inoculation on crop growth

Growth condition	Crop	Response	Reference
(a) Impact of bacterial inoculation on crop growth			
Lab studies	Rice (*Oryza sativa*)	Inoculation significantly increased germination, root and shoot length, and plant vigor	Vandana et al. (2018)
Lab studies		Inoculation suppressed economically important crop pathogens	Shakeel et al. (2015)
Field experiment		Improved all quality parameters	Choudhary et al. (2013)
Pot study	Maize (*Zea mays*)	Enhanced plant growth and yield parameters by suppressing fungal pathogens	Akhtar et al. (2018)
Lab and field conditions		Significantly affected root-system architecture	Vacheron et al. (2018)
Greenhouse and field conditions		Inoculation increased plant height and dry weight	Jarak et al. (2012)
Field and pot experiment		Improved shoot and root length, root dry weight, yield cob weight and P uptake	Baig et al. (2014)
Field experiments	Wheat (*Triticum aestivum*)	significantly increased macro and micro nutrients and grain total biomass yields	Turan et al. (2018)
Greenhouse conditions		Inoculation significantly increased plant biomass, at all stages	Nguyen et al. (2019)
Field experiment		Improved plant height, number of spikelets per spike and grain yield	Sial et al. (2018)
Lab conditions	Tomato (*Solanum lycopersicum*)	Significantly increased stem height and root mass	Cendales et al. (2017)
Growth chamber		Suppressed soil-borne diseases and promoted plant growth	Qiano et al. (2017)
Pot experiments		Improved fruit quality by increasing carbohydrates, sugar and ascorbic acid	Pishchik et al. (2018)
Field experiment under salt stress	Cotton (*Gossypium hirsutum* L.)	Increased seed germination, total shoot, root dry weight and yield under salt stress	Pulatov et al. (2016)
Field experiment		Enhanced yield	Alavo et al. (2015)
Field experiment		Improved plant growth by increasing phytohormone production	Pindi et al. (2014)

(continued)

Table 6.1 (continued)

Growth condition	Crop	Response	Reference
Pot study	Chickpea (*Cicer arietinum* L.)	Enhanced plant height, leaf area leaf and stem weight, pod number and weight	Gopalakrishnan et al. (2015)
Pot study		Increased fresh and dry biomass	Goswami et al. (2013)
Field study	Hungarian vetch (*Vicia pannonica*)	Increased dry matter, crude protein ADF, NDF macro and micro nutrients	Yolcu et al. (2012)
Greenhouse conditions	Radish (*Raphanus sativus*)	Increased photosynthetic pigments, free amino acids, proline, phytohormones and N, P, K^+ content	Mohamed and Gomaa (2012)
Culture media	Banana (*Musa acuminate*)	Enhanced fresh and dry weight, plant height, stem thickness and modified roots architecture	Martin et al. (2015)
Field study	Strawberry (*Fragaria ananassa*)	Increased yield per plant	Pirlak and Murat (2009)
Glass house conditions	Sorghum (*Sorghum bicolor*)	Bacterial strains showed enhanced plant growth parameters, chlorophyll, carbohydrates, phosphorus nitrogen and other nutrients	Kumar et al. (2012)
Control conditions	Thale cress (*Arabidopsis thaliana*)	Improved shoot length and rosette diameter	Schwachtje et al. (2012)
Pots study	Poinsettia (*Euphorbia pulcherrima*)	Increased number of leaves, leaf area and volume of roots	Zulueta-Rodriguez et al. (2014)
Field study	Castor (*Ricinus communis* L.)	Significantly increased leaf biomass, number of leaves, root and shoot length, stem base diameter and leaf moisture content	Sandilya et al. (2017)
(b) Impact of mycorrhizal inoculation on crop growth.			
Metal stress (Pot experiment)	Rice (*Oryza sativa*)	Decreased metal uptake	Zhang et al. (2011)
Pot experiment	Carrot (*Daucus* carota) sorghum (*Sorghum bicolor*)	Improved plant growth	Kim et al. (2017)
Field conditions	Artichoke (*Cynara cardunculus*)	Improved yield	Colonna et al. (2015)
Pot study	Red amaranth (*Amaranthus cruentus*) and spinach (*Spinacia oleracea*)	Improved growth and yield	Ghosh et al. (2017a, b)

(continued)

Table 6.1 (continued)

Growth condition	Crop	Response	Reference
Poly bags	Chilli (*Capsicum* annuum)	Increased plant growth and yield parameters, chlorophyll, and nutrient uptake	Elahi et al. (2012)
Field experiments	Melon (*Cucumis melo*) watermelon (*Citrullus lanatus*)	Significantly increased plant growth with higher nutrient content	Ortas (2012)
Green house	Finger millet (*Eleusine coracana*)	Highest growth parameters were observed.	Patil et al. (2013)
In vitro experiments	Banana (*Musa acuminata*)	Superior biocontrol potential for disease management	Ganesan et al. (2009)
In vivo experiment	Mango (*Mangifera indica)*	Inoculation significantly control stem end rot disease	Suhanna et al. (2013)
Field conditions	Brazilian fern tree (*Schizolobium parahyba*)	Increased wood yield	Cely et al. (2016)
Greenhouse and field conditions	Red sage (*Salvia miltiorrhiza*)	Improved root growth and boosts the secondary metabolism	Zhou et al. (2018)
Field study	Gum trees (*Eucalyptus sp.*)	Inoculation positively affected stem diameter, stem length, and the fresh and dry biomass	Vitorino et al. (2016)
Pot experiment	Cucumber (*Cucumis sativus*)	increased plant height, stem diameter, dry weight, and macro and micro nutrient	Chen et al. (2017)

6.6 Soil Microbes as Biocontrol Agent

Soil microorganisms act as biopesticides and protect plants from pathogens by producing a range of different metabolites (Table 6.2). They possess several mode of actions such as the production of antibiotics, biosurfactants, cell wall degrading enzymes, chitinase, glucanase, toxins, siderophores and induction of systemic resistance in plants (Perez-Montano et al. 2014; Kumar and Singh 2015). Disease-causing organisms including bacteria, fungi, and nematodes causing severe soil borne diseases negatively affect all crops. Among pathogenic microbes, fungi are responsible for huge losses in economically important crops (Perez-Montano et al. 2014).

In these conditions (PGPB) and fungi interact in a complex system. Bacterial strains often encountered in the rhizosphere that can act as biocontrol agents belong to different genera such as *Acetobacter*, *Azotobacter*, *Bacillus*, *Pseudomonas*, *Paenibacillus*, *Streptomyces* (Berg and Smalla 2009; Kumar and Singh 2015). Some strains such as *Pseudomonas aeruginosa*, *Ochrobactrum lupine*, *Novosphingobium* and *Pentaromativorans* spp. have shown a disease management capacity vs Tomato Cucumber mosaic and bacterial spot (Dashti et al. 2012; Hahm et al. 2012).

Table 6.2 Metabolites used in biocontrol produced by microbial strains

Species	Biocontrol potential	Reference
Bacteria		
Acinetobacter	Chitinase	Krithika and Chellaram (2016)
Achromobacter sp. *Achromobacter xylosoxidans, Achromobacter xylosoxidans*	HCN production	Ngoma et al. (2013)
Alcaligenes sp.	Siderophores	Patel et al. (2018)
Alcaligenes sp.	Siderophores	Sayyed and Patel (2011)
Pseudomonas strains	HCN, siderophores	Sandilya et al. (2017)
Pseudomonas sp.	Siderophores, HCN, lipase, protease	Ghodsalavi et al. (2013)
Pseudomonas sp.	Antibiotics 2,4 DAPG	Asadhi et al. (2013)
Pseudomonas aeruginosa	Antibiotic, siderophores, HCN production	Uzair et al. (2018)
Pseudomonas putida	HCN production, siderophores production	Vandana et al. (2018)
Bacillus cereus		
Pseudomonas brassicacearum	Secondary metabolites	Andersson (2012)
Pseudomonas fluorescens	Protease, chitinase, glucanases	Ruchi et al. (2012)
Pseudomonas fluorescens	Antibiotics 2,4-diacetylphloroglucinol	Weller et al. (2012)
Pseudomonas aureofaciens	Siderophores	Chaiharn et al. (2009)
Bacillus firmus		
Bacillus subtilis	Chitinase and HCN	Shakeela et al. (2017)
Bacillus cereus	Chitinase	Ajayi et al. (2016)
Bacillus anthracis	Siderophores, pectinase, chitinase	Pandya et al. (2015)
Paenibacillus taichungensis		
Paenibacillus xylanilyticus		
Bacillus thuringiensis	Siderophores, phenols, HCN production chitinase activity	Ahmed et al. (2014)
Pseudomonas fluorescens		
Pseudomonas poae		
Bacillus sp.	Chitinase activity	Han et al. (2014)
Paenibacillus sp.		
Bacillus cereus	Surfactin type lipopeptide	Jourdan et al. (2009)
Pseudomonas, Bacillus, Pantoea and *Serratia*	Siderophores production, protease, HCN production	Etminani and Harighi (2018)
Pseudomonas, Bacillus, Brevundimonas, Azotobacter, Enterobacter	HCN, Lipase, protease	Patel and Desai (2015)
Pseudomonas, Bacillus subtilis sp.	lipase, amylase, chitinase, HCN	Bhatt and Vyas (2014)
Klebsiella sp.		
Cronobacter malonaticus		

(continued)

Table 6.2 (continued)

Species	Biocontrol potential	Reference
Rhizobium nepotum	Siderophore	Ghorpade and Gupta (2016)
Rhizobium	Siderophore	Datta and Chakrabartty (2013)
Herbaspirillum seropedicae	Siderophore, chitinase	Trovero et al. (2018), Rosconi et al. (2015)
Streptomyces sp.	Antibiotics, volatile organic Compounds	Vurukonda et al. (2018)
Streptomyces sp.	Secondary metabolite.	Singh et al. (2016)
Streptomyces spp.	Siderophore, cellulase, lipase, protease, chitinase, hydrocyanic acid and β-1,3-glucanase	Sreevidya et al. (2016), Gopalakrishnan et al. (2014, 2015)
Streptomyces, *Bordetella*, *Achromobacter*	Antibiotics	Abbas et al. (2014)
Streptomyces spp.	Siderophores	Franco-Correa et al. (2010); Lee et al. (2012)
Klebsiella, *Enterobacter*, *Pantoea*	HCN, chitinase, ammonia, cellulose, pectinase	Rodrigues et al. (2016)
Fungi		
Trichoderma sp.	Siderophore	Srivastava et al. (2018)
Trichoderma sp.	Activated salicylic acid (SA) and jasmonic acid	Martínez-Medina et al. (2016)
Trichoderma sp.	Harzianic acid	Vinale et al. (2013)
Trichoderma spp.	Siderophore	Ghosh et al. (2017a, b)
Beauveria spp.		
Metarhizium spp.		
Trichoderma harzianum	Activated salicylic and jasmonic acids	Alkooranee et al. (2017)
Trichoderma brevicompactum *Trichoderma virens*	β-1,3-glucanase and β-1,4- glucanase	Ayoubi et al. (2014)
Trichoderma asperellum	Siderophore	Weizhen and Lei (2013)
Trichoderma harzianum, *T. reesei*	Siderophore	Lehner et al. (2013)
Trichoderma atroviride	Regulates salicylic and jasmonic acids, and ethylene	Salas-Marina et al. (2011)
Trichoderma virens, *Trichoderma atroviride*	Activate salicylic and jasmonic acids defense signalling pathway	Contreras-Cornejo et al. (2011)
Trichoderma viride, *Aspergillus flavus*, *Curvularia lunata*, *Rhizopus stolonifer*	Antibiotics	Makut and Owolewa (2011)
Penicillium sp.	Endoglucanase, β-glucosidase,	Santa-Rosa et al. (2017)

(continued)

Table 6.2 (continued)

Species	Biocontrol potential	Reference
Penicillium echinulatum	Endoglucanases, xylanases, and β-glucosidases	Schneider et al. (2014)
Penicillium echinulatum	Cellulases, xylanases	Ritter et al. (2013)
Penicillium simplicisssum, *Acremonium* sp.	Chitinase, β-1, 3-glucanase, amylase, Siderophore	Potshangbam et al. (2017)
Aspergillus terreus	Chitinase	Krishnaveni and Ragunathan (2014)
Phoma sp.	Volatile compounds	Naznin et al. (2014)
Cladosporium sp.		
Ampelomyces sp.		
Phoma sp.	Volatile compounds	Naznin et al. (2013)
Ralstonia solanacearum	Volatile compounds	Tahir et al. (2017)
Pseudozyma aphidis	Extracellular metabolites	Barda et al. (2015)
Beauveria Bassiana, *Metarhizium anisopliae*, *Paecilomyces* sp.	Amylases, cellulases, esterases, lipases, proteases, gelatin, caseinase, pectinase	Fernandes et al. (2012)
Talaromyces wortmannii FS2	Volatile compounds	Yamagiwa et al. (2011)

Pseudomonas spp. have been reported for their potential against *Rhizoctonia solani* and *Phytophthora capsici* (Arora et al. 2008). *Trichoderma* and *Sebacinales* spp. are well known to control foliar, fruit and root pathogens, or even invertebrates such as nematodes (Shoresh et al. 2010). Endophytic fungi also have antimicrobial activities and play an important role in the regulation, or even suppression, of plant diseases, i.e. powdery mildew in wheat (Xiang et al. 2016).

PGPB release different lytic enzymes to compete with pathogens such as chitinases, glucanases, proteases (Viterbo et al. 2002). They produce chitinases and glucanases that degrade the fungal cell wall (Kumar et al. 2010). Fungi from several groups, including *Acremonium* sp. *Hansfordia pulvinata*, *Sarocladium implicatum*, *Simplicillium lanosoniveum* and *Lecanicillium lecanii* are efficient in controlling other phytopathogenic fungi by producing enzymes, proteases, lipases and antifungal metabolites that inhibit the germination of pathogens' propagules.

Soil microbes also produce antibiotics such as guanidylfungin A, nigericin, geldanamycin, controlling other disease-causing species (Trejo-Estrada et al. 1998). They can also produce alkaloids which are highly reactive and active, including alkaloids, festuklavine and elimoklavine (Bekemakhanova and Shemshura 2001) Antibiotics are low molecular weight organic compounds produced by specific groups of microorganisms. Beneficial *Pseudomonas* spp. can inhibit disease development by producing DAPG (diacetyl-phloroglucinol) and HCN (Hydrogen cayanide) (Junaid et al. 2013). Microbes can even produce hydrogen cyanide to protect plants from pathogens, as reported by Martinez-Viveros et al. (2010). Various antibiotics produced by microbes have been reported, such as amphisin A, kanosamine, A, zwittermicin, oomycin A, oligomycin, cyclic lipopeptides, 2,4-diacetyl phloroglucinol, hydrogen cyanide, pyoluteorin, phenazine, pyrrolnitrin, and xanthobaccin (Hitendra et al. 2017).

Siderophores are iron chelating compounds produced by microorganisms in iron deficient conditions. These siderophores chelate iron by converting it into complexed forms that cannot be used by other microorganisms in Fe deficient conditions. This is an important mechanisms in biological control by microorganisms, as it deprives other competing plant pathogenic fungi and bacteria. There are three different groups of siderophores which have been reported, namely catecholate siderophores, hydroxamate siderophores, and mixed siderophores (Vellasamy et al. 2015).

PGPB also trigger the plant defense mechanisms prior to infection, and in this way reduce the disease incidence. The induction of a systematic resistance results by the modulation of salicylic acid, jasmonic acid and ethylene pathways in the plants. PGPB from several genera including *Pseudomonas*, *B. amyloliquifaciens*, *B. subtilis*, *B. pasteurii*, *B. cereus*, *B. pumilus*, *B. mycoides*, and *B. sphaericus* reduce the severity of various diseases on many plants (Choudhary et al. 2007).

Some cyanobacteria and algal extracts were also found as efficient biocontrol agents because they produce antibacterial and antifungal metabolites. They can be applied to activate plant resistance mechanisms such as induced systematic resistance (Shunmugam et al. 2015).

Finally, also beneficial fungi may improve plant growth and contribute in controlling their diseases. Species from genera *Aspergillus*, *Penicillium*, *Phoma*, *Piriformospora*, *Fusarium*, and *Trichoderma* have been reported for inducing systematic resistance in plants (Hossain et al. 2017), and systematic resistance and suppression of anthracnose in cucumber (Elsharkawy et al. 2015).

References

Abbas, S., Senthilkumar, R., & Arjunan, S. (2014). Isolation and molecular characterization of microorganisms producing novel antibiotics from soil sample. *European Journal of Experimental Biology, 4*, 149–155.

Abdel-Lateif, K., Bogusz, D., & Hocher, V. (2012). The role of flavonoids in the establishment of plant roots endosymbioses with Arbuscular mycorrhiza fungi, rhizobia and Frankia bacteria. *Plant Signaling & Behavior, 7*, 636–641.

Abdul-Kareem. (2012). *Parasitism and disease development* (1st ed.). Peshawar: Khyber Pakhtunkhwa Agricultural University.

Ahmad, I., Ahmad, F., & John, P. (2011). *Microbes and microbial technology: Agricultural and environmental applications*. New York: Springer. https://doi.org/10.1007/978-1-4419-7931-5-1.

Ahmed, E. A., Enas, A. H., Tobgy, K. M. K. E., & Ramadan, E. M. (2014). Evaluation of rhizobacteria of some medicinal plants for plant growth promotion and biological control. *Annals of Agricultural Sciences, 5*, 273–280.

Aira, M., Gómez-Brandón, M., Lazcano, C., Bååth, E., & Domínguez, J. (2010). Plant genotype strongly modifies the structure and growth of maize rhizosphere microbial communities. *Soil Biology and Biochemistry, 42*, 2276–2281.

Ajayi, A. A., Onibokun, E. A., George, F. O. A., & Atolagbe, O. M. (2016). Isolation and characterization of chitinolytic bacteria for chitinase production from the African catfish, *Clarias gariepinus*. *Research Journal of Microbiology, 11*, 119–125.

Akhtar, N., Naveed, M., Khalid, M., Nisar, A., Rizwan, M., & Siddique, S. (2018). Effect of bacterial consortia on growth and yield of maize grown in *Fusarium* infested soil. *Soil and Environment, 37*, 35–44.

Akiyama, K., Matsuzaki, K., & Hayashi, H. (2005). Plant sesquiterpenes induce hyphal branching in Arbuscular mycorrhizal fungi. *Nature, 435*, 824–827.

Alavo, T. B. C., Boukari, S., Fayalo, D. G., & Bochow, H. (2015). Cotton fertilization using PGPR *Bacillus amyloliquefaciens* FZB42 and compost: Impact on insect density and cotton yield in North Benin. *Cogent Food & Agriculture, 1*, 1–8.

Alkooranee, J. T., Aledan, T. R., Ali, K. A., Lu, G., Zhang, X., Wu, J., et al. (2017). Detecting the hormonal pathways in oilseed rape behind induced systemic resistance by *Trichoderma harzianum* TH12 to *Sclerotinia sclerotiorum*. *PLoS One*. https://doi.org/10.1371/journal.pone.0168850.

Andersson, P. F. (2012). *Secondary metabolites associated with plant disease, plant defense and biocontrol* (Doctoral Thesis Swedish University of Agricultural Sciences).

Andrade, S. A. L., Domingues, A. P., & Mazzafera, J. P. (2015). Photosynthesis is induced in rice plants that associate with arbuscular mycorrhizal fungi and are grown under arsenate and arsenite stress. *Chemosphere, 134*, 141–149.

Arafat, Y., Wei, X., Jiang, Y., Chen, T., Saqib, H. S. A., Lin, S., & Lin, W. (2017). Spatial distribution patterns of root-associated bacterial communities mediated by root exudates in different aged ratooning tea monoculture systems. *International Journal of Molecular Sciences*. https://doi.org/10.3390/ijms18081727.

Arora, N. K., Khare, E., & Oh, J. H. (2008). Diverse mechanisms adopted by *Pseudomonas fluorescent* PGC2 during the inhibition of Rhizoctonia solani and Phytophthora capsici. *World Journal of Microbiology & Biotechnology, 24*, 581–585.

Asadhi, S., Reddy, B. V. B., Sivaprasad, Y., Prathyusha, M., Krishna, T. M., & Kumar, K. V. K. (2013). Characterization, genetic diversity and antagonistic potential of 2, 4-diacetylphloroglucinol producing Pseudomonas fluorescens isolates in groundnut-based cropping systems of Andhra Pradesh, India. *Archives of Phytopathology and Plant Protection*. https://doi.org/10.1080/03235408.2013.782223.

Ayoubi, N., Doustmorad, Z., & Mirabolfathy, M. (2014). Evaluation of β-1,3-glucanase and β-1,4-glucanase enzymes production in some *Trichoderma* species. *Archives of Phytopathology and Plant Protection*. https://doi.org/10.1080/03235408.2013.862457.

Badri, D. V., Zolla, G., Bakker, M. G., Manter, D. K., & Vivanco, J. M. (2013). Potential impact of soil microbiomes on the leaf metabolome and on herbivore feeding behavior. *New Phytologist, 198*, 264–273.

Baig, K. S., Arshad, M., Khalid, A., Hussain, S., Abbas, M. N., & Imran, M. (2014). Improving growth and yield of maize through bioinoculants carrying auxin production and phosphate solubilizing activity. *Soil and Environment, 33*, 159–168.

Bais, H. P., Weir, T. L., Perry, L. G., Gilroy, S., & Vivanco, J. M. (2006). The role of root exudates in rhizosphere interactions with plants and other organisms. *Annual Review of Plant Biology, 57*, 233–266.

Barda, O., Shalev, O., Alster, S., Buxdorf, K., Gafni, A., & Levy, M. (2015). *Pseudozyma aphidis* induces salicylic-acid-independent resistance to clavibacter michiganensis in tomato plants. *The American Phytopathological Society, 99*, 621–626.

Bekemakhanova, N. E., & Shemshura, O. N. (2001). *Alkaloids of microscopic fungi for plant protection. Bioactive fungal metabolites. Impact and exploitation*. In: International Symposium by British Mycological Society.

Berg, G., & Smalla, K. (2009). Plant species and soil type cooperatively shape the structure and function of microbial communities in the rhizosphere. *FEMS Microbiology Ecology, 68*, 1–13.

Bhatt, P. V., & Vyas, B. R. M. (2014). Screening and characterization of plant growth and health promoting rhizobacteria. *International Journal of Current Microbiology and Applied Sciences, 3*, 139–155.
Birhane, E., Sterck, F. J., Fetene, M., & Bongers, F. (2012). Arbuscular mycorrhizal fungi enhance photosynthesis, water use efficiency, and growth of frankincense seedlings under pulsed water availability conditions. *Oecology, 169*, 895–904.
Biswas, K., Kumar, R., Prakash, V., Tarafdar, A., & Kumar, S. (2016). Soil microbes and their interaction with plants. In Rituparna & A. B. Mitra (Eds.), *Plant pathogen interaction: Recent trends* (1st ed.). New York: Springer.
Bogino, P. C., Oliva, M. L. M., Sorroche, F. G., & Giordano, W. (2013). The role of bacterial biofilms and surface components in plant-bacterial associations. *International Journal of Molecular Sciences, 14*, 15838–15859.
Brahmbhatt, D. N. H., & Kalasariya, H. S. (2015). Effect of algae on seedling growth of "Queen of Forages". *International Journal of Engineering Research and General Science, 3*, 827–833.
Bulgarelli, D., Rott, M., Schlaeppi, K., van Themaat, V. L. E., Ahmadinejad, N., Assenza, F., et al. (2012). Revealing structure and assembly cues for *Arabidopsis* root inhabiting bacterial microbiota. *Nature, 488*, 91–95.
Carvalhais, L. C., Dennis, P. G., Badri, D. V., Tyson, G. W., Vivanco, J. M., & Schenk, P. M. (2013). Activation of the jasmonic acid plant defence pathway alters the composition of rhizosphere bacterial communities. *PloS ONE, 8*. https://doi.org/10.1371/journal.pone.0056457.
Cely, M. V., Siviero, M. A., Emiliano, J., Spago, F. R., Freitas, V. F., Barazetti, A. R., et al. (2016). Inoculation of *Schizolobium parahyba* with mycorrhizal fungi and plant growth-promoting rhizobacteria increases wood yield under field conditions. *Frontiers in Plant Science*. https://doi.org/10.3389/fpls.2016.01708.
Cendales, C. T., Rodriguez González, C. A., Villota Cuásquer, C. P., Tapasco Alzate, O. A., & Hernández, R. A. (2017). *Bacillus* effect on the germination and growth of tomato seedlings (*Solanum lycopersicum* L). *Acta Biológica Colombiana, 22*, 37–44.
Chaiharn, M., Chunhaleuchanon, S., & Lumyong, S. (2009). Screening siderophore producing bacteria as potential biological control agent for fungal rice pathogens in Thailand. *World Journal of Microbiology & Biotechnology, 25*, 1919–1928.
Chaparro, J. M., Badri, D. V., & Vivanco, J. M. (2013). Rhizosphere microbiome assemblage is affected by plant development. *ISME Journal, 8*, 790–803.
Chen, S., Zhao, H., Zou, C., Li, Y., Chen, Y., Wang, Z., et al. (2017). Combined inoculation with multiple Arbuscular mycorrhizal fungi improves growth, nutrient uptake and photosynthesis in cucumber seedlings. *Frontiers in Microbiology*. https://doi.org/10.3389/fmicb.2017.02516.
Chen, H., Wu, H., Yan, B., Zhao, H., Liu, F., Zhang, H., et al. (2018). Core microbiome of medicinal plant *Salvia miltiorrhiza* seed: A rich reservoir of beneficial microbes for secondary metabolism. *International Journal of Molecular Sciences*. https://doi.org/10.3390/ijms19030672.
Choudhary, D. K., Anil, P., & Johri, B. N. (2007). Induced systemic resistance (ISR) in plants: Mechanism of action. *Indian Journal of Microbiology, 47*, 289–297.
Choudhary, R. L., Kumar, D., Shivay, Y. S., Anand, A., & Nain, L. (2013). Yield and quality of rice (*Oryza sativa*) hybrids grown by SRI method with and without plant growth promoting rhizobacteria. *Indian Journal of Agronomy, 58*, 430–433.
Chowdhury, S. P., Schmid, M., Hartmann, A., & Tripathi, A. K. (2009). Diversity of 16S-rRNA and nifH genes derived from rhizosphere soil and roots of an endemic drought tolerant grass. *European Journal of Soil Biology, 45*, 114–122.
Colonna, E., Rouphael, Y., De Pascale, S., & Barbieri, G. (2015). Effects of mycorrhiza and plant growth promoting rhizobacteria on yield and quality of artichoke. *Acta Horticulturae, 1147*, 43–50.
Contreras-Cornejo, H. A., Macías-Rodríguez, L., Beltrán-Peña, E., Herrera-Estrella, A., & López-Bucio, J. (2011). Trichoderma-induced plant immunity likely involves both hormonal- and camalexin-dependent mechanisms in Arabidopsis thaliana and confers resistance against necrotrophic fungus Botrytis cinerea. *Plant Signaling & Behavior, 6*, 1554–1563.

Dashti, N. H., Ali, N. Y., Cherian, V. M., & Montasser, M. S. (2012). Application of plant growth promoting rhizobacteria (PGPR) in combination with a mild strain of Cucumber mosaic virus (CMV) associated with viral satellite RNAs to enhance growth and protection against a virulent strain of CMV in tomato. *Canadian Journal of Plant Pathology, 34*, 177–186.

Datta, B., & Chakrabartty, P. K. (2013). Siderophore biosynthesis genes of *Rhizobium* sp. isolated from *Cicer arietinum* L. *Biotechnology, 4*, 391–401.

Dennis, P. G., Miller, A. J., Clark, I. M., Taylor, R. G., Valsami-Jones, E., & Hirsch, P. R. (2008). A novel method for sampling bacteria on plant root and soil surfaces at the microhabitat scale. *Journal of Microbiological Methods, 75*, 12–18.

Dennis, P. G., Miller, A. J., & Hirsch, P. R. (2010). Are root exudates more important than other sources of rhizodeposits in structuring rhizosphere bacterial communities? *FEMS Microbiology Ecology, 72*, 313–327.

Dotaniya, M., & Vasudev, M. (2015). Rhizosphere effect on nutrient availability in soil and its uptake by plants: A Review. *Proceedings of the National Academy of Sciences, India Section B: Biological Sciences, 85*. https://doi.org/10.1007/s40011-013-0297-0.

Dotaniya, M. L., Meena, H. M., Lata, M., & Kumar, K. (2013). Role of phytosiderophores in iron uptake by plants. *Agricultural Science Digest, 33*, 73–76.

El Kiyum, S. S. S. (2017). Influence of effective microorganisms on qualities of tomatoes (*Lycopersicon esculentum*) grown on tropical loam soil. *Journal of Natural Sciences Research, 7*, 53–56.

Elahi, I. F. E., Mridha, M. A. U., & Aminuzzman, F. M. (2012). Role of AMF on plant growth, nutrient uptake arsenic toxicity and chlorophyll content of chili grown in arsenic amended soil. *Bangladesh Journal of Agricultural Research, 37*, 635–644.

Elsharkawy, M., Shivanna, M., Meera, M., & Mitsuro, H. (2015). Mechanism of induced systemic resistance against anthracnose disease in cucumber by plant growth-promoting fungi. *Acta Agriculturae Scandinavica, Section B — Soil & Plant Science, 65*, 1–13.

Etminani, F., & Harighi, B. (2018). Isolation and identification of endophytic bacteria with plant growth promoting activity and biocontrol potential from wild pistachio trees. *Plant Pathology Journal, 34*, 208–217.

Fernandes, E. G., Valerio, H. M., Feltrin, T., & Van Der Sand, S. T. (2012). Variability in the production of extracellular enzymes by entomopathogenic fungi grown on different substrates. *Brazilian Journal of Microbiology, 43*, 827–833.

Franco-Correa, M., Quintana, A., Duque, C., Suarez, C., Rodriguez, M. X., & Barea, J. M. (2010). Evaluation of actinomycete strains for key traits related with plant growth-promotion and mycorrhiza helping activities. *Applied Soil Ecology, 45*, 209–217.

Ganesan, S., Swaminathan, U., & Muthukumar, A. (2009). Biocontrol with *Trichoderma* species for the management of postharvest crown rot of banana. *Phytopathologia Mediterranea, 48*, 214–225.

Gao, M., Yao, S., Liu, Y., Yu, H., Xu, P., Sun, W., Pu, Z., Hou, H., & Bao, Y. (2018). Transcriptome analysis of tomato leaf spot pathogen fusarium proliferatum: De novo assembly, expression profiling, and identification of candidate effectors. *International Journal of Molecular Sciences, 9*(1), 31.

Ghodsalavi, B., Ahmadzadeh, M., Soleimani, M., Brokanloui, P. M., & Taghizad-Farid, R. (2013). Isolation and characterization of rhizobacteria and their effects on root extracts of *Valeriana officinalis*. *AJCS, 7*, 338–344.

Ghorpade, V. M., & Gupta, S. G. (2016). Siderophore production by *Rhizobium nepotum* isolated from "Stem nodule of *Aeschynomene indica*". *International Journal of Advanced Research in Biological Sciences, 3*, 105–108.

Ghosh, A., Arif, T. U., Chamely, S. G., Rezwanul, H. M., Rahman, M., & Bhuiyan, A. H. (2017a). Comparative effect of arbuscular mycorrhiza, cowdung and phosphorus on growth and yield contributing characters of red amaranth (*Amaranthus tricolor* L.) and Indian spinach (*Basella alba* L.). *Tropical Plant Research, 4*, 254–263.

Ghosh, K. S., Banerjee, S., & Sengupta, C. (2017b). Bioassay, characterization and estimation of siderophores from some important antagonistic fungi. *Journal of Biopesticides, 10*, 105–112.
Gopalakrishnan, S., Vadlamudi, S., Bandikinda, P., Sathya, A., Vijayabharathi, R., Rupela, O., et al. (2014). Evaluation of *Streptomyces* strains isolated from herbal vermicompost for their plant growth-promotion traits in rice. *Microbiological Research, 169*, 40–48.
Gopalakrishnan, S., Srinivas, V., Alekhya, G., Prakash, B., Kudapa, H., & Varshney, R. K. (2015). Evaluation of broad-spectrum *Streptomyces* sp. for plant growth promotion traits in chickpea (*Cicer arietinum* L.). *Philippine Agricultural Scientist, 98*, 270–278.
Goswami, D., Vaghela, H., Parmar, S., Dhandhukia, P., & Thakker, J. N. (2013). Plant growth promoting potentials of *Pseudomonas* spp. strain OG isolated from marine water. *Journal of Plant Interactions, 8*, 281–290.
Guiñazú, L. B., Andres, J. A., DelPapa, M. F., Pistorio, M., & Rosas, S. B. (2010). Response of alfalfa (Medicago sativa L.) to single and mixed inoculation with phosphate-solubilising bacteria and Sinorhizobium meliloti. *Biology and Fertility of Soils, 46*, 185–190.
Hahm, M. S., Sumayo, M., Hwang, Y. J., Jeon, S. A., Park, S. J., Lee, J. Y., et al. (2012). Biological control and plant growth promoting capacity of rhizobacteria on pepper under greenhouse and field conditions. *Journal of Microbiology, 50*, 380–385.
Han, K. I., Patnaik, B. B., Kim, Y. H., Kwon, H. J., Han, Y. S., & Han, M. D. (2014). Isolation and characterization of chitinase-producing *Bacillus* and *Paenibacillus* strains from salted and fermented shrimp. *Journal of Food Science*. https://doi.org/10.1111/1750-3841.12387.
Hartmann, A., Rothballer, M., Schmid, M., & Hiltner, L. (2008). A pioneer in rhizosphere microbial ecology and soil bacteriology research. *Plant and Soil, 312*, 7–14.
Hartmann, A., Schmid, M., Van Tuinen, D., & Berg, G. (2009). Plant-driven selection of microbes. *Plant and Soil, 321*, 235–257.
Hassan, F., Qayyum, A., Malik, W., Maqbool, A., Hassan, M., Rehman, M. A., Shoaib, M., et al. (2016). Cotton leaf curl virus (CLCuV) disease in Pakistan: A critical review. *Applied Sciences and Business Economics, 3*, 8–14.
Hayat, S., Faraz, A., & Faizan, M. (2017). Root exudates: Composition and impact on plant-microbe interaction. In I. Ahmad & F. M. Husain (Eds.), *Biofilms in plant and soil health*. Hoboken: John Wiley & Sons Ltd. https://doi.org/10.1002/9781119246329.ch10.
Hitendra, J., Sohel, S., & Riyaz, S. (2017). Role of hydrolytic enzymes of rhizoflora in biocontrol of fungal phytopathogens. An overview. In S. Mehnaz (Ed.), *Rhizotrophs: Plant growth promotion to bioremediation* (Microorganisms for sustainability) (Vol. 2, pp. 183–203). Singapore: Springer.
Horiuchi, J., Prithiviraj, B., Bais, H., Kimball, B., & Vivanco, J. (2005). Soil nematodes mediate positive interactions between legume plants and rhizobium bacteria. *Planta, 222*, 848–857.
Hossain, M. M., Sultana, F., & Islam, S. (2017). Plant growth-promoting fungi (PGPF): Èhytostimulation and induced systemic resistance. In D. Singh, H. Singh, & R. Prabha (Eds.), *Plant-microbe interactions in agro-ecological perspectives* (pp. 135–191). Singapore: Springer. https://doi.org/10.1007/978-981-10-6593-4-6.
Hou, M. P., & Babalola, O. O. (2013). Evaluation of plant growth promoting potential of four rhizobacterial species for indigenous system. *Journal of Central South University, 20*, 164–171.
Huang, X. F., Chaparro, J., Reardon, K., Zhang, R., Shen, Q., & Jorge, M. V. (2014). Rhizosphere interactions: Root exudates, microbes, and microbial communities. *Botany, 92*. https://doi.org/10.1139/cjb-2013-0225.
Igiehon, N. O., & Babalola, O. O. (2017). Biofertilizers and sustainable agriculture: Exploring arbuscular mycorrhizal fungi. *Applied Microbiology and Biotechnology, 101*, 4871–4881.
Igiehon, N. O., & Babalola, O. O. (2018). Rhizosphere microbiome modulators: Contributions of nitrogen fixing bacteria towards sustainable agriculture. *International Journal of Environmental Research and Public Health*. https://doi.org/10.3390/ijerph15040574.
Jacoby, R., Peukert, M., Succurro, A., Koprivova, A., & Kopriva, S. (2017). The role of soil microorganisms in plant mineral nutrition-current knowledge and future directions. *Frontiers in Plant Science*. https://doi.org/10.3389/fpls2017.01617.

Jarak, M., Nastasija, M., Dragana, B., Dragana, J., Timea, H. J., & Dragana, S. (2012). Effects of plant growth promoting rhizobacteria on maize in greenhouse and field trial. *African Journal of Microbiology Research, 6*, 5683–5690.

Jeffrey, B., John, R. T., Daniel, R., Inga, A. Z., & Jude, L. (2010). Factors affecting soil microbial community structure in tomato cropping systems. *Soil Biology and Biochemistry, 42*, 831–841.

Jones, D. L., Nguyen, C., & Finlay, R. D. (2009). Carbon flow in the rhizosphere: Carbon trading at the soil-root interface. *Plant and Soil, 321*, 5–33.

Jourdan, E., Henry, G., Duby, F., Dommes, J., Barthelemy, J. P., Thonart, P., & Ongena, M. (2009). Insights into the defense-related events occurring in plant cells following perception of surfactin-type lipopeptide from *Bacillus subtilis*. *MPMI, 22*, 456–468.

Junaid, J. M., Dar, N. A., Bhat, T. A., Bhat, A. H., & Bhat, M. A. (2013). Commercial biocontrol agents and their mechanism of action in the management of plant pathogens. *International Journal of Plant, Animal and Environmental Sciences, 1*, 39–57.

Kaiser, C., Koranda, M., Kitzler, B., Fuchslueger, L., Schnecker, J., Schweiger, P., et al. (2010). Belowground carbon allocation by trees drives seasonal patterns of extracellular enzyme activities by altering microbial community composition in a beech forest soil. *New Phytologist. 187*, 843–858.

Kapoor, N., & Sachdeva, S. (2013). Rhizosphere: Its structure, bacterial diversity and significance. *Reviews in Environmental Science and Biotechnology, 13*, 63–77.

Kim, K. R., Owens, G., Naidu, R., & Kwon, S. L. (2010). Influence of plant roots on rhizosphere soil solution composition of long-term contaminated soils. *Geoderma, 155*, 86–92.

Kim, S. J., Eo, J. K., Lee, E. H., Park, H., & Eom, A. H. (2017). Effects of Arbuscular Mycorrhizal fungi and soil conditions on crop plant growth. *Mycobiology, 45*, 20–24.

Kollmann, J., Banuelos, M. J., & Nielsen, S. L. (2007). Effects of virus infection on growth of the invasive alien *Impatiens glandulifera*. *Preslia, 79*, 33–44.

Krishnaveni, B., & Ragunathan, R. (2014). Chitinase production from marine wastes by *Aspergillus terreus* and its application in degradation studies. *International Journal of Current Microbiology and Applied Sciences, 3*, 76–82.

Krithika, S., & Chellaram, C. (2016). Isolation, screening, and characterization of chitinase producing bacteria from marine wastes. *International Journal of Pharmacy and Pharmaceutical Sciences, 8*, 34–36.

Kumar, A., & Singh, V. (2015). Plant growth promoting rhizobacteria (PGPR). A promising approach for disease management. In S. Singh & D. P. Singh (Eds.), *Microbes and environmental management* (pp. 195–209). New Delhi: Studium Press.

Kumar, H., Bajpai, V. K., & Dubey, R. C. (2010). Wilt disease management and enhancement of growth and yield of *Cajanus cajan* (L) var. Manak by bacterial combinations amended with chemical fertilizer. *Crop Protection, 29*, 591–598.

Kumar, G. P., Desai, S., Leo, D. A. E., Hask, M., & Reddy, G. (2012). Plant growth promoting *Pseudomonas* spp. from diverse agro-ecosystems of India for *Sorghum bicolor* L. *Journal of Biofertilizers and Biopesticides, S7*, 001. https://doi.org/10.4172/2155-6202.S7-001.

Kuppireddy, V., Uversky, V., Toh, S., Tsai, M. C., Beckerson, W., Cahill, C., et al. (2017). Identification and initial characterization of the effectors of an anther smut fungus and potential host target proteins. *International Journal of Molecular Sciences, 8*(11), 2489. https://doi.org/10.3390/ijms18112489.

Lee, J., Postmaster, A., Soon, H. P., Keast, D., & Carson, K. C. (2012). Siderophore production by *Actinomycetes* isolates from two soil sites in Western Australia. *Biometals, 25*, 285–296.

Lehner, S. M., Atanasova, L., Neumann, N. K. N., Krska, R., Lemmens, M., Druzhinina, I. S., & Schuhmachera, R. (2013). Isotope-assisted screening for iron-containing metabolites reveals a high degree of diversity among known and unknown siderophores produced by *Trichoderma* spp. *Applied and Environmental Microbiology, 79*, 18–31.

Li, S., Peng, M., Liu, Z., & Shah, S. S. (2017). The role of soil microbes in promoting plant growth. *Molecular Microbiology Research, 7*, 30–37.

Liang, J. F., An, J., Gao, J. Q., Zhang, X. Y., & Yu, F. H. (2018). Effects of arbuscular mycorrhizal fungi and soil nutrient addition on the growth of *Phragmites australis* under different drying rewetting cycles. *PLoS ONE, 13*(1). https://doi.org/10.1371/journal.pone.0191999.

Ling, S. K., Rugang, L., & Zhang, W. (2014). First report of cucumber green mottle mosaic virus infecting greenhouse cucumber in Canada. *Plant Disease, 98*, 701–701.

Liu, H. W., Luo, L. X., Li, J. Q., Liu, P. F., Chen, X. Y., & Hao, J. J. (2013). Pollen and seed transmission of Cucumber green mottle mosaic virus in cucumber. *Plant Pathology, 63*, 72–77.

Lobmann, M. T., Vetukuri, R. R., Zinger, L. D., Alsanius, B. W., Grenville-Briggs, L. J., & Walter, A. J. (2016). The occurrence of pathogen suppressive soils in Sweden in relation to soil biota, soil properties, and farming practices. *Applied Soil Ecology, 107*, 57–65.

Makut, M., & Owolewa, O. A. (2011). Antibiotic-producing fungi present in the soil environment of Keffi metropolis, Nasarawa state, Nigeria. *Trakia Journal of Sciences, 9*, 33–39.

Martin, C. M., Eilyn, M., García, S. C., Berkis, R., SA, M., Tatiana, P., Mora, L. M., & Yelenys, A. C. (2015). The effects of plant growth promoting Bacillus pumilus CCIBP- C5 on 'Grande naine' (Musa AAA) plants in acclimatization stage. *Biotecnología Vegetal, 15*, 151–156.

Martínez-Medina, A., Fernandez, I., Lok, G. B., Pozo, M. J., Pieterse, C. M. J., Saskia, C., & Van Wees, S. C. (2016). Shifting from priming of salicylic acid to jasmonic acid regulated defences by *Trichoderma* protects tomato against the root knot nematode *Meloidogyne incognita*. *New Phytologist, 3*(3), 1363–1377.

Martinez-Viveros, O., Jorquera, M. A., Crowley, D. E., Gajardo, G., & Mora, M. L. (2010). Mechanisms and practical considerations involved in plant growth promotion by rhizobacteria. *Journal of Soil Science and Plant Nutrition, 10*, 293–319.

Mendes, R., Kruijt, M., de Bruijn, I., Dekkers, E., van der Voort, M., Schneider, J. H., et al. (2011). Deciphering the rhizosphere microbiome for disease suppressive bacteria. *Science, 332*, 1097–1100.

Miyasaki, Y., Rabenstein, J. D., Rhea, J., Crouch, M. L., Mocek, U. M., Kittell, P. E., et al. (2013). Isolation and characterization of antimicrobial compounds in plant extracts against multidrug-resistant *Acinetobacter baumannii*. *PLoS ONE, 8*, e61594.

Moe, L. A. (2013). Amino acids in the rhizosphere: From plants to microbes. *American Journal of Botany, 100*, 1692–1705.

Mohamed, H. I., & Gomaa, E. Z. (2012). Effect of plant growth promoting *Bacillus subtilis* and *Pseudomonas fluorescens* on growth and pigment composition of radish plants (*Raphanus sativus*) under NaCl stress. *Photosynthetica, 50*, 263–272.

Mohammadi, K., Khalesro, S., Sohrabi, Y., & Heidari, G. (2011). A review: Beneficial effects of the mycorrhizal fungi for plant growth. *Journal of Applied Environmental and Biological Sciences, 1*, 310–319.

Montalbán, B., Thijs, S., Lobo, M. C., Weyens, N., Ameloot, M., Vangronsveld, J., & Pérez-Sanz, A. (2017). Cultivar and metal-specific effects of endophytic bacteria in Helianthus tuberosus exposed to Cd and Zn. *International Journal of Molecular Sciences, 18*, E2026. https://doi.org/10.3390/ijms18102026.

Moriones, E., Praveen, S., & Chakraborty, S. (2017). Tomato leaf curl New Delhi virus: An emerging virus complex threatening vegetable and fiber crops. *Viruses, 9*, E264. https://doi.org/10.3390/v9100264.

Muller, M. S., Scheu, S., & Jousset, A. (2013). Protozoa drive the dynamics of culturable biocontrol bacterial communities. *PLoS One, 8*, 1–6.

Muller, D. B., Vogel, C., Bai, Y., & Vorholt, J. A. (2016). The plant microbiota: Systems-level insights and perspectives. In N. M. Bonini (Ed.), *Annual review of genetics* (Vol. 50, pp. 211–234). Palo Alto: Annual Reviews.

Nassal, D., Spohn, M., Eltlbany, N., Jacquiod, S., Smalla, K., Marhan, S., & Kandeler, E. (2018). Effects of phosphorus-mobilizing bacteria on tomato growth and soil microbial activity. *Plant and Soil, 427*, 17–37.

Naznin, H. A., Kimura, M., Miyazawa, M., & Hyakumachi, M. (2013). Analysis of volatile organic compounds emitted by plant growth-promoting fungus *Phoma* sp. GS8-3 for growth promotion effects on tobacco. *Microbes and Environments, 28*, 42–49.

Naznin, H. A., Kiyohara, D., Kimura, M., Miyazawa, M., Shimizu, M., & Hyakumachi, M. (2014). Systemic resistance induced by volatile organic compounds emitted by plant growth-promoting fungi in Arabidopsis thaliana. *PLoS One, 9*. https://doi.org/10.1371/journal.pone.0086882.

Neumann, G., George, T. S., & Plassard, C. (2009). Strategies and methods for studying the rhizosphere-the plant science toolbox. *Plant and Soil, 321*, 43–456.

Ngoma, L., Esau, B., & Babalola, O. O. (2013). Isolation and characterization of beneficial indigenous endophytic bacteria for plant growth promoting activity in Molelwane Farm, Mafikeng, South Africa. *African Journal of Biotechnology, 12*, 4105–4114.

Nguyen, M. L., Spaepen, S., Jardin, P. D., & Delaplace, P. (2019). Biostimulant effects of rhizobacteria on wheat growth and nutrient uptake depend on nitrogen application and plant development. *Archives of Agronomy and Soil Science, 65*, 58–73.

Nihorimbere, V., Ongena, M., Smargiassi, M., & Thonart, P. (2011). Beneficial effect of the rhizosphere microbial community for plant growth and health. *Biotechnology, Agronomy, Society and Environment, 15*, 327–337.

Niu, Y. F., Chai, R. S., Jin, G. L., Wang, H., Tang, C. X., & Zhang, Y. S. (2012). Responses of root architecture development to low phosphorus availability: A review. *Annals of Botany, 112*, 391–408.

Ortas, I. (2012). The effect of mycorrhizal fungal inoculation on plant yield, nutrient uptake and inoculation effectiveness under long-term field conditions. *Field Crops Research, 125*, 35–48.

Pandya, M., Rajput, M., Rajkumar, S., Najmeh, A., Doustmorad, Z., & Mansoureh, M. (2015). Exploring plant growth promoting potential of non rhizobial root nodules endophytes of Vigna radiate. *Microbiology, 84*, 80–89.

Patel, T. S., & Desai, P. B. (2015). Isolation and screening of PGPR from rhizospheric and non rhizospheric soil of Bt-cotton. *Indo-American Journal of Agricultural and Veterinary Sciences, 3*, B1110–B1120.

Patel, P. R., Shaikh, S. S., & Sayyed, R. Z. (2018). Modified chrome azurol S method for detection and estimation of siderophores having affinity for metal ions other than iron. *Environmental Sustainability, 1*, 81–87.

Patil, G. B., Lakshman, L. H., Romana, M., & Agadi, B. S. (2013). Effect of co-inoculation of AM fungi and two beneficial microorganisms on growth and nutrient uptake of *Eleusine coracana Gaertn*. (*Finger millet*). *Asian Journal of Plant Science & Research, 3*, 26–30.

Pazarlar, S., Gümüş, M., & Öztekin, G. B. (2013). The effects of tobacco mosaic virus infection on growth and physiological parameters in some pepper varieties (*Capsicum annuum* L.). *Notulae Botanicae Horti Agrobotanici Cluj-Napoca, 41*, 427–443.

Perez-Montano, F., Alias-Villegas, C., Bellogín, R. A., del Cerro, P., Espuny, M. R., Jiménez-Guerrero, I., et al. (2014). Plant growth promotion in cereal and leguminous agricultural important plants from microorganism capacities to crop production. *Microbiological Research, 169*, 325–336.

Philippot, L., Raaijmakers, J. M., Lemanceau, P., & van der Putten, W. H. (2013). Going back to the roots: The microbial ecology of the rhizosphere. *Nature Reviews Microbiology, 11*, 789–799.

Pindi, P. K., Sultana, T., & Vootla, P. K. (2014). Plant growth regulation of Bt-cotton through Bacillus species. *Biotechnology, 4*, 305–315.

Pirlak, L., & Murat, K. (2009). Effects of plant growth promoting rhizobacteria on yield and some fruit properties of strawberry. *Journal of Plant Nutrition, 32*, 1173–1184.

Pishchik, V., Vorobyev, N., Ostankova, Y., Semenov, A., Areg, A., Popov, A., et al. (2018). Impact of *Bacillus subtilis* on tomato plants growth and some biochemical characteristics under combined application with humic fertilizer. *International Journal of Plant & Soil Science, 22*, 1–12.

Potshangbam, M., Devi, S. I., Sahoo, D., & Strobel, G. A. (2017). Functional characterization of endophytic fungal community associated with *Oryza sativa* L. and *Zea mays* L. *Frontiers in Microbiology*. https://doi.org/10.3389/fmicb.2017.00325.

Prakash, O. M., Sharma, R., Sharma, P., & Kerthikeyan, N. (2015). Role of microorganisms in plant nutrition and health. In A. Rakshit, H. B. Singh, & A. Sen (Eds.), *Nutrient use efficiency: From basics to advances*. New Delhi: Springer.

Pulatov, A. A. S., Khudayberdi, N., Maksud, A., & Botir, K. (2016). Effect of biofertilizers on growth and yield of cotton in different soil conditions. *Cotton Genomics and Genetics, 7*, 1–7. https://doi.org/10.5376/cgg.2016.07.0001.

Qiano, J., Xiang, Y., Liang, X., Liu, Y., Borriss, R., & Liu, Y. (2017). Addition of plant growth promoting *Bacillus subtilis* PTS-394 on tomato rhizosphere has no durable impact on composition of root microbiome. *Microbiology, 117*, 131. https://doi.org/10.1186/s12866-017-1039-x.

Rajendiran, S., Coumar, M. V., Kundu, S., Dotaniya, M. L., & Rao, A. S. (2012). Role of phytolith occluded carbon of crop plants for enhancing soil carbon sequestration in agroecosystems. *Current Science, 103*, 911–920.

Rawat, S. (2015). Food spoilage: Microorganisms and their prevention. *Asian Journal of Plant Science & Research, 5*, 47–56.

Rillig, M. C., Caldwell, B. A., Wosten, H. A. B., & Sollins, P. (2007). Role of proteins in soil carbon and nitrogen storage: Control of persistence. *Biogeochemistry, 85*, 25–44.

Ritter, C., Camassola, M., Zampieri, D., Silveira, M. M., & Dillon, A. J. P. (2013). Cellulase and xylanase production by *Penicillium echinulatum* in submerged media containing cellulose amended with sorbitol. *Enzyme Research, 2013*, 1–9. https://doi.org/10.1155/2013/240219.

Rodrigues, A. A., Forzani, M. V., Soares, R. D. S., Sibov, S. T., & Vieira, J. D. G. (2016). Isolation and selection of plant growth-promoting bacteria associated with sugarcane. *Pesquisa Agropecuária Tropical, 46*(2), 149–158.

Rosconi, F., Trovero, M. M., Souza, E., & Fabiano, E. (2015). Serobactins mediated iron acquisition systems optimize competitive fitness of Herbaspirillum seropedicae inside rice plants. *Environmental Microbiology, 18*, 2523–2533. https://doi.org/10.1111/1462-2920.13202.

Ruchi, R. K., Kumar, A., Kumar, A., Patil, S., Thapa, S., & Kaur, M. (2012). Evaluation of plant growth promoting attributes and lytic enzyme production by *Fluorescent Pseudomonas* diversity associated with Apple and Pear. *International Journal of Scientific and Research Publications, 2*, 1–8.

Salas-Marina, M. A., Silva-Flores, M. A., Uresti-Rivera, E. E., Castro-Longoria, E., Herrera-Estrella, A., & Casas-Flores, S. (2011). Colonization of Arabidopsis roots by *Trichoderma atroviride* promotes growth and enhances systemic disease resistance through jasmonate and salicylate pathways. *European Journal of Plant Pathology, 131*, 15–26.

Sanchez-Lopez, A., Pintelon, I., Stevens, V., Imperato, V., Timmermans, J. P., Gonzalez-Chavez, C., et al. (2018). Seed endophyte microbiome of *Crotalaria pumila* unpeeled: Identification of plant-beneficial methylobacteria. *International Journal of Molecular Sciences, 19*, E291. https://doi.org/10.3390/ijms19010291.

Sandilya, S. R. S. P., Bhuyan, P. M., Nageshappa, V., Gogoi, D. K., & Kardong, D. (2017). Impact of *Pseudomonas aeruginosa* MAJ PIA03 affecting the growth and phytonutrient production of castor, a primary host-plant. *Journal of Soil Science and Plant Nutrition, 17*, 499–5150.

Santa-Rosa, P., Souza, A. L., Roque, R. A., Andrade, V. E., Filho, S. A., Adolfo, M., & Silva, N. (2017). Production of β-glucosidase and CMCase thermostables by *Penicillium Sp.* LMI01 isolated in the Amazon region. *Electronic Journal of Biotechnology, 31*, 84–92. https://doi.org/10.1016/j.ejbt.2017.11.005.

Sanwal, S. K., Venkataravanappa, V., & Singh, B. (2016). Resistance to bhendi yellow vein mosaic disease: A review. *Indian Journal of Agricultural Sciences, 86*, 835–843.

Sayyed, R. Z., & Patel, P. R. (2011). Biocontrol potential of siderophore producing heavy metal resistant Alcaligenes sp. and *Pseudomonas* aeruginosa rzs3 vis-à-vis organophosphorus fungicide. *Indian Journal of Microbiology, 51*, 266–272.

Schirawski, J., & Perlin, M. H. (2018). Plant microbe interaction 2017-The good, the bad and the diverse. *International Journal of Molecular Sciences, 9*, 1374. https://doi.org/10.3390/ijms19051374.

Schneider, W. D. H., dos Reis, L., Camassola, M., & Dillon, A. J. P. (2014). Morphogenesis and production of enzymes by *Penicillium echinulatum* in response to different carbon sources. *BioMed Research International, 2014*, 254863. https://doi.org/10.1155/2014/254863.

Schwachtje, J., Karojet, S., Kunz, S., Brouwe, S., & van Dongen, J. T. (2012). Plant-growth promoting effect of newly isolated rhizobacteria varies between two *Arabidopsis* ecotypes. *Plant Signaling & Behavior, 7*, 623–627.

Shakeel, M., Afroz, R., Muhammad, H., & Fauzia, H. (2015). Root associated *Bacillus* sp. improves growth, yield and zinc translocation for basmati Rice (*Oryza sativa*) varieties. *Frontiers in Microbiology, 6*, 780. https://doi.org/10.3389/fmicb.2015.01286.

Shakeela, S., Padder, S A., & Bhat, Z. A. (2017). Isolation and characterization of plant growth promoting rhizobacteria associated with medicinal plant *Picrorhiza Kurroa*. *Journal of Pharmacognosy and Phytochemistry, 6*, 157–168.

Shariatmadari, Z., Hossein, R., Seyed, M., Hashtroudi, S., Ghassempour, M., & Aghashariatmadary, A. (2013). Plant growth promoting cyanobacteria and their distribution in terrestrial habitats of Iran. *Soil Science and Plant Nutrition, 59*, 535–547.

Sharma, N., Sudarshan, Y., Sharma, R., & Singh, G. (2008). RAPD analysis of soil microbial diversity in western Rajasthan. *Current Science, 94*, 1058–1061.

Shoresh, M., Harman, G. E., & Mastouri, F. (2010). Induced systemic resistance and plant responses to fungal biocontrol agents. *Annual Review of Phytopathology, 48*, 1–23.

Shukla, M., Patel, R. H., Verma, R., Deewan, P., & Dotaniya, M. L. (2013). Effect of bio-organics and chemical fertilizers on growth and yield of chickpea (*Cicer arietinum* L.) under middle Gujarat conditions. *Vegetos, 26*, 183–187.

Shunmugam, S., Muralitharan, G., & Thajuddin, N. (2015). Cyanobacteria and algae: Potential sources of biological control agents used against phytopathogenic bacteria. In S. Shunmugam, G. Muralitharan, & N. Thajuddin (Eds.), *Sustainable approaches to controlling plant pathogenic bacteria*. New York: CRC Press. https://doi.org/10.1201/b18892-13.

Sial, N. L., Ahmed, A. S., Abbas, M., Irfan, M., & Depar, N. (2018). Growth and yield of wheat as affected by phosphate solubilizing bacteria and phosphate fertilizer. *Pakistan Journal of Biotechnology, 15*, 475–479.

Singh, A. A., Gupta, R., Srivastava, M., Gupta, M. M., & Pandey, R. (2016). Microbial secondary metabolites ameliorate growth, in planta contents and lignification in *Withania somnifera* (L.) Dunal. *Physiology and Molecular Biology of Plants, 22*, 253–260.

Singh, D. P., Patil, J. H., Prabha, R., & Yandigeri, M. S. (2018). Actinomycetes as potential plant growth-promoting microbial communities. In Prasad et al. (Eds.), *New and future developments in microbial biotechnology and bioengineering: Crop improvement through microbial biotechnology* (pp. 27–38). New York: Elsevier.

Somers, E., Vanderleyden, J., & Srinivasan, M. (2004). Rhizosphere bacterial signalling: A love parade beneath our feet. *Critical Reviews in Microbiology, 304*, 205–240.

Spanic, V., Marcek, T., Abicic, I., & Sarkanj, B. (2018). Effects of fusarium head blight on wheat grain and malt infected by *Fusarium culmorum* toxins. *Toxins (Basel), 10*, 17. https://doi.org/10.3390/toxins10010017.

Spence, C., Alff, E., Johnson, C., Ramos, C., Donofrio, N., Sundaresan, V., & Bais, H. (2014). Natural rice rhizospheric microbes suppress rice blast infections. *BMC Plant Biology, 14*, 130. https://doi.org/10.1186/1471-2229-14-130.

Sreevidya, M., Gopalakrishnan, S., Kudapa, H., & Varshney, R. K. (2016). Exploring plant growth-promotion actinomycetes from vermicompost and rhizosphere soil for yield enhancement in chickpea. *Brazilian Journal of Microbiology, 47*, 85–95.

Srivastava, M. P., Gupta, S., & Sharm, Y. K. (2018). Detection of siderophore production from different cultural variables by CAS-agar plate assay. *Asian Journal of Pharmacy and Pharmacology, 4*, 66–69.

Su, A. Y., Niu, S. Q., Liu, Y. Z., He, A. L., Zhao, Q., Pare, P., et al. (2017). Synergistic effects of *Bacillus amyloliquefaciens* (GB03) and water retaining agent on drought tolerance of perennial ryegrass. *International Journal of Molecular Sciences, 18*, 2651. https://doi.org/10.3390/ijms1812265.

Suhanna, H. A., Norhanis, Y., & Hartinee, A. (2013). Application of Trichoderma spp. to control stem end rot disease of mango var. *Journal of Tropical Agriculture and Food Science, 41*, 159–168.

Tahir, A. S., Qin, G., Huijun, W. U., Waseem, R., Safdar, A., Huang, Z., et al. (2017). Effect of volatile compounds produced by *Ralstonia solanacearum* on plant growth promoting and systemic resistance inducing potential of *Bacillus volatiles*. *BMC Plant Biology, 17*, 133. https://doi.org/10.1186/s12870-017-1083-6.

Trejo-Estrada, R., Paszczynski, S. A., & Crawford, D. (1998). Antibiotics and enzymes produced by the biocontrol agent *Streptomyces violaceusniger* YCED-9. *Journal of Industrial Microbiology and Biotechnology, 21*, 81–90.

Trovero, M. F., Scavone, P., Platero, R., de Souza, E. M., Fabiano, E., & Rosconi, F. (2018). *Herbaspirillum seropedicae* differentially expressed genes in response to iron availability. *Frontiers in Microbiology, 9*, 1430. https://doi.org/10.3389/fmicb.2018.01430.

Turan, M., Medine, G., Cakmakci, R., Taskin, O., & Fikrettin, S. (2018). The effect of PGPR strain on wheat yield and quality parameters. In: World Congress of Soil Science, pp. 140–143.

Ueno, D., Rombola, A. D., Iwashita, T., Nomoto, K., & Ma, J. F. (2007). Identification of two novel phytosiderophores secreted by perennial grasses. *New Phytologist, 174*, 304–310.

Uren, N. C. (2007). Types, amounts and possible functions of compounds released into the rhizosphere by soil-grown plants. In R. Pinton, Z. Varanini, & P. Nannipieri (Eds.), *The rhizosphere: Biochemistry and organic substances at the soil-plant interface* (2nd ed., pp. 1–21). Boca Raton: CRC Press.

Uzair, B., Kausar, R., & Bano, S. A. (2018). Isolation and molecular characterization of a model antagonistic *Pseudomonas aeruginosa* divulging in vitro plant growth promoting characteristics. *BioMed Research International, 2018*, 6147380. https://doi.org/10.1080/03235408.2013.

Vacheron, J., Desbrosses, G., Renoud, S., Padilla, R., Walker, V., Muller, D., & Prigent-Combaret, C. (2018). Differential contribution of plant-beneficial functions from *Pseudomonas kilonensis* F113 to root system architecture alterations in *Arabidopsis thaliana* and *Zea mays*. *MPMI, 31*, 212–223.

Vandana, U. K., Chopra, A., Choudhury, A., Adapa, D., & Mazumder, P. B. (2018). Genetic diversity and antagonistic activity of plant growth promoting bacteria, isolated from tea-rhizosphere: A culture dependent study. *Biomedical Research, 29*, 853–864.

Vellasamy, S., Nithya, K., Hariharan, H., Jayaprakashvel, M., & Balasubramanian, N. (2015). Biocontrol Mechanisms of siderophores against bacterial plant pathogens. pp. 167–189. In V. R. Kannan & K. K. Bastas (Eds.), *Sustainable approaches to controlling plant pathogenic bacteria* (pp. 167–186). Boca Raton: CRC Press. https://doi.org/10.1201/b18892-9.

Venkataravanappa, V., Kodandaram, M., Lakshminarayana, N., Reddy, C., Shankarappa, K. S., & Reddy, M. K. (2017). Comparative transmission of Bhendi yellow vein mosaic virus by two cryptic species of the whitefly, *Bemisia tabaci* (Hemiptera: Aleyrodidae). *Biotechnology, 7*, 331. https://doi.org/10.1007/s13205-017-0970-8.

Vinale, F., Nigro, M., Sivasithamparam, K., & Flematti, G. (2013). Harzianic acid: A novel siderophore from *Trichoderma harzianum*. *FEMS Microbiology Letters, 347*, 123–129.

Viterbo, A., Ofir, R., Leonid, C., & Ilan, C. (2002). Significance of lytic enzymes from *Trichoderma* spp. in the biocontrol of fungal plant pathogens. *Antonie van Leeuwenhoek, 81*, 549–556.

Vitorino, L. C., Bessa, L. A., Carvalho, L. G., & Silva, F. G. (2016). Growth promotion mediated by endophytic fungi in cloned seedlings of *Eucalyptus grandis* × *Eucalyptus urophylla* hybrids. *African Journal of Biotechnology, 15*, 2729–2738.

Vurukonda, S. S. K. P., Giovanardi, D., & Stefani, E. (2018). Plant growth promoting and biocontrol activity of *Streptomyces* spp. as endophytes. *International Journal of Molecular Sciences, 19*, E952. https://doi.org/10.3390/ijms19040952.

Walker, T. S., Bais, H. P., Grotewold, E., & Vivanco, J. M. (2003). Root exudation and rhizosphere biology. *Plant Physiology, 132*, 44–51.

Weidner, S., Latz, E., Agaras, B., Valverde, C., & Jousset, A. (2017). Protozoa stimulate the plant beneficial activity of rhizospheric *pseudomonads*. *Plant and Soil, 410*, 509–515.

Weizhen, Q., & Lei, Z. (2013). Study of the siderophore-producing *Trichoderma asperellum* Q1 on cucumber growth promotion under salt stress. *Journal of Basic Microbiology, 53*, 355–364. https://doi.org/10.1002/jobm.201200031.

Weller, D. M., Mavrodi, D. V., van Pelt, J. A., Pieterse, C. M., van Loon, L. C., & Bakker, P. A. (2012). Induced systemic resistance in *Arabidopsis thaliana* against *Pseudomonas syringae* pv. tomato by 2,4-diacetylphloroglucinol-producing *Pseudomonas fluorescens*. *Phytopathology, 102*, 403–412.

Xiang, L., Gong, S., Yang, L., Hao, J., Xue, M. F., Zeng, F. S., et al. (2016). Biocontrol potential of endophytic fungi in medicinal plants from Wuhan Botanical Garden in China. *Biological Control, 94*, 47–55.

Yamagiwa, Y., Inagaki, Y., Ichinose, Y., Toyoda, K., & Hyakumachi, M. (2011). *Talaromyces wortmannii* FS2 emits β-caryophyllene, which promotes plant growth and induces resistance. *Journal of General Plant Pathology, 77*, 336–341.

Yolcu, H., Gunes, A., Gullap, M. K., & Cakmakci, R. (2012). Effects of plant growth-promoting rhizobacteria on some morphologic characteristics, yield and quality contents of Hungarian vetch. *Turkish Journal of Field Crops, 17*, 208–214.

Zak, D. R., Pregitzer, K. S., King, J. S., & Holmes, W. E. (2000). Elevated atmospheric CO_2 fine roots and the response of soil microorganism: A review and hypothesis. *New Phytologist, 147*, 201–222.

Zhang, X. H., Wang, Y. S., & Lin, A. J. (2011). Effects of Arbuscular mycorrhizal colonization on the growth of upland rice (*Oryzal Sativa* L.) in soil experimentally contaminated with Cu and Pb. *Journal of Clinical Toxicology, S3*, 003. https://doi.org/10.4172/2161-0495.S3-00.

Zhang, Q., Gao, X., Ren, Y., Ding, X., Qiu, J., Li, N., et al. (2018). Improvement of Verticillium wilt resistance by applying arbuscular mycorrhizal fungi to a cotton variety with high symbiotic efficiency under field conditions. *International Journal of Molecular Sciences, 19*, E241. https://doi.org/10.3390/ijms19010241.

Zhou, L. S., Tang, K., & Guo, S. X. (2018). The plant growth-promoting fungus (PGPF) *Alternaria* sp. A13 markedly enhances *Salvia miltiorrhiza* root growth and active ingredient accumulation under greenhouse and field conditions. *International Journal of Molecular Sciences, 19*, E270. https://doi.org/10.3390/ijms19010270.

Zulueta-Rodriguez, R., Cordoba-Matson, M. V., Hernandez-Montiel, L. G., Murillo-Amador, B., Rueda-Puente, E., & Lara, L. (2014). Effect of *Pseudomonas putida* on growth and anthocyanin pigment in two poinsettia (*Euphorbia pulcherrima*) cultivars. *The Scientific World Journal*, 1–6. https://doi.org/10.1155/2014/810192.

Chapter 7
Conventional and Modern Technologies for the Management of Post-Harvest Diseases

Rashad Qadri, Muhammad Azam, Imran Khan, Yaodong Yang, Shaghef Ejaz, Muhammad Tahir Akram, and M. Arslan Khan

Abstract Postharvest losses mostly occur due to senescence, microbial decay and pathogen attack, which greatly affect the quantity and quality of food. Number of techniques are used to minimize the postharvest losses and diseases, by treating products with several physical, biochemical and biological means, directly controlling pathogen infestation and extends products shelf life. Numerous physical techniques (refrigeration, cold atmosphere storage, low pressure storage and modified atmosphere storage) used to control postharvest diseases are either curative or preventive, aiming at halting disease spreading. Among physical techniques, heat treatment is considered the most effective technique especially to manage fungal diseases, which are the most common in postharvest (chilling injury). Moreover, UV treatments (UV-C, UV-B and UV-A) are used to sterilize commodities, reducing the decay due to microorganisms, helping in extending shelf life and to maintain fruits and vegetables quality. Recently, exogenous application of calcium based chemicals helped in stabilizing plant cell wall, maintaining quality of fruits and vegetables. Postharvest biological control agents have been extensively studied. By introducing natural enemies of the pathogen to be targeted its population may be reduced by restricting normal growth or activity. Additionally, volatile compounds are usually applied on a commercial scale for flavoring and seasoning agents in

---

R. Qadri (✉) · M. Azam · M. T. Akram · M. A. Khan
Institute of Horticultural Sciences, University of Agriculture Faisalabad, Faisalabad, Punjab, Pakistan

I. Khan
Institute of Soil and Environmental Sciences, University of Agriculture Faisalabad, Faisalabad, Punjab, Pakistan

Y. Yang
Hainan Key Laboratory of Tropical Oil Crops Biology, Coconut Research Institute, Chinese Academy of Tropical Agricultural Sciences, Wenchang, Hainan, China

S. Ejaz
Department of Horticulture, Bahauddin Zakariya University, Multan, Pakistan

© Springer Nature Switzerland AG 2020

I. Ul Haq, S. Ijaz (eds.), *Plant Disease Management Strategies for Sustainable Agriculture through Traditional and Modern Approaches*, Sustainability in Plant and Crop Protection 13, https://doi.org/10.1007/978-3-030-35955-3_7

foods, that strongly reduce the incidence of microbial pathogens. These volatile compounds have various properties such as antiprotectants, antimicrobial, are less harmful to mammalians, are environment friendly, and could be used as alternatives for chemical fungicides. Plants represent a huge reservoir of natural compounds harboring fungicidal activities with potential to replace synthetic fungicides. Many species produce volatile substances and essential oils that could serve as antifungal or antimicrobial preservatives for fruits and other harvested commodities. Thus, combining various treatment options may offer a more consistent, durable, practical, and sustainable solution to stakeholders and producers for postharvest control of infections. This chapter will highlight the importance of conventional and modern technologies used to control pathogens infestation, postharvest disorders to maintain quality of fruit and vegetables.

Keywords Post harvest technologies · Essential oils · Biofilm · Organic volatile compounds · Antagonistic microorganisms

7.1 Introduction

Globally, world population is increasing day by day and it is estimated that world food demand will be increased by 50–70% in the middle of this century. There are, however, many food production constraints such as land degradation, changing climate, water scarcity and food loss and weight (FLW). The term "FLW" refers to the loss of edible, available food for humans. Food losses take place due to preharvest and postharvest factors whereas food waste occurs at the last stage of the supply chain, during retail and consumption. FLW losses are categorized as qualitative and quantitative. Qualitative losses include faded color, distorted shape and poor taste, while quantitative losses include physical, mechanical and pathological damages to the commodity. According to Food and Agriculture Organization (FAO), about 30–50% of food produced in the world is lost, without being consumed.

These losses occur mostly in the postharvest phase due to senescence, microbial decay and pathogens attack. Postharvest pathogens greatly affect quantity and quality of food. Number of techniques are used to minimize postharvest losses. They are adopted to minimize postharvest diseases by treating products with several physical, biochemical and biological means, by directly controlling pathogens infestation and extending their shelf life by acting on the fruit physiology. In these techniques, optimization of water, temperature, carbon assimilation, different doses of fungicides and coating materials are used to control postharvest infestations and extend shelf life. In this chapter we discuss in detail all conventional and modern technologies to control postharvest diseases.

7.2 Physical Agents

There are numerous techniques based on physical agents that are being used since more than 100 years to control postharvest diseases and infestations and to provide safe and sound food to consumers. The physical techniques are either curative or preventive (Tripathi and Giri 2014). Physical agents mostly include refrigeration, cold atmosphere storage, low pressure storage and modified atmosphere storage, which prevent pathogen activity and disease spreading. Physical techniques kill insects and increase the shelf life of food. Further, they are also helpful in preventing postharvest disorders and maintain food quality.

7.2.1 *Low Pressure Storage*

Low pressure storage, also called hypobaric storage, is used since many years to extend shelf life of vegetables, fruit and other metabolic active compounds. This technique is used to fulfil the consumers' demand of microbe-free and high quality food, with extended shelf life. In this technique the oxygen level is reduced, and heat and harmful gases are expelled out of the room or container rapidly. In modern low pressure systems, high humidity and moisture are also maintained to prevent water loss and wilting. Low pressure reduces fruit ripening by lowering down its respiration rate and ethylene production (Stenvers and Stork 1977). Moreover, it inactivates the enzymes activity and microbial activity, which results in the extension of shelf life.

7.2.2 *Low Storage Temperature*

Fruits and vegetables are highly perishable commodities are . Among the several postharvest techniques used, low temperature is the most common. It is used to slow down the internal metabolism of fruits and vegetables, and to delay their ripening. Low temperature storage is an effective technology which is used to conserve nutritional value and quality of products after harvest. Precooling treatment during hot weather is very effective and beneficial to slow down the respiration rate of harvested product. Normally, the produce is kept at 10–13 °C during hot weather after harvesting, or when its shipment is delayed. For fresh horticultural products, refrigeration is a wide and most common technology used to delay postharvest ripening and deterioration. Further cold storage rooms are developed to preserve commodities for long durations. Every fruit and vegetable has its specific temperature at which it can be stored up to a maximum time period. This storage temperature varies with the stage of the plant, i.e. eggplant fruit and orange green fruit can be stored for maximum duration at 10 °C, avocado and red fruit can be stored at 8 °C while green fruit can be stored at 15 °C for maximum duration (Tripathi et al. 2013). The fluctuations in temperature or storage below specific temperature creates many problems such as deterioration and postharvest disorders. Among these, chilling injury is the most common disorder of fruit and vegetables, as many horticultural products are sensitive to chilling.

7.2.3 Modified Controlled Atmosphere for Storage

This technique is applied to preserve food by maintaining its quality, and to extend its storage life. In this technique food storage life is extended by modifying the surrounding temperature. This modified atmosphere seizes the respiration rate of food, slowing down the chemical decomposition, fruit softening, ripening, ethylene production, also reducing the activity of insects and other microorganisms present. As a consequence, the food shelf life increases. There are two terms used to identify this method: modified atmosphere (MA) and controlled atmosphere (CA) that are almost interchangeable. Both differ based on the atmospheric composition. In MA storage, there is a modification in gases (O_2 and CO_2 concentrations) along with the respiration rate of food and with the packing structure or film around the food. In CA storage O_2 and CO_2 are continuously controlled throughout the period of storage. The efficacy of both CA and MA storage can be sustained with temperature and humidity which result in maintaining quality and increasing shelf life.

Tropical fruits are mostly stored at less than 1% O_2 level with more than 12% CO_2 level. Similarly, temperate fruit i.e. loquat (*Eriobotrya japonica*) can be stored in MA for 2 months by using 20 μm polyethylene (PE) bags and maintaining temperature at 5 °C, with 5 kPa CO_2 and 4 kPa O_2 level (Ding et al. 2002). For subtropical fruits such as grapefruit or lemons, treatments with 20–45 kPa CO_2 reduce rind pitting and delay de-greening (Jayas and Jeyamkondan 2002; Hatton and Spalding 1990). Similarly, CO_2 treatment in MA storages at 8–20 kPa also suppress fungal decay (Ahmadi et al. 1999). This technology retards fruit softening, decay, flesh disorders and unpleasant smell. It is also observed that this technology is also helpful to control diseases such as anthracnose, during transport.

7.2.4 Heat Treatments

Due to chemicals harmful side effects, physical techniques are considered as the substitutes to control postharvest diseases. Among them, heat treatment is considered the most effective especially vs fungal diseases, which are the most common in postharvest. This technique is applied as a preventive measure to control insects and pests. Heat treatments are also useful to control chilling injury in fruits during cold storage.

There are three different heat treatment methods commonly used to control post-harvest diseases. They include hot water, vapor heat and hot air treatments, the latter also known as forced air treatment. Among them, hot water treatment is one of the most common and easy technique, used worldwide to control fungi. The fungal spores are mostly present on the upper cell layer of peel, especially in fruits and vegetables. These are mostly dipped in hot water for a short period, depending on the commodity. Mostly, fruits are dipped into hot water for 90 min at 46 °C. However, some fruits and vegetables are dipped into hot water for 10 min at 50–60 °C (Barkai-Golan and Phillips 1991).

Fungicides such as thiabendazole and imazalil are mixed in hot water to control fungal diseases of citrus. Similarly, ethanol, sulfur dioxide and sodium carbonate are added to hot water at 45 °C to control citrus green mold (Smilanick et al. 1997). Hot water treatment is helpful against insects as well. It is applied to different tropical and subtropical fruit to control fruit fly and other species. It is considered the most effective technique as uniform temperature can be maintained in the bath. The effectiveness of fungicides can be also increased by using hot water.

Vapor heat treatment is another technique mostly used to control insects. In this technique hot vapors are applied on the skin of fruits and vegetables to kill the insect larvae (Lurie 1998). This technique was used to kill i.e. the Mexican and Mediterranean fruit fly. However, due to the development of cheap fumigants its use became less frequent as it is expensive as compared to fumigants. However, it is becoming commercially convenient again due to the organic food demand by consumers, and the ban of certain toxic fumigants. In modern techniques hot air is introduced into the chamber by pallets. This technique is fast as compared to other such as air circulation . Vapor heat treatment is helpful to control fungi as well. It is applied on tomato to control *Botrytis cinerea* and on apples to control *Penicillium expansum*.

7.2.5 Ultraviolet Light Treatment (UV Treatment)

UV irradiation is used to sterilize commodities killing exposed microorganism. This treatment is effective to control postharvest diseases. In the last few years, this technology has been used to reduce decay by microorganism, and to maintain quality of fruits and vegetables, extending their shelf life. UVs are called non ionizing radiations and contain higher energy as compared to visible light. These radiations are further divided into three categories; UV-C, UV-B and UV-A depending on the light frequency. UV-C have short wavelength and their frequency is in the range from 200 to 280 nm. UV-B have medium wavelength and their frequency lies between 280 to 320 nm. UV-A have long wavelength, their frequency ranging from 320 to 400 nm.

The eradication and destruction of microorganisms depend on the dose of UV radiation. In postharvest, UV-C radiations are applied by UV lamps at 254 nm wavelength for a specific period of time to destroy the DNA of microorganism (Bintsis et al. 2000). In fruits and vegetables, UV doses depend on the commodities, but in most of cases its range lies between 0.2 and 20 kJ m^{-2}. While in horticultural products it depends on ripening stage and harvesting season. UV-C radiations have been effectively used to control postharvest decay in fruits (apples, grape, grapefruit, tangerine, mango, peaches and strawberry) and vegetables (tomato and pepper) (Droby et al. 1993). The most common postharvest pathogens that are controlled by UV include *B. cinerea, Alternaria alternata, Colletotrichum gloeosporioides, Rhizopus stolonifera* and *P. italicum.*

UV radiations induce physiological changes in a commodity. Inappropriate UV doses may affect color, pigments, firmness, ethylene production, respiration rate, flavor, aroma, antioxidants, carotenoids, ascorbic acids and chilling injury. Chlorophyll degradation in fruits and vegetables results in loss of quality. It is observed that UV treatments reduce chlorophyll degradation in tomato while UV treatment at 10 kJ m^{-2} was found to be helpful to maintain chlorophyll in broccoli (Chairat et al. 2013). In plant tissues, cell wall is the main barrier that resists pathogen infestation. It helps to maintain turgor pressure and cell enlargement, which ultimately helps to maintain the tissues firmness. UV radiations are helpful to maintain firmness in fruits and vegetables. Radiations at 4.1 kJ m^{-2} reduced firmness loss in strawberry (Marquenie et al. 2003; Marquenie et al. 2002). Similarly, UV irradiations may also delay firmness losses in tomato and pepper at of 3.6 and 7 kJ m^{-2}, respectively (Barka 2001). UV radiations are also helpful to slow down ethylene production and respiration rate of horticultural products, ultimately extending their shelf life and delaying insect pest infestation on perishable commodities.

7.3 Chemical Agents

In postharvest, multiple activities may be interrelated with each other. Since 1960, chemicals are being used to reduce postharvest diseases. This has been the main method to control diseases, but due to human health safety concern, several polices have been developed on their use. Therefore, fungicide treatments on fruits and vegetables are monitored by regulating agencies and the chemical treatment option is restricted to few authorized fungicides. The type and number of registered fungicides depend on each country rule and regulation. The fungicide banned in one country might not be banned in other one. For example, four fungicides are used in USA to control *P. expansum* in postharvest, including pyrimethanil, fludioxonil, thiabendazole (TBZ) and captan. In China only thiabendazole (TBZ) is instead allowed. However, all these four chemicals are strictly prohibited in the EU.

The chemicals' mode of action is similar to other physical agents, as they slow down the respiration rate and physiological decay. The chemicals help to stabilize the fruit surface and prevent degradation. Among chemicals, some of them are reported to induce resistance in plant diseases. Some react to pathogen metabolism that cause diseases while some other seize the activity of the pathogens. However, chemicals used to control postharvest diseases should be environment friendly and safe for humans. The most abundant chemicals used to control postharvest diseases are calcium chloride ($CaCl_2$), sodium bicarbonate and chitosan. They are environment friendly and non-toxic for human health.

7.3.1 Calcium Chloride ($CaCl_2$)

Calcium has a basic role in maintaining quality of fruits and vegetables (Chepngeno et al. 2016). It contributes in links to pectic components, which increase the cell wall cohesion, and is helpful in reducing senescence and fruit decay. It allows bonds between the cell wall and the middle lamella, which affect firmness. The exogenous application of calcium helps in stabilizing the plant cell wall by protecting it from cell wall degrading enzymes. Calcium chloride ($CaCl_2$) is a non-toxic fungicide that is used worldwide. The application of calcium chloride is helpful in reducing fruit ripening, moreover it delays senescence (Chéour and Souiden 2015). It is also helpful to induce resistance against fungal attacks. $CaCl_2$ is also used to delay fruit softening in horticultural products (Kirkby and Pilbeaam 1984). It is used in several horticultural products including loquat, mango, apples, tomato, cherries, chilies and strawberries.

It was observed that postharvest calcium dipping increases calcium levels, without causing fruit injury, as compared to pre-harvest treatments. The concentration of $CaCl_2$ varies along with the horticultural product and duration of storage. In strawberry, 1% $CaCl_2$ at 3 °C is used to prevent fungal attacks and fruit ripening, up to 10 days. While with similar concentrations of $CaCl_2$, kiwi fruit can be stored for 12 weeks at 0 °C storage. The postharvest $CaCl_2$ application is helpful in maintaining membrane integrity by delaying its catabolism (Chen et al. 2011). Furthermore, it is also helpful to maintain tissue firmness and cell turgor pressure by extending shelf life.

7.3.2 Sodium Carbonate and Bicarbonate

Carbonic acid salts are extensively used all over the world due to their non-toxicity. They are sodium carbonate (Na_2CO_3) and sodium bicarbonate ($NaHCO_3$) that are commonly known as soda ash and baking soda, respectively. Both compounds are frequently used in the world as food additives and have no restrictions policies in America and Europe (Multon 1988). These two compounds are further classified as safe food and are labeled as “organic food” by the United States Department of Agriculture (USDA). Sodium bicarbonate is used as an alternate to borax and is used to control post-harvest diseases such as blue mold and green mold. Similarly, sodium carbonate is also used on some commodities to control postharvest diseases.

Both these two salts have several advantages as they are least toxic, less expensive, easily available in market and give minimum fruit injury (Palou et al. 2001). Both are used as synthetic fungicides to control *P. italicum* and *P. digitatum*. Sodium bicarbonate is used to control green mold in oranges. However, its influence to control the diseases varies with temperature, dipping period and storage conditions of the commodity (Palou et al. 2002). It is one of the most extensive compound that is used abundantly on fruits such as citrus, peaches, grapes, apples, banana and papayas to control postharvest storage rots, green molds and blue molds.

7.3.3 Chitosan

Biodegradable natural compounds derived from plants and animals are considered for use in organic agriculture as they have no harmful effect on human health. Chitosan is derived from the outer shell of shrimps, craps and krills. It is a polysaccharide having chemically similarity with cellulose. It is a sub form of chitin consisting of glucosamine, 2-amino-2-deoxy-β-D-glucose (Freepons 1991). Chitosan has a polycationic nature conferring antifungal properties. It is also believed that this compound has certain fungal enzymes which change fungal morphology, as i.e. *Trichoderma longibrachiatum* cell wall becomes very thin when treated. Similarly, *F. oxysporum* and *R. stolonifera* showed abnormal shapes, size reduction and swollen hyphae when treated with chitosan.

In postharvest, chitosan is applied to inhibit postharvest mycelium growth and spore germination. It is also observed that it enhances resistance in plant cells, by inducing defense mechanism in tissues. Chitosan application in oranges increases defensive enzymes such β-1,3- glucanase and chitinase (Gutiérrez-Martínez et al. 2012). The application of chitosan enhanced the defensive system in banana, mango and jack fruit, as confirmed through molecular techniques. Chitosan also reduces the tissue browning by inhibiting the activity of peroxidase. It is also effective against bacterial and has antibacterial properties as well. Application of chitosan at postharvest inhibits the growth of *Agrobacterium tumefaciens* and *Erwinia amylovora* strains by disrupting cell membrane, causing cellular leakage. In postharvest storage chitosan coating reduced gray mold on strawberry and Rhizopus rot in raspberries. Similarly, in temperate fruits such as apple, peaches and pear, chitosan reduced storage rot problems. It is also effective to control anthracnose in papaya during postharvest storage. Its application on fruits or vegetables forms a semi permeable membrane around the commodity reducing respiration and delaying ripening. It is also reported that chitosan application on tomato and pears reduced the internal CO_2 formation, increasing shelf life.

7.4 Organic Volatile Compounds and Fungicides

In plants there are many compounds that have naturally antifungal properties (Talibi et al. 2012). Volatile antifungal compounds derived from plants are nerolidol, jasmonates, acetaldehydes, ethyl formate, benzaldehyde, ethyl benzoate, ethyl formate and 2-nonanone. They have antifungal properties and are effective against *P. digitatum* and *P. italicum*. Aromatic plants contain several volatiles that have antifungal properties such as C_{10} and C_{15} terpenes of citrus. Similarly, among citrus essential oils, citral is a monoterpene having antifungal properties and is effective to control *P. digitatum* in pre- and postharvest (Wuryatmo et al. 2014). Similarly, a phenolic compound derived from grapefruit peel, 7-geranoxy coumarin, reduced decay and was very effective against *P. digitatum* (Droby et al. 2008). Several phenolic and volatile compounds are extracted from plant tissues and are converted into gels, powders and aqueous solutions. The extracts taken from eucalyptus, aloe vera, acacia and garlic have antifungal properties against *P. digitatum* and *P. italicum* and are frequently used in postharvest to reduce decay due to fungi (Chen et al. 2019).

Besides natural volatile compounds, there are several organic fungicides that are available in the market such as EcoCarb®. Its bioactive ingredient is potassium bicarbonate. Its mode of action by changing the pH of leaves severely affecting fungal spores. It is used to control powdery mildew in grapes and roses. Another bio stimulant, Aminogro®, consisting of several amino acids, stimulates the plant immune system and provides a defensive mechanism against several fungal diseases. A product named Bion® contains acibenzolar-S-methyl as active ingredient, which effectively controls postharvest diseases of mango (Zainuri 2006). Similarly, another product, named naturalGreen® is registered as organic product in Germany. Its composition consists of $CaCO_3$ (79%), $MgCO_3$ (4.6%), silicon and several trace elements (Smith et al. 2011). It improves yields and plants defensive system against several diseases. Another product, Scholar®, contains fludioxonil as active ingredient. When is applied at postharvest it reduces anthracnose and stem end rots in mango, blue and green molds in citrus and brown rot in stone fruits (Smith et al. 2011).

7.5 Biological Agents

During and after harvest, mechanical injuries, improper hygienic conditions and physiological decay are the leading causes of postharvest fungal infestations in horticultural commodities, which result in substantial losses of fresh products. Even newly developed storage, distribution and transportation facilities could not control pathogens populations once they infested a damaged product. Therefore, the traditional use of fungicides, in field or after harvesting, is the most accepted and widely used method for postharvest disease control. However, postharvest use of fungicides has been banned in some countries as the use of synthetic chemicals in agricultural food production system is being discouraged, mainly owing to toxicity to humans and environment, and resistance development in pathogens(Adaskaveg and Förster 2010). Moreover, preference of people for organically produced commodities led to limiting the use of fungicides, at any stage. Therefore, development of a safe and efficient technique such as biological agents provides a better and much needed opportunity to control diseases at the postharvest stage, when bruises and cuts caused by harvesting, handling, packing and transportation can be protected from pathogenic attacks (Liu et al. 2013).

7.5.1 Biocontrol

7.5.1.1 Antagonistic Microorganism

For the past few decades, postharvest biological control based on microbial antagonists has been extensively investigated. The strategy relies on the introduction of natural enemies of the pathogen to be targeted, to reduce its population by restricting normal growth or activity. Many antagonists have been used to control postharvest diseases in apple (Calvo et al. 2007; Mikani et al. 2008), citrus (Canamas et al. 2008; Kinay and Yildiz 2008), grapefruit (Hershkovitz et al. 2013), strawberry (Wei et al. 2014), kiwifruit (Sui and Liu 2014) and tomato (Ma et al. 2015).

7.5.1.2 Source of Antagonistic Microorganism

Mostly, the antagonists so far isolated and used to developed different commercial formulations have been isolated and screened from pre-existing microbiomes, present on the surface of fruit and vegetables (Droby and Wisniewski 2018). This decade old strategy worked very well in isolating few of the very earliest antagonists (Wilson et al. 1993).

Besides, other sources include phyllosphere (above-ground portions of plants, i.e. leaves, flowers, seeds etc.), roots, sea and soils. Kalogiannis et al. (2006) isolated the yeast *Rhodotorula glutinis* from tomato phyllosphere. The yeast was antagonistic to *B. cinerea* the causal agent of grey mold on tomato. Another yeast, *Kloeckera apiculata*, was isolated from the epiphytes of citrus roots by Long et al. (2005) and tested against *P. italium* and *B.cinerea* responsible for the postharvest rot of citrus and grape fruits, respectively. Wild plants provide a better opportunity as a source of antagonists because their microbiome have developed better tolerance to pathogens. Recently, endophytic bacteria have been isolated from the seeds of *Paullinia cupana* plants growing in Amazonas and Bahia regions (Santos et al. 2016). These endophytic bacteria showed antagonistic activity against *C. gloeosporioides*. The marine environment has been also searched for antagonists as many studies showed the existence of complex bacterial communities in the sea.

Rhodosporidium paludigenum (a marine yeast) obtained from the East China Sea showed an inhibitory capacity vs *P. expansum* on pear fruits (Wang et al. 2010). Hernandez-Montiel et al. (2010) argued that yeasts isolated from the sea may be more suitable as antagonists, rather than those isolated from fruit microbiome, mainly owing to their greater tolerance against osmotic stress.

Agricultural or natural soils are also a diverse sources of antagonists. One of the first microbial antagonist isolated from agricultural soil was *Bacillus subtilis* strain B3, active against brown rot of peach, showing an efficient inhibition of the causal organism *Monilia fructicola* (Janisiewicz and Korsten 2002a, b).

7.5.1.3 Development and Production of Antagonistic Microorganism

For any postharvest disease, development of a potent antagonist requires time and funding, as several steps are involved from isolation to product registration. Generally, the whole procedure has two main stages: (1) isolation and identification and (2) commercial product development (Nunes 2012). The development of an antagonist starts with its isolation and screening as a biocontrol agent. However, certain points need to be considered before isolating an antagonist including pathogen, host, disease epidemiology, constitutive or induced host resistance, and the environmental conditions where the biocontrol agent will be used (Nunes 2012). Isolation and efficacy analysis of an antagonist determine its ability to successfully control the targeted pathogen. These trials are carried out in laboratory, packing houses and storage facilities to observe mode of action and interaction and ensure compatibility of the antagonist with host, environment and postharvest procedures. After optimizing biocontrol capacity and growth condition for sustainable production, the microorganism is ready to be used in a bioformulation for market commercialization. Economical production of an antagonist on a large scale is key criterium to develop a commercial biocontrol product. Droby et al. (2016) described two main steps necessary for mass production of an antagonist. First, a cultural medium should be developed that is economical, less complicated to prepare and maintain, providing adequate nutrients and energy for growth and propagation. Second, the growth conditions including temperature, pH, gaseous concentration and agitation should be optimized for proper growth and population stability.

7.5.1.4 Mechanism of Biocontrol

Antagonistic activity relies on multiple factors working separately or synergistically to suppress growth and activity of a pathogen. The introduction of -omics has helped researchers to develop an in-depth understanding of different mechanisms an antagonist may use for inhibition. Major mechanisms include competition for nutrients and space, synthesis of hydrolytic enzymes and antibiotics, parasitism and induction of resistance in the host.

7.5.1.4.1 Competition for Nutrients and Space

An antagonist survives on macro- and micro-nutrients, carbohydrates, minerals, amino acids and vitamins and these are also the key essential elements for pathogens. Moreover, as the colony size increases, more space is required to either the antagonist or pathogen. Thus, an effective antagonist assimilates most nutrient resources and space for its multiplication, thus suppressing the pathogen by limiting its spore germinating and the host infesting capability (Hernandez-Montiel et al. 2018).

This reduction in rate of spore germination due to nutrient inaccessibility was also shown by bacteria, i.e. *Pseudomonas putida* against *P. digitatum* (Yu and Lee 2015). Similarly, as for competition for space, if the rate of multiplication of an antagonist is higher it may form defensive barriers such as an extracellular polysaccharide matrix, at the site of injury (Andrews et al. 1994).

7.5.1.4.2 Siderophores

Iron is one of the critical elements for growth of microorganism, especially fungi. In presence of O_2 and H_2O, iron is oxidized to ferric ions to form a stable iron oxide complex. These molecules are sequestered by siderophores produced by microbial antagonists, hence making them unavailable for growth and germination of pathogens (Carmona-Hernandez et al. 2019). Yeasts also take a competitive advantage by low iron and oxygen microenvironments, by producing siderophores (Spadaro and Droby 2016).

7.5.1.4.3 Parasitism

In a parasitic relationship, an antagonist may attack hyphae and produce cell wall hydrolytic enzymes (Dukare et al. 2018). This involves direct contact of the antagonist with the pathogen, recognition and then secretion of lytic enzymes resulting in either its complete killing or damaging of propagules (Talibi et al. 2014). The constituents of the fungal cell wall include chitin, chitosan and ß-glucans, associated with cell wall proteins. The antagonist hydrolytic enzymes, especially chitinases, glucanases, cellulases and proteases, may disintegrate the structure of the pathogen cell wall (Spadaro and Droby 2016). The antagonistic efficiency of *Bacillus* and *Pseudomonas* spp. is greatly attributed to such a biological activity, originating from the direct action of chitinase (Yu et al. 2008).

7.5.1.4.4 Antibiosis

One of the mechanisms used by antagonists to kill or reduce growth and activity of a pathogen is the production of antibiotics. Some examples of antibiotic producing antagonistic bacteria are *B. subtilis* (producing iturin, a potent antifungal peptide), *Ps. cepacia* (producing pyrrolnitrin), whereas among fungi *Myrothecium roridum* produces trichothecenes (Torres et al. 2014). Besides, other bacterial genera such as *Streptomyces*, *Burkholderia*, *Pantoea*, *Lysobacter* and *Enterobacter* also produce and secrete antibiotics active in antibiosis (Dukare et al. 2018). Antibiotics suppress the growth and development of fungal pathogens through a variety of modes of actions. The most recognized mechanisms include prohibiting the synthesis and further repair of the cell wall, degradation and disintegration of cell membrane structure and disruption of protein synthesis, by inhibiting the formation of initiation complexes on the smaller ribosomal subunits (De Souza et al. 2003).

7.5.1.4.5 Induced Resistance

Inducing resistance in host fruit against invading fungi is one of the mechanisms of biological control conferred by some antagonists. The latter induce a defense reaction in the host, active against the pathogen. The application of yeast antagonists (i.e. *Cryptococcus laurentii*) in combination with methyl jasmonic acid on peaches stimulated the activity of defensive enzymes such as chitinase, glucanase, phenylalanine ammonia lyase and peroxidase (Yao and Tian 2005).

Apart from production of enzymes, reactive oxygen species (ROS) have a signaling and a direct antimicrobial role, and are associated with induced resistance in many hosts. Further, the closure of stomata induced by an antagonist may be also considered as a host defense response. Yeasts acting as antagonists may provoke ROS production and, hence, defense signaling in fruit tissues, thus stimulating both antioxidant gene expression and antioxidant enzyme activity, as observed in peach fruit tissues (Xu and Tian 2008).

7.5.1.4.6 Production of Antifungal Volatile Compounds

Microbial antagonists synthesize, among several antifungal metabolites, volatile low molecular weight, lipophilic compounds. These volatiles were found to affect the growth of fungal pathogens (Mari et al. 2016). *In-vitro* and *in-vivo* studies have shown that *Aureobasidium pullulans* produces volatile compounds, including phenethyl alcohol, methyl butanol and methyl propanol, as a possible mode of action against *B. cinerea*, *C. acutatum* and *Penicillium* spp. (Di Francesco et al. 2014). All categories of microbial antagonists including bacteria, fungi and yeasts produce volatile compounds controlling pathogenic diseases at postharvest stage (Morath et al. 2012; Zheng et al. 2013; Di Francesco et al. 2015).

7.5.1.4.7 Formation of Biofilms

In another biocontrol strategy antagonists form mechanical barriers, generally known as biofilms, between the pathogen cells and the host lesion surface. Microcolonies in the biofilm are enclosed in an extracellular matrix consisting of proteins, nucleic acids and high molecular weight polysaccharides, produced by the antagonistic microbes (Carmona-Hernandez et al. 2019). Vero et al. (2013) proposed that this is one of the possible mechanisms that yeasts used in biocontrol. Similarly, in apples, the yeast *Saccharomyces cerevisiae* showed such biofilm forming ability against *P. expansum* inducing rot (Scherm et al. 2003).

7.6 Qualities of a Potent Antagonist

Protocols aiming at identifying and isolating potential antagonists used the same method as adopted to investigate biocontrol agents in foliage and soil-borne diseases. Mostly in this strategy, a single potent antagonist is screened out. Generally, an isolated antagonist targets wound pathogens that typically infest a food product, at the site of injury. Pathogen spores germinate within 24 hours of infection and colonize the host tissues, due to free availability of nutrients and carbohydrates (Droby et al. 2016). Therefore, the most important feature of an antagonist is to suppress the pathogen population by competing for space, through increased growth and colony formation, at a quick pace. Wilson et al. (1993) used this criterion to devise a method for screening and identifying successful antagonists of *Botrytis* and *Penicillium* rots of apple. They collected potential antagonists from microbial populations on fruit surfaces of apple, oranges, and tomato, introduced them to the wounds on fresh apple surface and infected these wounds with postharvest rot pathogens. Afterwards, the antagonists from wounds exempt of rot symptoms were cultured on nutrient agar and tested.

Some microbial secondary metabolites are assumed to act as antibiotics, and can become a hurdle in the registration process. Therefore, antagonists producing such secondary metabolites may be excluded.

Antagonists should also have higher survival and growth rates, to outpace the pathogen population growth . This survival and growth should last in the variable conditions of the supply chain (storage, transportation, shelf-life etc.). An antagonist should in fact survive in a variety of microclimates e.g. different ranges of temperatures and humidity and varying levels of O_2, CO_2 and ethylene, during various phases of the supply chain. Furthermore, the antagonist population should be sustainable during various kinds of postharvest operations. Considering the microbial ecology and diversity in fruit surface microbiome and the diversity in physiology of fruits and vegetables, an antagonist developed for commercial purposes is expected to perform on a variety of fruit or vegetable species. A single product should be hence applicable to several pathogens and commodities. This will reduce the cost and skills required of developing and maintaining an antagonist and will also simplify the use of products on a consumer's scale. This is the reason explaining why, despite developing certain products, their use has been restricted to a limited scale. For a successful and widespread acceptance of a product, it needs to be as nearly efficient as a chemical fungicide in controlling the disease, up to 98–100% (Droby et al. 2016).

7.7 Commercial Antagonists

Generally, a biological control formulation consists of a microorganism (an antagonist identified as an active ingredient), a carrier (an inert material) and an adjuvant (a nutrient and compounds enhancing survival) (Droby et al. 2016). In the recent past, several microbial bioformulations have been developed and commercially registered (Zhang et al. 2018). The biggest challenge for a commercialized product has been to sustain its efficacy in field conditions. Most biotic and abiotic factors, such as temperature, humidity, gaseous levels and host surface microbiome, markedly vary during the postharvest phase. Moreover, these factors also interact with the physiology and genetics of hosts, demanding for a product that remains efficient in very diverse conditions.

Numerous commercial microbial products, based on bacteria, fungi and yeast, have been developed for phytopathogens. These products are considered as first generation for preharvest and postharvest applications (Carmona-Hernandez et al. 2019). Table 7.1 shows several commercialized products that have been used, worldwide.

7.8 Natural Plant-Based Antimicrobial Substances

7.8.1 Volatile Aromatic Components

Volatile compounds are naturally present in a wide variety of fruits and vegetables, and in herb and spice essential oils. They are usually applied on commercial scale as flavoring and seasoning agents in foods and strongly reduce the incidence of microbial and fungal pathogens (Tripathi et al. 2004). It has been found that ripe fruits produce aroma volatiles that are main sensory attributes of fruits as well as have some efficient role in plants defense (Archbold et al. 2000).

Table 7.1 Commercial antagonistic microbe-based biocontrol products for controlling postharvest rots in fruits and vegetables

	Product	Antagonist	Target pathogen	Fruit	Country
Bacteria-based products	Biosave	*Pseudomonas syringae*	*Mucor, Botrytis, Penicillium*	Pome, citrus fruit, cherry, potato, sweet potato	United States
	Pantovital	*Pantoea agglomerans*	*Monilinia, Botrytis, Penicillium*	Citrus fruit, pome fruit	Spain
	Amylo-X	*Bacillus amyloliquefaciens*	*Botrytis cinerea*	Grapes	Italy
	Serenade	*B. subtilis*	*Botrytis cinerea*	Strawberry	Germany
	Avogreen (Preharvest)	*B. subtilis*	*Cercospora, Colletotrichum*	Avocado	South Africa
Yeast/fungi-based products	Yield plus	*Cryptococcus albidus*	*Mucor, Botrytis, Penicillium*	Pome fruit	South Africa
	Shemer	*Metschnikowia fructicola*	*Rhizopus, Aspergillus, Botrytis, Penicillium*	Table grape, strawberry, sweet potato	Netherlands
	Nexy	*Candida oleophila*	*Botrytis, Penicillium*	Pome fruit	Belgium, EU
	Aspire	*C. Oleophila*	*Botrytis, Penicillium*	Citrus fruit, pome fruit	United States
	Candifruit	*C. sake*	*Rhizopus, Penicillium, Botrytis*	Pome fruit	Spain
	Boni protect (Preharvest)	*Aureobasisium pullulans*	*Penicillium, Botrytis, Monilinia*	Pome fruit	Austria

Adapted from Dukare et al. 2018 and Zhang et al. 2018

7.8.2 Acetaldehyde, Six Carbon (C_6) Aldehydes, Benzaldehyde

Acetaldehyde (AA) is a natural product which performs a vital role in producing aroma and flavor in ripening fruits. AA is produced from pyruvic acid by activity of pyruvate decarboxylase enzyme (Pesis 2005). El Sheikh et al. (2000) reported that peach fruit rots caused by different fungi (*Rhizopus stolonifera*, *Monilinia fructicola, B. cinerea*) was studied with application of different aldehydes and biopesticides. Abd Alla et al. (2008) investigated that AA vapor treatments decreased growth of mycelium. However, AA application could not control the incidence of some serious potato postharvest diseases such as *Pectobacterium atroseptic*, *C. coccodes* and *Helminthosporium solani* (Wood et al. 2013). Prasad and Stadelbacher (1974) reported that AA fumigation on strawberry cv. Midway and tow grapes cultivar (Perlette and Sultanina) significantly decreased gray mold and *Rhizopus* during storage periods. Application of AA significantly delayed and remarkably decreased the bacterial postharvest disease (*E. carotovora*) in plastic polyethylene containers on bell pepper (*Capsicum annum*) cv. Bell Tower. Several researchers found that AA applied at time-dependent concentration and exposure, could play an important role in pathogens control during the storage periods (Utama et al. 2010). Utama et al. (2002) also explored growth inhibition of different fungi (*Erwinia carotovora*, *P. digitatum* and *C. musae*) with minimum growth inhibition value ranging from 0.88 to 0.91 mmol dish^{-1}. AA volatiles applied at 0.56 mL L^{-1} concentration decreased the growth of *C. acutatum* during one-week storage at 23 °C (Auras et al. 2007).

7.8.3 Acetic Acid

Acetic acid has unique antibacterial and antifungal properties. It is usually used in food industry to inhibit microbial growth and is generally considered as a safe compound for industry (Barkai-Golan 2001). Moreover, it has no toxic residues during shelf life enhancement of fruit and is effective in controlling postharvest decay. Acetic acid plays a role in destroying the spores of *B. cinerea* and *P. expansum* spreading infection inside the fruit packing, and is capable of decreasing the effect of chilling injury. The general mechanism of acetic acid involves the cell membrane by disturbing the metabolites transport and membrane potential. Protons entering the membrane normally destroy the membrane and reduce their activity. The researcher also investigated acetic acid fumigation, in the inhibition and elimination of different fungal diseases on cherries, apricots, and peaches (Sholberg and Gaunce 1996). Radi et al. (2010) showed that the use of a hot acetic acid solution inhibited *P. expansum* infection on apples during storage of 14 or 28 days. Acetic acid application at 2 and 3% (50 °C) were the most effective treatments, respectively.

7.8.4 Jasmonates

Jasmonates are signaling molecules (plant-specific elicitors) which trigger numerous essential processes during plant growth and development (physiological and developmental). Their biosynthesis is induced by wounding, as pathogen attacks play an important role activating defense responses in plants (Delaunois et al. 2014). Methyljasmonate (MeJA) universally distributed in the plant kingdom, was primarily detected from *Jasminum grandiflorum* flowers as a fragrant volatile compound. Jasmonate is synthesized from α-linolenic acid via the octadecanoid pathway. They are essential plant regulators active in plant growth and development, pathogens resistance, and diverse environmental stresses (Wasternack and Hause 2002). Ding et al. (2001) found that MeJA application upregulated genes which participate in cellular compartmentation, secondary metabolism, JA biosynthesis, defense and stress proteins release. MeJA is a multifunctional signaling molecule (elicitor or signaling agent), and has been widely applied in fruits and vegetables to enhance their shelf life and marketing, during storage (Tian and Zhang 2016). Additionally, recent studies revealed that MeJA significant affected quality and biochemical properties of fruit and vegetables (Karaman et al. 2013; Fung et al. 2006).

Gonzalez-Aguilar et al. (2000) found that postharvest MeJA treatments significantly improved fruit color and quality, reduced symptoms and maintained biochemical properties in the mango cv. Kent. While, MJ fumigation (30 ppm) could not significantly decreased anthracnose or severity of stem end rot during storage at 13 °C in the mango cv. Nam Dok Mai. However, MeJA treatments in low temperature storage significantly improved ethylene production, with higher β-carotene in fruit pulp of cv. Nam Dok Mai (Boonyaritthongchai et al. 2016). MeJA applications induced ethylene biosynthesis which improved fruit maturing and enzymes related to the synthesis of aroma compounds.

Lalel et al. (2003) showed that higher levels of MeJA effectively increased ethanol production in fruit pulp and reduced cis-3-hexenol, regardless of ripening period. Reyes-Diaz et al. (2016) reported that activities of antioxidant enzymes also increased, improving nutritional value. MeJA application on grapevine enhanced resistance to environmental stresses e.g. foliar application of MeJA on cuttings reduced powdery mildew, enhancing the activity of phenolic and aroma-related compounds (Belhadj et al. 2006; Ruiz-García et al. 2014; Ruiz-García et al. 2012).

7.8.5 Hexenal and Hexanal

Plants naturally produce hexanal through the lipoxygenase pathway, after disruption of tissues. Diverse hexanal play an important role in extending fruit quality, by reducing activity of phospholipase-D, the main enzyme involved in spoilage (Paliyath and Murr 2007). As a natural plant volatile, hexanal (C6), has antimicrobial properties enhancing the postharvest quality of fruits (Tiwari and Paliyath 2011) and vegetables (Cheema et al. 2014).

Hexanal is safe and approved as food additive from FDA (EAFUS 2006). Numerous attempts have been performed to investigate the effects of various concentration of hexanal/hexane as pre- and postharvest sprays, dips and as vapors during maturing (Sharma et al. 2010; Cheema et al. 2014; Paliyath et al. 2015; Gill et al. 2016). In previous studies, application of hexenal significantly improved the quality attributes, enhanced the postharvest shelf life and delayed the ripening and senescence process (Misran et al. 2015; Paliyath and Murr 2007; Paliyath et al. 2003).

Hexanal is volatile and can easily be uses as a vapor form. It has been reported in previous studies that hexanal vapor treatments has antimicrobial properties against different fungi such as *P. expansum, B. cinerea, A. alternata, Sclerotinia sclerotiorum, C. gloeosporioides*, and *M. fructicola* during postharvest storage (Song et al. 2007; Fan et al. 2006; Thavong et al. 2010; Utto et al. 2008). Its application also helps to enhance shelf life of longan (*Dimocarpus longan*) and sweet cherry fruits during storage (Thavong et al. 2010; Sharma et al. 2010). Hexanal application in a modified atmosphere (70% N_2 and 30% CO_2) reduced the incidence of browning and stimulated the aroma production in apple slices of Jonagold and Golden Delicious cvs (Song et al. 1998; Lanciotti et al. 1999).

7.8.6 Glucosinolates

Glucosinolates (GLs) are found in at least 16 higher plant families and consist of a large group of thioglucoside-*N*-hydroxysulphate anionic compounds. A single GL is mainly present in cruciferous species (Fahey et al. 2001; Fenwick et al. 1983). Myrosinase (Myr) hydrolyse GLs producing several compounds such as isothiocyanates (ITCs), in combination with various nitriles, thiocyanates and oxazolidine thiones, depending on the GL chemical nature and hydrolysis conditions. GLs belong to a diverse group of secondary metabolites which are mainly produced in vegetables such as *Brassica* spp. (broccoli). They show different chemical groups such as d-thioglucose, a sulphonated oxime group and amino acids, sharing a similar chemical structure (Moreno et al. 2006). It has been reported that GLs derived ITCs are active against different postharvest fungal diseases on e.g. pear, and peaches or nectarines such as *P. expansum* and *M. laxa,* respectively (Mari et al. 2002, 2008).

7.8.7 Essential Oils

Essential oils (EO) are volatile compounds produced from secondary metabolism in many plant species and characterized by a strong odor. They consist of 20 to 60 compounds present with various amounts in many plant species (Bakkali et al. 2008). These volatile compounds have various properties acting as antiprotectants or antimicrobials, are less harmful to mammalians, environment friendly, and could be used as alternatives to chemical fungicides (Isman 2000; Kalemba and Kunicka 2003; Burt 2004). Properties showed a main role in plant defense mechanisms, anti-repellency and antigermination (Bakkali et al. 2008, Grande-Tovar et al. 2016). EOs are generally recognized as safe for the environment and human health (Hyldgaard et al. 2012; Adorjan and Buchbauer 2010; Barbosa et al. 2008; Edris 2007) and may play a pivotal role in postharvest disease control (Caccioni and Guizzardi 1994).

EOs have not residual effects on fresh produce, possess antimicrobial and biodegradable properties and are natural antioxidants (Kalemba and Kunicka 2003). Being volatile, applications in low quantities are safe for humans as well. Moreover, EOs are commonly used in cooking purposes and accepted easily by the consumers being also used as flavoring components in foods. As GRAS (generally regarded as safe) compounds they are generally applied as biopesticides to control different pests and diseases and ensure food safety. EOs have antifungal properties, and are commercially used as biofumigation agents in the form of vapors to control postharvest diseases enhancing postharvest life and fruit quality, reducing decay percentage and inhibiting the incidence of microbial deterioration (Guerreiro et al. 2015). Due to their antimicrobial activity EOs application may be regarded as a method more advanced than chemical fungicides (Maqbool et al. 2011). They also increased resistance of fruit by inhibiting spore germination and mycelial growth (Regnier et al. 2010), disturbing the cellular metabolism of microorganisms (Sivakumar and Bautista-Baños 2014). EOs have been used to control *M. laxa* in stone fruit, and postharvest diseases e.g. molds, food-borne and various bacteria in citrus and tomato (Neiri et al. 2007; Banihashemi and Abivardi 2011; Macias et al. 2007). For other applications see Sect. 7.11 in this Chapter.

7.8.8 Plant Extracts

Plants extracts (PEs) are mainly known as therapeutic and immunization compounds for human use since prehistoric times (Dellavalle et al. 2011), due to phytochemical and chemical properties. They have been also used as organic antifungal compounds with a reduced toxicity and improved consumer acceptance (Tomazoni et al. 2016). PEs include a wide group of chemicals due to the presence of different secondary substances with antifungal activity and a broader spectrum (Nerio et al. 2010; Da Cruz-Cabral et al. 2013). Chavez-Quintal et al. (2011) found that papaya leaves and seed extracts, and Moringa (*Moringa oleifera*) Leaf extract (MLE), exhibited antifungal activity against fungal disease (John et al. 2013).

MLE have diverse phytochemicals compounds (such as sitosterol, niazin A, stigma sterol, kaempferol and quercetin) and antimicrobial properties (Rao et al. 2001). MLE was used for extension and quality improvement of avocado cv. Fuerte (Yousef et al. 2015). Tesfay and Magwaza (2017) found that application of 1% carboxymethyl cellulose (w/v) containing MLE on the avocado cvs Hass and Fuerte enhanced the fruits postharvest life and quality.

Other Pes include phenolics from *Hedera helix* (ivy) increasing resistance to phytopathogenic fungi (Parvu et al. 2015), and pomegranate extracts, that exhibited higher resistance against fungal diseases (Li Destri Nicosia et al. 2016). Karim et al. (2015) reported that incidence of citrus sour rot was effectively controlled with application of plant extracts from *Cistus populifolius* and *C. ladanifer*. PEs of *Orobanche crenata* and *Sanguisorba minor* exhibited significant control against fungi (Gatto et al. 2013). Gatto et al. (2011) showed that nine wild edible herbaceous species were used among which *O. crenata* and *S. minor* exhibited significant antifungal responses, and were effective in decreasing *in vitro* germination of *Monilina laxa* conidia and fungal diseases in fruits.

7.8.9 *Propolis*

Propolis is a resinous material comprised of a large amount of constituents and various biological properties, collected by honey bees from plant exudates (Sforcin and Bankova 2011). It consists of waxes, vitamins essential oils, and key compounds such as flavonoids, phenolics and their esters with antimicrobial properties (Juliano et al. 2007; Siqueira et al. 2009). Alone or in combination with other chemicals propolis significantly improved the shelf life of fresh horticultural commodities. Propolis also exhibited activity against yeast, several bacteria and fungi (Umthong et al. 2009; Silici et al. 2005; Campana et al. 2009). Flavonoids occurrence were supposed to increase the antimicrobial activities of propolis through synergistic effects with phenolic compounds (Vardar-Unlu et al. 2008).

Numerous studies indicated that use of propolis extracts produced a eco-friendly film which improved the exchange of gases on various fruits (Zahid et al. 2013; Ali et al. 2014, 2015; Passos et al. 2016). However, there are many factors which affect the biological action of propolis, such as plant sources, time of collection and environment (Passos et al. 2016). Recently, it has been used, alone or combined with other coatings, to improve and increase the storage period of fruits such as banana, oranges, apples, pomegranate, grapes, dragon (*Hylocereus undatus*) and cherries (Passos et al. 2016; Kamel et al. 2015; Zahid et al. 2013; El-Badawy et al. 2012; Pastor et al. 2011; Ozdemir et al. 2010). Mattiuz et al. (2015) reported that dipping of Kent mango in 1.5% ethanolic extract of propolis exhibited lower anthracnose as compared to chitosan. Zahid et al. (2013) reported that 0.5% ethanolic extract of propolis increased fruit quality and postharvest life of dragon berries.

7.9 Antimicrobial Substances from Soil

7.9.1 *Fusapyrone and Deoxyfusapyrone*

Several naturally-occurring products have been identified enhancing plant resistance vs microbes, particularly fungal pathogens. They include (but are not limited to) β-aminobutyric acid, ethephon, fusapyrone and deoxyfusapyrone (Tripathi et al. 2013; Shuping and Eloff 2017; Babychan et al. 2017). These products have been applied to improve crop product quality and agricultural productivity.

Fusapyrone ($C_{34}H_{54}O_8$) and deoxyfusapyrone ($C_{34}H_{54}O_8$) are very important in the soil environment as antifungal metabolites. Fusapyrone and deoxyfusapyrone have been identified, isolated and purified from rice crop soil. Their main source in soil is *Fusarium semitectum*. Fusapyrone and deoxyfusapyrone were tested in several bioassays for their antibiotic activities and showed considerable antifungal activity against several fruit and/or plant pathogenic fungi (Tripathi and Shukla 2007; Sharma and Pongener 2010; Tripathi et al. 2013).

Extracts of fungal cultures containing fusapyrone and deoxyfusapyrone showed antibiotic activities towards *Geotrichum candidum* (Evidente et al. 1994). Both *in vitro* and *in vivo* assays on grapes showed the inhibitory effect of fusapyrone on growth of *B. cinerea* (Altomare et al. 2004). A considerable suppression of *B. cinerea* conidia germination has been reported on grapes treated with fusapyrone (100 μg mL^{-1}) that significantly inhibited growth and development of grey mold on spoiled grapes (Altomare et al. 1998). Low phyto- and zoo-toxicity of fusapyrone and deoxyfusapyrone promoted their application in controlling grapes (i.e. *B. cinerea*) and other crop diseases (Altomare et al. 2000, 2004; Sharma and Pongener 2010). Due to their selective action and low toxicity towards animals and plants, their use in biocontrolling postharvest fruits and plant diseases has been recommended (Altomare et al. 2004).

7.10 Effect of Other Materials on Biocontrol

Biocontrol of post-harvest decay of fruits and vegetables by microbial antagonists is at an advanced stage and proved successful during the last decade (Tripathi et al. 2013; Babychan et al. 2017). For biocontrol application, it is very important to determine its feasibility which mainly depends on the cost and reliability of the system (Janisiewicz et al. 1992). To maintain the population of an antagonist on fruits, some chemicals called enhancers may be also used, to provide the antagonists with nutrition and other supporting material . As they could also be detrimental to the disease causing microbes, they also improve overall quality and efficacy of the biocontrol system.

L-proline and L-asparagine are known to enhance biological control of *Ps. syringae* against *Pe. expansum* (Janisiewicz et al. 1992; Babychan et al. 2017). Glycochitosan has been reported to improve biocontrol efficacy of *Candida saitoana* on citrus and apple fruits for postharvest decay (El-Ghaouth et al. 2000c). Similarly, use of calcium salts improved efficacy of yeast biocontrol agents against *B. cinerea* and *P. expansum* (McLaughlin et al. 1992). Sometimes, the success of any enhancers is concentration dependent (McLaughlin et al. 1992; Tripathi et al. 2013). For instance, addition of calcium chloride was only effective in improving biocontrol by *Ca. oleophila* on apple fruits when applied at concentrations higher than 90 mM (Wisnieweski et al. 1995).Increased amounts of CaCl were needed to improve biocontrol on pears than on apple fruits. The antagonist *Ca. saitoana* and glycolchitosan (2%) together were more successful in controlling blue and gray and molds of apple fruits, and green mold of lemons and orange fruits induced by *P. digitatu*m (El-Ghaouth et al. 2000a, b; Babychan et al. 2017).

Different fruit coatings were also useful in further controlling the decay, particularly when used with biocontrol agents (Bancroft 1995). Pre-treatment with sodium bicarbonate has been reported to further improve control of green mold on lemon fruits (El-Ghaouth et al. 2000a). The TAL Pro-long (a fruit coating mixture of sucrose esters of fatty acids, sodium salts of carboxy-methyl-cellulose, soap and sucroglycerides) significantly decreased the postharvest decay of pome fruits (Bancroft 1995). Although the TAL-Pro-long action mechanism is not very clear, it is known to extend resistance against pathogen invasion and greatly reduced the fruit ripening in storage (Bancroft 1995).

Sodium bicarbonate, ethanol and sodium carbonate are known as GRAS, regarded as safe substances, and can significantly decrease conidial germination of *P. digitatum* causal agent of citrus green mold disease (Smilanick et al. 1997, 1999). The antagonist alone is sometimes incapable of controlling the disease, particularly when fruits are inoculated with the pathogen 1 day before the antagonist application. On the other hand, carbonate salts could successfully control such infections (Smilanick et al. 1997). However, carbonate salts fail to provide long-term protection after treatment in case of re-infection. In contrast, the antagonist could persist for a longer time after application and thus successfully protect fruits from any re-infection. Similarly, a combination of ethanol-resistant *S. cerevisiae* strains and ethanol (10%) has been reported to decrease incidence of gray mold disease on apple or lemon fruits (Smilanick et al. 1997; Mari and Carati 1998).

The potential of biocontrol in postharvest decay has not been fully exploited and less efforts have been made in formulations preparation and subsequent commercialization (Babychan et al. 2017). In addition, biocontrol management by farmers, particularly in developing countries, remains limited due to the lack of awareness and unavailability of information on benefits.

7.11 Botanicals

Due to the environmental concern and consumer awareness regarding synthetic chemical additives, the use of natural additives (e.g., biodegradable and environmentally safe) in food have become popular (Jhalegar et al. 2013). In the recent years, research on natural food additives having a broad spectrum of antimicrobial activity has gained much attention (Marino et al. 2001; Jhalegar et al. 2013; Khatoon et al. 2018). Plants represent a huge reservoir of natural compounds harboring fungicidal activities with potential to replace synthetic fungicides.

Historically, EOs have been used as flavoring agents in several food products and many have excellent antimicrobial or antifungal activities (Kim et al. 1995; Alzoreky and Nakahara 2002; Borges et al. 2018). Moreover, the bioactivity of essential oils in the vapor phase make them attractive fumigants for protection of food in storage. Some EOs have shown greater potential for inhibiting post-harvest decay caused by fungi, enhancing the storage life of fruits and vegetables (Hidalgo et al. 2002; Sharma and Verma 2004). Recently, natural products with antimicrobial potential are being used in packaging materials. With this recent advancement, it has now become possible to stop decay causing organisms at the surface of the product (Han 2003).

In sweet cherries, the post-harvest grey mold rot caused by *B. cinerea* and brown rot caused by *M. fructicola* may be now controlled by thymol (30 mg L^{-1}) fumigation

(Chu et al. 1999, 2001). Thymol was also very effective in controlling brown rot on apricots, and its fumigation (even at low concentrations i.e. 2 mg L^{-1}) could greatly reduce the postharvest decay in plums, with no sign of phytotoxicity (Liu et al. 1993). Carvone, a product of the herb *Carum carvi* essential oil, is effective in inhibiting the potato sprouting during storage with a greater antifungal activity for protection of tubers from rotting, causing no toxicity to the consumers (Hartmans et al. 1995).

Furthermore, EOs of *Mentha arvensis*, *Zingiber officinale* and other plants were effective in controlling blue mold in oranges (Tripathi 2001).

The potential application of antimicrobial substances for controlling the postharvest decay and improving the quality of the food products will mainly depend on their; (1) cost effectiveness, (2) safety for human consumption, (3) dispersion in harvested tissue and biological activity, and (4) effectiveness on target organisms at low or non-phytotoxic levels.

Although several research studies have shown the effectiveness and potential of natural fungicides (plant-based) in controlling postharvest losses and decay of vegetables and fruits (Tripathi et al. 2013; Borges et al. 2018; Khatoon et al. 2018), more in depth studies are required to further explore their potential and improve their effectiveness, and to fully understand the mechanism to improve existing biocontrol systems. In order to further develop and accelerate research in natural fungicides for postharvest decay, it would be imperative to; (1) explore and develop relationship between chemical structure and biological activity, resulting in screening of promising compounds, (2) devise simple, efficient and reproducible biological assays, revealing the fungicidal potential and activity of natural fungicides, and their possible toxicity towards the plants and other living organisms, and (3) develop a database of literature relevant to already known plant derived natural fungicides.

7.12 Integrated Management

Among different postharvest treatment options, treatment methods such as physical, biological or thermal may result effective to only some extent, particularly when used alone. For example, although hot air treatment can eradicate or reduce the decay of apple induced by *P. expansum*, it may fail to completely eradicate the decay caused by *C. acutatum* (Leverentz et al. 2000; Janisiewicz et al. 2003). Similarly, control of decay by biological means, with narrower spectrum of activity, may not be as fast or effective as other means. Also, effectiveness of some chemical treatment methods, when used alone, may result limited. For example, sodium bicarbonate cannot provide protection of fruits if infection happened after treatment (Smilanick et al. 1999). A combination or an integrative use of different treatments options may hence result more effective than any control option alone (Tripathi et al. 2013).

Integrative use of UV illumination and radiation was reported to be effective against *Colletotrichum* spp. (Barkai-Golan 2001). The integrative application of sodium bicarbonate, fungicides and hot water showed higher efficacy in reducing postharvest decay (Barkai-Golan 2001; Conway et al. 2005). Likewise, combined use of radiation and fungicides at low doses effectively provided cumulative protection against postharvest decay (Barkai-Golan 2001; Korsten 2006). Antagonists in combination with edible waxes showed successful control on mango postharvest decay (Korsten et al. 1993). Moreover, natural waxes in combination with fungicides enhanced protection efficacy compared to their application alone (Tripathi et al. 2013). Also, wrapping of fruits with plastic or shrinking after temperature treatments proved to be an efficient integrated option (Barkai-Golan 2001).

Biocontrol could effectively be integrated with chemicals at low concentration or with disinfectants. For instance, use of *B. subtilis* with chitosan enhanced postharvest biocontrol effectiveness of *Penicillium* spp. on citrus (Obagwu and Korsten 2003). Similarly, low doses of fungicides in different combinations with various biocontrol agents were more efficacious against many postharvest diseases than when applied alone (Korsten et al. 1993; Tripathi et al. 2013). Some other specific combinations of sodium bicarbonate and calcium salts with biocontrol agents have also been proven to be very effective in controlling postharvest decay (Barkai-Golan 2001; Janisiewicz and Korsten 2002a, b; Conway et al. 2004). An effective and enhanced control of oranges and grapefruits decay caused by *P. italicum* and *P. digitatum* was obtained when sodium bicarbonate was used in combination with *B. subtilis* or other antagonists (Porat et al. 2003; Obagwu and Korsten 2003). The apple fruits decay caused by *C. acutatum* or *P. expansum* was completely eradicated by combining heat with heat tolerant antagonists (yeasts). Antagonist or heat alone could only reduce decay induced by *C. acutatum*. An integrated application, however, was needed to completely eradicate the decay (Conway et al. 2004; Tripathi et al. 2013).

Although radiation, sodium carbonate and heat treatments can control fungal spores at the application time, they often fail to provide protection against possible future decay or infections. On the other hand, antagonists can successfully provide protection against future infection, although they often fail to provide protection against inborn infections. Therefore, it can be concluded that the combined use of these treatment options is essential to achieve a better control of a specific postharvest infection. Integration is generally considered indeed as more effective than any individual treatment option. Thus, combining various options may offer a more consistent, durable, practical, and sustainable solution to stakeholders and producers for control of postharvest diseases.

References

Abd Alla, M. A., El-Sayed, H. Z., & Riad, S. E. M. (2008). Control of rhizopus rot disease of apricot fruits (*Prunus armeniaca* L.) by some plant volatiles aldehydes. *Research Journal of Agriculture and Biological Sciences, 4*, 424–433.

Adaskaveg, J. E., & Förster, H. (2010). New developments in postharvest fungicide registrations for edible horticultural crops and use strategies in the United States. In D. Prusky & M. L.

Gullino (Eds.), *Post-harvest pathology series: plant pathology in the 21st century* (Vol. 2, pp. 107–111). Dordrecht: Springer.

Adorjan, B., & Buchbauer, G. (2010). Biological properties of essential oils: An updated review. *Flavour and Fragrance Journal, 25*, 407–426.

Ahmadi, H., Biasi, W. V., & Mitcham, E. J. (1999). Control of brown rot decay of nectarines with 15% carbon dioxide atmospheres. *Journal of the American Society for Horticultural Science, 124*, 708–712.

Ali, A., Cheong, C. K., & Zahid, N. (2014). Composite effect of propolis and gum Arabic to control postharvest anthracnose and maintain quality of papaya during storage. *International Journal of Agriculture and Biology, 16*, 1117–1122.

Ali, A., Wei, Y. Z., & Mustafa, M. A. (2015). Exploiting propolis as an antimicrobial edible coating to control post-harvest anthracnose of bell pepper. *Packaging Technology and Science, 28*, 173–179.

Altomare, C., Pengue, R., Favilla, M., Evidente, A., & Visconti, A. (2004). Structure-activity relationships of derivatives of fusapyrone, an antifungal metabolite of *Fusarium semitectum*. *Journal of Agricultural and Food Chemistry, 52*, 2997–3001.

Altomare, C., Perrone, G., Stornelli, C., & Bottalico, A. (1998). Quaderni della Scuola di specializzazione in Viticoltura ed. Enologia. *University of Torino, Italy, 22*, 59–66.

Altomare, C., Perrone, G., Zonno, M. C., Evidente, A., Pengue, R., Fanti, F., & Polonelli, L. (2000). Biological characterization of fusapyrone and deoxyfusapyrone, two bioactive secondary metabolites of *Fusarium*. *Journal of Natural Products, 63*, 1131–1135.

Alzoreky, N. S., & Nakahara, K. (2002). Antimicrobial activity of extracts from some edible plants commonly consumed in Asia. *International Journal of Food Microbiology, 80*, 223–230.

Andrews, J. H., Harris, R. F., Spear, R. N., Lau, G. W., & Nordheim, E. V. (1994). Morphogenesis and adhesion of *Aureobasidium pullulans*. *Canadian Journal of Microbiology, 40*, 6–17.

Archbold, D. D., Hamilton-Kemp, T. R., & Fallik, E. (2000). Aroma volatiles as modulators of postharvest mold development on fruit: *in vivo* role and fumigation tools. *Acta Horticulturae, 518*, 87–92.

Auras, R., Almenar, E., Wharton, P., Rubino, M., & Harte, B. (2007). Release of acetaldehyde from b-cyclodextrins inhibits postharvest decay fungi in vitro. *Journal of Agricultural and Food Chemistry, 55*, 7205–7212.

Babychan, M., Jojy, E. T., & Syriac, G. M. (2017). Bio-control agents in management of post-harvest diseases. *Life Sciences International Research Journal, 4*(1), 51–55.

Bakkali, F., Averbeck, S., Averbeck, D., & Idaomar, M. (2008). Biological effects of essential oils – A review. *Food and Chemical Toxicology, 46*, 446–475.

Bancroft, R. D. (1995). The use of a surface coating to ameliorate the rate of spread of post-harvest fungal diseases of top fruit. *International Biodeterioration and Biodegradation, 36*, 385–405.

Banihashemi, Z., & Abivardi, C. (2011). Evaluation of fungicidal and fungistatic activity of plant essential oils towards plant pathogenic and saprophytic fungi. *Phytopathologia Mediterranea, 50*, 245–256.

Barbosa, L. C. A., Teixeira, R. R., & Montanari, R. M. (2008). Phytotoxic natural products as models for the development of crop protection agents. In F. Epifano (Ed.), *Current trends in Phytochemistry* (pp. 21–59). Trivandrum: Research Signpost.

Barka, E. A. (2001). Protective enzymes against reactive oxygen species during ripening of tomato (*Lycopersicum esculentum*) fruits in response to low amounts of UV-C. *Australian Journal of Plant Physiology, 28*, 785–791.

Barkai-Golan, G. R. (2001). *Postharvest disease of fruit and vegetables: Development and control*. Amsterdam, NL: Elsevier Science B.V. 418 pp.

Barkai-Golan, R., & Phillips, D. J. (1991). Postharvest heat treatment of fresh fruits and vegetables for decay control. *Plant Disease, 75*, 1085–1089.

Belhadj, A., Saigne, C., Telef, N., Cluzet, S., Bouscaut, J., Corio-Costet, M. F., & Merillon, J. M. (2006). Methyl jasmonate induces defense responses in grapevine and triggers protection against Erysiphe necator. *Journal of Agricultural and Food Chemistry, 54*, 9119–9125.

Bintsis, T., Litopoulou-Tzanetaki, E., & Richard, K. R. (2000). Existing and potential applications of UV light in the food industry. *Journal of the Science of Food and Agriculture, 80*, 637–645.

Boonyaritthongchai, P., Chimvaree, C., Buanong, M., Uthairatanakij, A., & Jitareerat, P. (2016). Effect of methyl jasmonate on physical and chemical properties of mango fruit cv. Nam Dok Mai. *Horticulture, 3*, 18.

Borges, D. F., Lopes, E. A., Moraes, F. A. R., Soares, M. S., Visôtto, L. E., Oliveira, C. R., & Valente, V. N. M. (2018). Formulation of botanicals for the control of plant-pathogens: A review. *Crop Protection, 110*, 135–140.

Burt, S. (2004). Essential oils: Their antibacterial properties and potential applications in foods-a review. *International Journal of Food Microbiology, 94*, 223–253.

Caccioni, D. R. L., & Guizzardi, M. (1994). Inhibition of germination and growth of fruit and vegetable postharvest pathogenic fungi by essential oil components. *Journal of Essential Oil Research, 6*, 173–179.

Calvo, J., Calvente, V., de Orellano, M. E., Benuzzi, D., & de Tosetti, M. I. S. (2007). Biological control of postharvest spoilage caused by *Penicillium expansum* and *Botrytis cinerea* in apple by using the bacterium *Rahnella aquatilis*. *International Journal of Food Microbiology, 113*, 251–257.

Campana, R., Patrone, V., Franzini, I. T., Diamantini, G., Vittoria, E., & Baffone, W. (2009). Antimicrobial activity of two propolis samples against human *Campylobacter jejuni*. *Journal of Medicinal Food, 12*, 1050–1056.

Canamas, T. P., Vinas, I., Usall, J., Torres, R., Anguera, M., & Teixido, N. (2008). Control of post-harvest diseases on citrus fruit by preharvest applications of biocontrol agent *Pantoea agglomerans* CPA-2. Part II. Effectiveness of different cell formulations. *Postharvest Biology and Technology, 49*, 96–106.

Carmona-Hernandez, S., Reyes-Pérez, J. J., Chiquito-Contreras, R. G., Rincon-Enriquez, G., Cerdan-Cabrera, C. R., & Hernandez-Montiel, L. G. (2019). Biocontrol of postharvest fruit fungal diseases by bacterial antagonists: A review. *Agronomy, 9*, 121.

Chairat, B., Nutthachai, P., & Varit, S. (2013). Effect of UV-C treatment on chlorophyll degradation, antioxidant enzyme activities and senescence in Chinese kale (*Brassica oleracea* var. alboglabra). *International Food Research Journal, 20*, 623–628.

Chávez-Quintal, P., González-Flores, T., Rodríguez-Buenfil, I., & Gallegos-Tintoré, S. (2011). Antifungal activity in ethanolic extracts of *Carica papaya* L. cv. Maradol leaves and seeds. *Indian Journal of Microbiology, 51*, 54–60.

Cheema, A., Padmanabhan, P., Subramanian, J., Blom, T., & Paliyath, G. (2014). Improving quality of greenhouse tomato (*Solanum lycopersicum*: L) by pre- and postharvest applications of hexanal containing formulations. *Postharvest Biology and Technology, 95*, 13–19.

Chen, F., Liu, H., Yang, H., Lai, S., Cheng, X., Xin, Y., & Deng, Y. (2011). Quality attributes and cell wall properties of strawberries (*Fragaria annanassa* Duch.) under calcium chloride treatment. *Food Chemistry, 126*, 450–459.

Chen, J., Yuting, S., Chuying, C., & Chunpeng, W. (2019). Inhibition of key citrus postharvest fungal strains by plant extracts *in vitro* and *in vivo*: A review. *Plants, 8*, 26.

Chéour, F., & Souiden, Y. (2015). Calcium delays the postharvest ripening and related membrane-lipid changes of tomato. *Journal of Nutrition & Food Sciences, 5*, 1–6.

Chepngeno, J., Owino, W., Kinyuru, J., & Nenguwo, N. (2016). Effect of calcium chloride and hydro cooling on postharvest quality of selected vegetables. *Journal of Food Research, 5*, 23–40.

Chu, C. L., Liu, W. T., & Zhou, T. (2001). Fumigation of sweet cherries with thymol and acetic acid to reduce post-harvest brown rot and blue mold rot. *Fruits, 56*, 123–130.

Chu, C. L., Liu, W. T., Zhou, T., & Tsao, R. (1999). Control of post-harvest gray mold rot of modified atmosphere packaged sweet cherries by fumigation with thymol and acetic acid. *Canadian Journal of Plant Science, 79*, 685–689.

Conway, W. S., Leverentz, B., Janisiewicz, W. J., Blodgett, A. B., Saftner, R. A., & Camp, M. J. (2004). Integrating heat treatments, biocontrol and/ or sodium bicarbonate to control post-

harvest decay of apple fruit *Colletotrichum acutatum* and *Penicillium expansum*. *Postharvest Biology and Technology, 34*, 11–20.
Conway, W. S., Leverentz, B., Janisiewicz, W. J., Saftner, R. A., & Camp, M. J. (2005). Improving biocontrol using antagonistic mixtures with heat and/ or sodium bicarbonate to control postharvest decay of apple fruit. *Postharvest Biology and Technology, 36*, 235–244.
Da Cruz-Cabral, L., Pinto, V. F., & Patriarca, A. (2013). Application of plant derived compounds to control fungal spoilage and mycotoxin production in foods. *International Journal of Food Microbiology, 166*, 1–14.
De Souza, J. T. A., Arnould, C., Deulvot, C., Lemanceau, P., Gianinazzi-Pearson, V., & Raaijmakers, J. M. (2003). Effect of 2, 4-diacetylphloroglucinol on *Pythium*: Cellular responses and variation in sensitivity among propagules and species. *Phytopathology, 93*, 966–975.
Delaunois, B., Farace, G., Jeandet, P., Clément, C., Baillieul, F., Dore, S., & Cordelier, S. (2014). Elicitors as alternative strategy to pesticides in grapevine? Current knowledge on their mode of action from controlled conditions to vineyard. *Environmental Science and Pollution Research, 21*, 4837–4846.
Dellavalle, P. D., Cabrera, A., Alem, D., Larrañaga, P., Ferreira, F., & Dalla Rizza, M. (2011). Antifungal activity of medicinal plant extracts against phytopathogenic fungus *Alternaria* spp. *Chilean Journal of Agricultural Research, 71*, 231.
Di Francesco, A., Roberti, R., Martini, C., Baraldi, E., & Mari, M. (2015). Activities of *Aureobasidium pullulans* cell filtrates against *Monilinia laxa* of peaches. *Microbiological Research, 181*, 61–67.
Di Francesco, A., Ugolini, L., Lazzeri, L., & Mari, M. (2014). Production of volatile organic compounds by *Aureobasidium pullulans* as a potential mechanism of action against postharvest fruit pathogens. *Biological Control, 81*, 8–14.
Ding, C. K., Chachin, K., Ueda, Y., Imahori, Y., & Wang, C. Y. (2002). Modified atmosphere packaging maintains postharvest quality of loquat fruit. *Postharvest Biology and Technology, 24*, 341–348.
Ding, K., Wang, C. Y., Gross, K. C., & Smith, D. L. (2001). Reduction of chilling injury and transcript accumulation of heat shock proteins in tomato fruit by methyl jasmonate and methyl salicylate. *Plant Science, 161*, 1153–1159.
Droby, S., Chalutz, E., Horev, B., Cohen, L., Gaba, V., Wilson, C. L., & Wisniewski, M. E. (1993). Factors affecting UV induced resistance in grapefruit against the green mould decay caused by *Penicillium digitatum*. *Phytopathology, 42*, 418–424.
Droby, S., Eick, A., Macarisin, D., Cohen, L., Rafael, G., Stange, R., et al. (2008). Role of citrus volatiles in host recognition, germination and growth of *Penicillium digitatum* and *Penicillium italicum*. *Postharvest Biology and Technology, 49*, 386–396.
Droby, S., & Wisniewski, M. (2018). The fruit microbiome: A new frontier for postharvest biocontrol and postharvest biology. *Postharvest Biology and Technology, 140*, 107–112.
Droby, S., Wisniewski, M., Teixidó, N., Spadaro, D., & Jijakli, M. H. (2016). The science, development, and commercialization of postharvest biocontrol products. *Postharvest Biology and Technology, 122*, 22–29.
Dukare, A. S., Paul, S., Nambi, V. E., Gupta, R. K., Singh, R., Sharma, K., & Vishwakarma, R. K. (2018). Exploitation of microbial antagonists for the control of postharvest diseases of fruits: A review. *Critical Reviews in Food Science and Nutrition, 16*, 1–16.
EAFUS. (2006). *A food additive database center of food safety and applied nutrition*. U.S. Food and Drug Administration (FDA).
Edris, E. A. (2007). Pharmaceutical and therapeutic potentials of essential oils and their individual volatile constituents: A review. *Phytotherapy Research, 21*, 308–323.
El Sheikh, A. M. M., Baraka, M. A., & Abbass, A. G. E. S. (2000). The effectiveness of fumigants and biological protection of peach against fruit rots. *Assiut Journal of Agricultural Sciences, 31*, 19–31.

El-Badawy, H. E. M., Baiea, M. H. M., Eman, A. A., & Abd El-Moneim, E. A. A. (2012). Efficacy of propolis and wax coatings in improving fruit quality of Washington navel orange under cold storage. *Research Journal of Agriculture and Biological Sciences, 8*, 420–428.

El-Ghaouth, A., Smilanick, J. L., & Wilson, C. L. (2000a). Enhancement of the performance of *Candida saitona* by the addition of glycolchitosan for the control of postharvest decay of apple and citrus fruit. *Postharvest Biology and Technology, 19*, 103–110.

El-Ghaouth, A., Smilanick, J. L., Wisniewski, M., & Wilson, C. L. (2000c). Improved control of apple and citrus fruit decay with a combination of *Candida saitoana* and 2-deoxy-d-glucose. *Plant Disease, 84*, 249–253.

Evidente, A., Conti, L., Altomare, C., Bottalico, A., Sindona, G., Segre, A. L., & Logrieco, A. (1994). Fusapyrone and deoxyfusapyrone, two antifungal αpyrones from *Fusarium semitectum*. *Natural Toxins, 2*, 4–13.

Fahey, J. W., Zalcmann, A. T., & Talalay, P. (2001). The chemical diversity and distribution of glucosinolates and isothiocyanates among plants. *Phytochemistry, 56*, 5–51.

Fan, L., Song, J., Beaudry, R. M., & Hildebrand, P. D. (2006). Effect of hexanal vapor on spore viability of *Penicillium expansum*, lesion development on whole apples and fruit volatile biosynthesis. *Journal of Food Science, 71*, 105–109.

Fenwick, G. R., Heaney, R. K., & Mullin, W. J. (1983). Glucosinolates and their breakdown products in food and food plants. *Critical Reviews in Food Science and Nutrition, 1*, 123–201.

Freepons, D. (1991). Chitosan, does it have a place in agriculture? In *Proceedings of Plant Growth Regulation Society of America* (pp. 11–19).

Fung, R. W., Wang, M., Smith, C. T., Gross, L. G., Tao, K. C., & Tian, M. (2006). Characterization of alternative oxidase (AOX) gene expression in response to methyl salicylate and methyl jasmonate pre-treatment and low temperature in tomatoes. *Journal of Plant Physiology, 163*, 1049–1060.

Gatto, M. A., Ippolito, A., Linsalata, V., Cascarano, N. A., Nigro, F., Vanadia, S., & Di Venere, D. (2011). Activity of extracts from wild edible herbs against postharvest fungal diseases of fruit and vegetables. *Postharvest Biology and Technology, 61*, 72–82.

Gatto, M. A., Sanzani, S. M., Tardia, P., Linsalata, V., Pieralice, M., Sergio, L., & Di Venere, D. (2013). Antifungal activity of total and fractionated phenolic extracts from two wild edible herbs. *Natural Science, 5*, 895–902.

Gill, K. S., Dhaliwal, H. S., Mahajan, B. V. C., Paliyath, G., & Boora, R. S. (2016). Enhancing postharvest shelf life and quality of guava (*Psidium guajava* L.) cv. Allahabad Safeda by pre-harvest application of hexanal containing aqueous formulation. *Postharvest Biology and Technology, 112*, 224–232.

Gonzalez-Aguilar, G. A., Fortiz, J., Cruz, R., Bae, R. M., & Wan, C. Y. (2000). Methyl jasmonate reduces chilling injury and maintains postharvest quality of mango fruit. *Journal of Agricultural and Food Chemistry, 48*, 515–519.

Grande-Tovar, C., Chaves-López, C., Viuda-Martos, M., Serio, A., Delgado, J., & Perez- Alvarez, J. A. (2016). Sublethal concentrations of Colombian *Austroeupatorium inulifolium* (H.B.K.) essential oil and their effects on fungal growth and production of enzymes. *Industrial Crops and Products, 87*, 315–323.

Guerreiro, A. C., Gago, C. M. L., Faleiro, M. L., Miguel, M. G. C., & Antunes, M. D. C. (2015). The use of polysaccharide-based edible coatings enriched with essential oils to improve shelf-life of strawberries. *Postharvest Biology and Technology, 110*, 51–60.

Gutiérrez-Martínez, P., Bautista-Baños, S., & Barrera-Necha, L. (2012). Uso potencial de extractos vegetales, aceites esenciales y quitosano para reducir el daño causado por hongos postcosecha en productos hortofrutícolas. In J. A. R. de León (Ed.), *Avances en ciencia y tecnología alimentaria en México* (Vol. 1, pp. 485–495).

Han, J. H. (2003). Antimicrobial food packaging. In R. Ahvenainen (Ed.), *Novel food packaging technologies* (pp. 50–70). Boca Raton: CRC Press.

Hartmans, K. J., Diepenhorst, P., Bakker, W., & Gorris, L. G. M. (1995). The use of carvone in agriculture, sprout suppression of potatoes and antifungal activity against potato tuber and other plant diseases. *Industrial Crops and Products, 4*, 3–13.

Hatton, T. T., & Spalding, D. H. (1990). Controlled atmosphere storage of some tropical fruits. In M. Calderon & R. Barkai-Golan (Eds.), *Food preservation by modified atmospheres* (pp. 301–314). Boca Raton: CRC Press.

Hernandez-Montiel, L. G., Gutierrez-Perez, E. D., Murillo-Amador, B., Vero, S., Chiquito-Contreras, R. G., & Rincon-Enriquez, G. (2018). Mechanisms employed by *Debaryomyces hansenii* in biological control of anthracnose disease on papaya fruit. *Postharvest Biology and Technology, 139*, 31–37.

Hernandez-Montiel, L. G., Ochoa, J. L., Troyo-Dieguez, E., & Larralde-Corona, C. P. (2010). Biocontrol of postharvest blue mold (*Penicillium italicum* Wehmer) on Mexican lime by marine and citrus *Debaryomyces hansenii* isolates. *Postharvest Biology and Technology, 56*, 181–187.

Hershkovitz, V., Sela, N., Taha-Salaime, L., Liu, J., Rafael, G., Kessler Aly, R., Wisniewski, M., & Droby, S. (2013). De-novo assemble and characterization of the transcriptome of *Metschnikowia fructicola* reveals differences in gene expression following interaction with *Penicillium digitatum* and grapefruit peel. *BMC Genomics, 14*, 168.

Hidalgo, P. J., Libera, J. L., Santos, J. A., La Front, F., Castellanos, C., Palomino, A., & Roman, M. (2002). Essential oils in *Culamintha sylvatica*. Bromf. ssp. *ascendens* (Jorden) P.W. Ball: Wild and cultivated productions and antifungal activity. *Journal of Essential Oil Research, 14*, 68–71.

Hyldgaard, M., Mygind, T., Meyer, R. L., & Debabov, D. (2012). Essential oils in food preservation: Mode of action, synergies, and interactions with food matrix components. *Frontiers in Microbiology, 3*, 1–24.

Isman, B. M. (2000). Plant essential oils for pest and disease management. *Crop Protection, 19*, 603–608.

Janisiewicz, W. J., & Korsten, L. (2002a). Biological control of postharvest diseases of fruit. *Annual Review of Phytopathology, 40*, 411–441.

Janisiewicz, W. J., & Korsten, L. (2002b). Control of postharvest diseases of vegetables by application of microbes. In J. Bredt & J. Bartz (Eds.), *Postharvest physiology of vegetables* (pp. 543–562). New York: Marcel Dekker.

Janisiewicz, W. J., Usall, J., & Bors, B. (1992). Nutritional enhancement of biocontrol of blue mold on apples. *Phytopathology, 82*, 1364–1370.

Janisiewicz, W. J., Leverentz, B., Conway, W. S., Saftner, R. A., Reed, A. N., & Camp, M. J. (2003). Control of bitter rot and blue mold of apples by integrating heat and antagonist treatments on 1- MCP treated fruit stored under controlled atmosphere condition. *Postharvest Biology and Technology, 29*, 129–143.

Jayas, D. S., & Jeyamkondan, S. (2002). Modified atmosphere storage of grains, meats, fruits and vegetables. *Biosystems Engineering, 82*, 235–251.

Jhalegar, M. J., Sharma, R. R., & Singh, D. (2013). Antifungal efficacy of botanicals against major postharvest pathogens of Kinnow mandarin and their use to maintain postharvest quality. *Fruits, 69*, 233–237.

John, R. N., Gala, V. C., & Sawan, C. S. (2013). Inhibitory effects of plant extracts of multi species dental biofilm formation in vitro. *International Journal of Pharma and Bio Sciences, 4*, 487–495.

Juliano, C., Pala, C. L., & Cossu, M. (2007). Preparation and characterization of polymeric films containing propolis. *Journal of Drug Delivery Science and Technology, 17*, 177–181.

Kalemba, D., & Kunicka, A. (2003). Antibacterial and antifungal properties of essential oils. *Current Medicinal Chemistry, 10*, 813–829.

Kalogiannis, S., Tjamos, S. E., Stergiou, A., Antoniou, P. P., Ziogas, B. N., & Tjamos, E. C. (2006). Selection and evaluation of phyllosphere yeasts as biocontrol agents against grey mould of tomato. *European Journal of Plant Pathology, 116*, 69–76.

Kamel, H. M., Zaki, Z. A., & Abd El-Moneim, E. A. A. (2015). The effect of propolis and sodium metabisulfite as postharvest treatments on pomegranate arils storage. *American-Eurasian Journal of Agricultural & Environmental Sciences, 15*, 1962–1973.

Karaman, S., Ozturk, B., Genc, N., & Celik, S. M. (2013). Effect of preharvest application of methyl jasmonate on fruit quality of plum (*Prunus salicina* Lindell cv. "Fortune") at harvest and during cold storage. *Journal of Food Processing & Preservation, 37*, 1049–1059.

Karim, H., Boubaker, H., Askarne, L., Talibi, I., Msanda, F., Boudyach, E. H., Saadi, B., & Ait Ben Aoumar, A. (2015). Antifungal properties of organic extracts of eight Cistus L. species against postharvest citrus sour rot. *Letters in Applied Microbiology, 62*, 16–22.

Khatoon, A., Mohapatra, A., & Satapathy, K. B. (2018). Botanicals to control post-harvest decay of Colocasia (*Colocasia esculenta* L.) tubers in Odisha, India. *Research Journal of Pharmacy and Technology, 11*, 1804–1809.

Kim, J., Marshall, M. R., & Wei, C. (1995). Antimicrobial activity of some essential oil components against five food borne pathogens. *Journal of Agricultural and Food Chemistry, 43*, 2839–2845.

Kinay, P., & Yildiz, M. (2008). The shelf life and effectiveness of granular formulations of *Metschnikowia pulcherrima* and *Pichia guilliermondii* yeast isolates that control postharvest decay of citrus fruit. *Biological Control, 45*, 433–440.

Kirkby, E. A., & Pilbeaam, D. J. (1984). Calcium as a plant nutrient. *Plant, Cell & Environment, 7*, 397–405.

Korsten, L., De Villiers, E. E., & Lonsdale, J. H. (1993). Biological control of mango postharvest diseases in the packhouse. *South African Mango Grower' Association Yearbook, 13*, 117–121.

Korsten, L. (2006). Advances in control of postharvest diseases in tropical fresh produce. *International Journal of Postharvest Technology and Innovation, 1*, 48–61.

Lalel, H. J. D., Singh, Z., & Tan, S. C. (2003). The role of methyl jasmonate in mango ripening and biosynthesis of aroma volatile compounds. *The Journal of Horticultural Science and Biotechnology, 78*, 470–484.

Lanciotti, R., Corbo, M. R., Gardini, F., Sinigaglia, M., & Guerzoni, M. E. (1999). Effect of hexanal on the shelf life of fresh apple slices. *Journal of Agricultural and Food Chemistry, 47*, 4769–4776.

Leverentz, B., Janisiewicz, W. J., Conway, W. S., Saftner, R. A., & Fuchs, Y. (2000). Combining yeasts or a bacterial biocontrol agent and heat treatment to reduce postharvest decay of 'Gala' apples. *Postharvest Biology and Technology, 21*, 87–94.

Li Destri Nicosia, M. G., Pangallo, S., Raphael, G., Romeo, F. V., Strano, M. C., Rapisarda, P., et al. (2016). Control of postharvest fungal rots on citrus fruit and sweet cherries using a pomegranate peel extract. *Postharvest Biology and Technology, 114*, 54–61.

Liu, J., Stevens, C., Khan, V. A., Lu, J. Y., Wilson, C. L., Adeyeye, O., et al. (1993). Application of ultraviolet-C light on storage rots and ripening of tomatoes. *Journal of Food Protection, 56*, 868–672.

Liu, J., Sui, Y., Wisniewski, M., Droby, S., & Liu, Y. (2013). Review: Utilization of antagonistic yeasts to manage postharvest fungal diseases of fruit. *International Journal of Food Microbiology, 167*, 153e160.

Long, C. A., Zheng, W., & Deng, B. X. (2005). Biological control of *Penicillium italicum* of citrus and *Botrytis cinerea* of grape by strain 34–9 of *Kloeckera apiculata*. *European Food Research and Technology, 211*, 197–201.

Lurie, S. (1998). Postharvest heat treatments. *Postharvest Biology and Technology, 14*, 257–269.

Ma, X., Wang, X., Cheng, J., Nie, X., Yu, X., Zhao, Y., & Wang, W. (2015). Microencapsulation of *Bacillus subtilis* B99-2 and its biocontrol efficiency against *Rhizoctonia solani* in tomato. *Biological Control, 90*, 31–41.

Macias, F. A., Molinillo, J. M. G., Varela, R. M., & Galindo, J. C. G. (2007). Allelopathy: A natural alternative for weed control. *Pest Management Science, 63*, 327–348.

Maqbool, M., Ali, A., Alderson, P., Mohamed, M. T. M., Siddiqui, Y., & Zahid, N. (2011). Postharvest application of gum Arabic and essential oils for controlling anthracnose and quality of banana and papaya during cold storage. *Postharvest Biology and Technology, 62*, 71–76.
Mari, M., & Carati, A. (1998). Use of *Saccharomyces cerevisiae* with ethanol in the biological control of grey mould on pome fruits. COST 914-COST 915 Joint Workshop. In *Non conventional methods for the control of postharvest diseases and microbial spoilage* (pp. 85–91). Luxembourg: Publ EC.
Mari, M., Leoni, O., Bernardi, R., Neri, F., & Palmieri, S. (2008). Control of brown rot on stone fruit by synthetic and glucosinolate-derived isothiocyanates. *Postharvest Biology and Technology, 47*, 61–67.
Mari, M., Leoni, O., Iori, R., & Cembali, T. (2002). Antifungal vapour-phase activity of allyl isothiocyanate against *Penicillium expansum* on pears. *Plant Pathology, 51*, 231–236.
Marino, M., Bersani, C., & Comi, G. (2001). Impedance measurement to study antimicrobial activity of essential oils from Lamiaceae and Compositae. *International Journal of Food Microbiology, 67*, 187–195.
Mari, M., Bautista-Baños, S., & Sivakumar, D. (2016). Decay control in the postharvest system: Role of microbial and plant volatile organic compounds. *Postharvest Biology and Technology, 122*, 70–81.
Marquenie, D., Michiels, C., Geeraerd, A., Schenk, A., Soontjens, C., Van Impe, J., & Nicolaï, B. (2002). Using survival analysis to investigate the effect of UV–C and heat treatment on storage rot of strawberry and sweet cherry. *International Journal of Food Microbiology, 73*, 187–196.
Marquenie, D., Michiels, C. W., Van Impe, J. F., Schrevens, E., & Nicolaï, B. M. (2003). Pulsed white light in combination with UV-C and heat to reduce storage rot of strawberry. *Postharvest Biology and Technology, 28*, 455–461.
Mattiuz, B. H., Ducamp-Collin, M. N., Machado, M. C. F., Vigneault, C., Magalhães, M. K., Sagoua, W., & Montet, D. (2015). Effect of propolis on postharvest control of anthracnose and quality parameters of 'Kent' mango. *Scientia Horticulturae, 184*, 160–168.
Mclaughlin, R. J., Wilson, C. L., Droby, S., Ben-Arie, R., & Chalutz, E. (1992). Biological control of postharvest diseases of grape, peach, and apple with the yeasts *Kloeckera apiculata* and *Candida guilliermondii*. *Plant Disease, 76*, 470–473.
Mikani, A., Etebarian, H. R., Sholberg, P. L., Gorman, D. T., Stokes, S., & Alizadeh, A. (2008). Biological control of apple gray mold caused by *Botrytis mali* with *Pseudomonas fluorescens* strains. *Postharvest Biology and Technology, 48*, 107–112.
Misran, A., Padmanabhan, P., Sullivan, J. A., Khanizadeh, S., & Paliyath, G. (2015). Composition of phenolics and volatiles in strawberry cultivars and influence of preharvest hexanal treatment on their profiles. *Canadian Journal of Plant Science, 95*, 115–126.
Morath, S. U., Hung, R., & Bennett, J. W. (2012). Fungal volatile organic compounds: A review with emphasis on their biotechnological potential. *Fungal Biology Reviews, 26*, 73–83.
Moreno, D. A., Carvajal, M., & García-Viguera, C. (2006). Chemical and biological characterization of nutraceutical compounds of broccoli. *Journal of Pharmaceutical and Biomedical Analysis, 41*, 1508–1522.
Multon, J. L. (1988). *Aditivos y auxiliares de fabricación en las industrias agro alimentarias*. Editorial Acribia, Zaragoza, Spain.
Neiri, F., Mari, M., Brigati, M., & Bertolini, P. (2007). Fungicidal activity of plant volatile compounds for controlling *Monilinia laxa* in stone fruit. *Plant Disease, 91*, 30–35.
Nerio, L. S., Olivero-Verbel, J., & Stashenko, E. (2010). Repellent activity of essential oils: A review. *Bioresource Technology, 101*, 372–337.
Nunes, C. A. (2012). Biological control of postharvest diseases of fruit. *The European Journal of Plant Pathology, 133*, 181–196.
Obagwu, J., & Korsten, L. (2003). Integrated control of citrus green and blue molds using *Bacillus subtilis* in combination with sodium bicarbonate or hot water. *Postharvest Biology and Technology, 28*, 187–194.

Ozdemir, A. E., Candr, E. E., Kaplankran, M., Soylu, E. M., Sahinler, N., & Gul, A. (2010). The effects of ethanol-dissolved propolis on the storage of grapefruit cv. Star ruby. *Turkish Journal of Agriculture and Forestry, 34*, 155–162.

Paliyath, G., & Murr, D. P. (2007). Compositions for the preservation of fruits and vegetables. US patent no. 7198811B2.

Paliyath, G., Murr, D. P., Yada, R. Y., & Pinhero, R. G. (2003). Inhibition of phospholipase D. US patent no. 6514914.

Paliyath, G., Padmanabhan, P., Mihindukulasurya, S., Janavi, G. J., Sharkawy, I., & Subramaniyan, S., et al. (2015). In *ACS national meeting and exposition. applied nanotechnology for food and agriculture*.

Palou, L., Smilanick, J. L., Usall, J., & Viñas, I. (2001). Control of postharvest blue and green molds of oranges by hot water, sodium carbonate, and sodium bicarbonate. *Plant Disease, 85* 371–376.

Palou, L., Usall, J., Munoz, J. A., Joseph, L., & Vin, S. I. (2002). Hot water, sodium carbonate, and sodium bicarbonate for the control of postharvest green and blue molds of clementine mandarins. *Postharvest Biology and Technology, 24*, 93–96.

Parvu, M., Vlase, L., Parvu, A. E., Rosca-Casian, O., Gheldiu, A. M., & Parvu, O. (2015). Phenolic compounds and antifungal activity of *Hedera helix* L. (Ivy) flowers and fruits. *Notulae Botanicae Horti Agrobotanici Cluj-Napoca, 43*, 53–58.

Passos, F. R., Mendes, F. Q., Cunha, M. C., Pigozzi, M. T., & Carvalho, A. M. X. D. (2016). Propolis extraction in postharvest conservation Banana 'Prata'. *Revista Brasileira de Fruticultura, 38*(2), p. e-931.

Pastor, C., Sanchez-Gonzalez, L., Marcilla, A., Chiralt, A., Chafer, M., & Gonzalez-Martinez, C. (2011). Quality and safety of table grapes coated with hydroxypropylmethylcellulose edible coatings containing propolis extract. *Postharvest Biology and Technology, 60*, 64–70.

Pesis, E. (2005). The role of the anaerobic metabolites, acetaldehyde and ethanol, in fruit ripening, enhancement of fruit quality and fruit deterioration. *Postharvest Biology and Technology, 37*, 1–9.

Porat, R., Daus, A., Weiss, B., Cohen, L., & Droby, S. (2003). Effect of combining hot water, sodium bicarbonate and biocontrol on postharvest decay of citrus fruit. *The Journal of Horticultural Science and Biotechnology, 77*, 441–445.

Prasad, K., & Stadelbacher, G. J. (1974). Effect of acetaldehyde on postharvest decay and market quality of fresh strawberries. *Phytopathology, 64*, 948–951.

Radi, M., Mohsen, R., Hassan, A. J., Gholamreza, M., Asgar, F., & Sedigheh, A. (2010). Effect of hot acetic acid solutions on postharvest decay caused by *Penicillium expansum* on red delicious apples. *Scientia Horticulturae, 126*, 421–425.

Rao, A. V., Devi, P. U., & Kamath, R. (2001). *In vivo* radioprotective effect of *Moringa oleifera* leaves. *Indian Journal of Experimental Biology, 39*, 858.

Regnier, T., Combrick, S., Du Plooy, W., & Botha, B. (2010). Evaluation of *Lippia scaberrima* essential oil and some pure terpenoids constituents as postharvest mycobiocides for avocado fruit. *Postharvest Biology and Technology, 57*, 176–182.

Reyes-Diaz, M., Lobos, T., Cardemil, L., Nunes-Nesi, A., Retamales, J., Jaakola, L., et al. (2016). Methyl jasmonate: An alternative for improving the quality and health properties of fresh fruits. *Molecules, 21*, 1–18.

Ruiz-García, Y., Romero-Cascales, I., Gil-Muñoz, R., Fernández-Fernández, J. I., Lopez- Roca, J. M., & Gomez-Plaza, E. (2012). Improving grape phenolic content and wine chromatic characteristics through the use of two different elicitors: Methyl jasmonate versus benzothiadiazole. *Journal of Agricultural and Food Chemistry, 60*, 1283–1290.

Ruiz-García, Y., López-Roca, J. M., Bautista-Ortín, A. B., Gil-Muñoz, R., & Gómez-Plaza, E. (2014). Effect of combined use of benzothiadiazole and methyl jasmonate on volatile compounds of Monastrell wine. *American Journal of Enology and Viticulture, 65*, 238–243.

Santos, S. M. C., Polonio, J. C., Quecine, M. C., Almeida, T. T., Bogas, A. C., Pamphile, J. A., et al. (2016). Endophytic cultivable bacteria community obtained from the *Paullinia cupana* seed in

Amazonas and Bahia regions and its antagonistic effects against *Colletotrichum gloeosporioides*. *Microbial Pathogenesis, 98*, 16–22.
Scherm, B., Ortu, G., Muzzu, A., Budroni, M., Arras, G., & Migheli, Q. (2003). Biocontrol activity of antagonistic yeasts against *Penicillium expansum* on apple. *Journal of Plant Pathology, 85*, 1–9.
Sforcin, J. M., & Bankova, V. (2011). Propolis: Is there a potential for the development of new drugs? *Journal of Ethnopharmacology, 133*, 253–260.
Sharma, M., Jacob, J. K., Subramanian, J., & Paliyath, G. (2010). Hexanal and 1-MCP treatments for enhancing the shelf life and quality of sweet cherry (*Prunus avium* L.). *Scientia Horticulturae, 125*, 239–247.
Sharma, N., & Verma, U. K. (2004). Bioactivity of *Hyptis suaveolens* on storage mycoflora. In *International conference on fumigation and controlled atmosphere*. Conrad Jupiters, Gold Coast, Australia, (Abstract), p. 55.
Sharma, R. R., & Pongener, A. (2010). Natural products for postharvest decay control in horticultural produce: A review. *Stewart Postharvest Review, 4*, 1–9.
Sholberg, P. L., & Gaunce, A. P. (1996). Fumigation of stone fruit with acetic acid to control postharvest decay. *Crop Protection, 15*, 681–686.
Shuping, D. S. S., & Eloff, J. N. (2017). The use of plants to protect plants and food against fungal pathogens: A review. *African Journal of Traditional, Complementary, and Alternative Medicines, 14*, 120–127.
Silici, S., Koc, N. A., Ayangil, D., & Canaya, S. (2005). Antifungal activities of propoli collected by different races of honeybees against yeasts isolated from patients with superficial mycoses. *Journal of Pharmaceutical Sciences, 99*, 39–44.
Siqueira, A. B., Gomes, B. S., Cambuim, I., Maia, R., Abreu, S., & Souza-Motta, C. M. (2009). *Trichophyton* species susceptibility to green and red propolis from Brazil. *Letters in Applied Microbiology, 48*, 90–96.
Sivakumar, D., & Bautista-Baños, S. (2014). A review on the use of essential oils for postharvest decay control and maintenance of fruit quality. *Crop Protection, 64*, 27–37.
Smilanick, J. L., Mackey, B. E., Reese, R., Usall, J., & Margosan, D. A. (1997). Influence of concentration of soda ash, temperature, and immersion period on the control of postharvest green mold of oranges. *Plant Disease, 81*, 379–382.
Smilanick, J. L., Margosan, D. A., Milkota, F., & Usall, J. (1999). Control of citrus green mold by carbonate and bicarbonate salts and influence of commercial postharvest practices on their efficacy. *Plant Disease, 83*, 139–145.
Smith, L. A., Dann, E. K., Leonardi, J., Dean, J. R., & Cooke, A. W. (2011). Exploring non-traditional products for management of postharvest anthracnose and stem end rot in avocado. In *Proceedings of VII world avocado congress 2011, Cairns, Australia*.
Song, J., Hildebrand, P. D., Fan, L., Forney, C., Renderos, W., Palmer, L., & Doucette, C. (2007). Effect of hexanal vapor on the growth of postharvest pathogens and fruit decay. *Journal of Food Science, 72*, 108–112.
Song, J., Leepipattanawit, R., Deng, W., & Beaudry, R. M. (1998). Hexanal vapor acts as residueless antifungal agent that enhances aroma biosynthesis in apple fruit. *Acta Horticulturae, 464*, 219–224.
Spadaro, D., & Droby, S. (2016). Development of biocontrol products for postharvest diseases of fruit: The importance of elucidating the mechanisms of action of yeast antagonists. *Trends in Food Science & Technology, 47*, 39e49.
Stenvers, N., & Stork, H. W. (1977). Growth, ripening and storage of tomato fruits. *Gartenbauwissenschaft, 42*, 66–70.
Sui, Y., & Liu, J. (2014). Effect of glucose on thermotolerance and biocontrol efficacy of the antagonistic yeast *Pichia guilliermondii*. *Biological Control, 74*, 59–64.
Talibi, I., Askarne, L., Boubaker, H., Boudyach, E. H., Msanda, F., Saadi, B., & Aoumar, A. (2012). Antifungal activity of Moroccan medicinal plants against citrus sour rot agent *Geotrichum candidum*. *Letters in Applied Microbiology, 55*, 155–161.

Talibi, I., Boubaker, H., Boudyach, E. H., & Ait Ben Aoumar, A. (2014). Alternative methods for the control of postharvest citrus diseases. *Journal of Applied Microbiology, 117*, 1–17.

Tesfay, S. Z., & Magwaza, L. S. (2017). Evaluating the efficacy of moringa leaf extract: Chitosan and carboxymethyl cellulose as edible coatings for enhancing quality and extending postharvest life of avocado (*Persea americana* mill.) fruit. *Food Packaging and Shelf Life, 11*, 40–48

Thavong, P., Archbold, D. D., Pankasemsuk, T., & Koslanund, R. (2010). Postharvest use of hexanal vapor and heat treatment on longan fruit decay and consumer acceptance. *Thammasat International Journal of Science and Technology, 15*, 54–63.

Tian, S., & Zhang, Z. (2016). Methyl jasmonate. In M. W. Siddiqui, J. F. Ayala Zavala, & C. A. Hwang (Eds.), *Postharvest management approaches for maintaining quality of fresh produce* (pp. 97–111). Cham: Springer.

Tiwari, K., & Paliyath, G. (2011). Microarray analysis of ripening-regulated gene expression and its modulation by 1-MCP and hexanal. *Plant Physiology and Biochemistry, 49*, 329–340.

Tomazoni, E. Z., Pansera, M. R., Pauletti, G. F., Moura, S., Ribeiro, R. T., & Schwambach, J. (2016). In vitro antifungal activity of four chemotypes of *Lippia alba* (Verbenaceae) essential oils against *Alternaria solani* (Pleosporeaceae) isolates. *Anais da Academia Brasileira de Ciências, 88*, 999–1010.

Torres, R., Solsana, C., Viñas, I., Usall, J., Plaza, P., & Teixidó, N. (2014). Optimization of packaging and storage conditions of a freeze dried *Pantoea agglomerans* formulation for controlling postharvest diseases in fruit. *Journal of Applied Microbiology, 117*, 173–184.

Tripathi, M. K., & Giri, S. K. (2014). Probiotic functional foods: Survival of probiotics during processing and storage. *Journal of functional foods, 9*, 225–241.

Tripathi, P., Dubey, N., Banerji, R., & Chansouria, J. (2004). Evaluation of some essential oils as botanical fungitoxicants in management of post-harvest rotting of citrus fruits. *World Journal of Microbiology and Biotechnology, 20*, 317–321.

Tripathi, A., Neeta, S., Vinay, S., & Afroz, A. (2013). A review on conventional and non-conventional methods to manage post-harvest diseases of perishables. *Research, 5*, 6–19.

Tripathi, P. (2001). *Evaluation of some plant products against fungi causing post-harvest diseases of some fruits. Ph.D. thesis*, Department of Botany, Banaras Hindu University, India.

Tripathi, P., & Shukla, A. K. (2007). Emerging non-conventional technologies for control of post-harvest diseases of perishables. In *Fresh produce, global science books*.

Umthong, S., Puthong, S., & Chanchao, C. (2009). *Trigona laeviceps* propolis from Thailand: Antimicrobial, antiproliferative and cytotoxic activities. *The American Journal of Chinese Medicine, 37*, 855–865.

Utama, I. M. S., Will, R. B. H., Ben-Yehoshua, S., & Kuek, C. (2002). *In vitro* efficacy of plant volatiles for inhibiting the growth of fruit and vegetable decay microorganisms. *Journal of Agricultural and Food Chemistry, 50*, 6371–6377.

Utama, I. M. S., Wills, R. B. H., Antara, I. N. S., Mandrini, I. A. B., & Kencana, P. K. D. (2010). The efficacy of acetaldehyde vapour against the growth of soft rot bacteria (*Erwinia carotovora*) inoculated on capsicum fruits. In *Proceedings of the national seminar on horticulture, Udayana University, Indonesia* (pp. 436–444).

Utto, W., Mawson, A. J., & Bronlund, J. E. (2008). Hexanal reduces infection of tomatoes by *Botrytis cinerea* whilst maintaining quality. *Postharvest Biology and Technology, 47*, 434–437.

Vardar-Unlu, G., Silici, S., & Unlu, M. (2008). Composition and in vitro antimicrobial activity of Populus buds and poplar-type propolis. *World Journal of Microbiology and Biotechnology, 24*, 1011–1017.

Vero, S., Garmendia, G., Gonzalez, N. B., Bentancur, O., & Wisniewski, W. (2013). Evaluation of yeasts obtained from Antarctic soil samples as biocontrol agents for the management of post-harvest diseases of apple (Malus x domestica). *FEMS Yeast Research, 13*, 189–199.

Wang, Y., Wang, P., Xia, J., Yu, T., Luo, B., Wang, J., & Zheng, X. (2010). Effect of water activity on stress tolerance and biocontrol activity in antagonistic yeast *Rhodosporidium paludigenum*. *International Journal of Food Microbiology, 143*, 103–108.

Wasternack, C., & Hause, B. (2002). Jasmonates and octadecanoids: Signals in plant stress responses and development. *Progress in Nucleic Acid Research and Molecular Biology, 72*, 165–221.

Wei, Y., Mao, S., & Tu, K. (2014). Effect of preharvest spraying *Cryptococcus laurentii* on postharvest decay and quality of strawberry. *Biological Control, 73*, 68–74.

Wilson, C. L., Wisniewski, M., Droby, S., & Chalutz, E. (1993). A selection strategy for microbial antagonist to control postharvest diseases of fruits and vegetables. *Scientia Horticulturae, 53*, 183–189.

Wisniewski, M., Droby, S., Chalutz, E., & Eilam, Y. (1995). Effects of Ca^{2+} and Mg^{2+} on *Botrytis cinerea* and *Penicillium expansum* in vitro and on the biocontrol activity of *Candida oleophila*. *Plant Pathology, 44*, 1016–1024.

Wood, E. M., Miles, T. D., & Wharton, P. S. (2013). The use of plant volatile compounds for the control of the potato postharvest diseases, black dot, silver scurf and soft rot. *Biological Control, 64*, 152–159.

Wuryatmo, E., Able, A. J., Ford, C. M., & Scott, E. S. (2014). Effect of volatile citral on the development of blue mould, green mould and sour rot on navel orange. *Australasian Plant Pathology, 43*, 403–441.

Xu, X., & Tian, S. (2008). Salicylic acid alleviated pathogen-induced oxidative stress in harvested sweet cherry fruit. *Postharvest Biology and Technology, 49*, 379–385.

Yao, H. J., & Tian, S. P. (2005). Effects of a biocontrol agent and methyl jasmonate on postharvest diseases of peach fruit and the possible mechanisms involved. *Journal of Applied Microbiology, 98*, 941–950.

Yousef, A. R. M., Abd El-Moniem, E. A. A., & Saleh, M. M. S. (2015). The effect of some natural products on storability and fruit properties of 'Fuerte' avocado. *International Journal of ChemTech Research, 8*, 1454–1462.

Yu, S. M., & Lee, Y. H. (2015). Genes involved in nutrient competition by Pseudomonas putida JBC17 to suppress green mold in postharvest Satsuma mandarin. *Journal of Basic Microbiology, 55*, 898–906.

Yu, T., Wang, L., Yin, Y., Wang, Y., & Zheng, X. (2008). Effect of chitin on the antagonistic activity of Cryotococcus laurentii against *Penicillium expansum* in pear fruit. *International Journal of Food Microbiology, 122*, 44–48.

Zahid, N., Ali, A., Siddiqui, Y., & Maqbool, M. (2013). Efficacy of ethanolic extract of propolis in maintaining postharvest quality of dragon fruit during storage. *Postharvest Biology and Technology, 79*, 69–72.

Zainuri. (2006). *Defence mechanisms and induced resistance in 'Kensington Pride' mango*. Brisbane: The University of Queensland.

Zhang, H., Mahunu, G. K., Castoria, R., Yang, Q., & Apaliya, M. T. (2018). Recent developments in the enhancement of some postharvest biocontrol agents with unconventional chemicals compounds. *Trends in Food Science & Technology, 78*, 180–187.

Zheng, M., Shi, J., Shi, J., Wang, Q., & Li, Y. (2013). Antimicrobial effects of volatiles produced by two antagonistic *Bacillus* strains on the anthracnose pathogen in postharvest mangos. *Biological Control, 65*, 200–206.

Chapter 8
Application of Nanotechnology for Integrated Plant Disease Management

Imran Ul Haq, Siddra Ijaz, and Nabeeha Aslam Khan

Abstract Nanotechnology is an innovative and emerging discipline in the field of science and technology. With its broad application, it is now becoming a key part of life sciences, including approaches to target phytopathogens for disease management. Agrochemicals application against phytopathogens is not sustainable anymore because of insufficient bioavailability of active and low-impact compounds. Hence, the nature of nanoparticles (NPs), nanoemulsions and nanoformulations make them efficient nanopesticides to target in a very efficient way, showing higher solubility, permeability and stability. This chapter will provide details on this technology as integrated in plant disease management. Antimicrobial action, potential and application of NPs and NPs-based nanopesticides for managing the plant diseases are described.

Keywords Nanobiotechnology · Metallic nanoparticles · Nanoparticles · Nanobarcodes · Nanofungicides

8.1 Introduction

Nanotechnology is an emerging multidisciplinary approach with applications in various spheres of science and technology. Many reports are available on the potential of nanotechnology in various important sectors including textile industry, health care, information technology, power and energy. Natural resources available for crop cultivation such as land, irrigation water and soil nutrients are limited, but

I. Ul Haq (✉) · N. A. Khan
Department of Plant Pathology, University of Agriculture Faisalabad, Faisalabad, Pakistan

S. Ijaz
Centre of Agricultural Biochemistry and Biotechnology (CABB),
University of Agriculture Faisalabad, Faisalabad, Pakistan

© Springer Nature Switzerland AG 2020

I. Ul Haq, S. Ijaz (eds.), *Plant Disease Management Strategies for Sustainable Agriculture through Traditional and Modern Approaches*, Sustainability in Plant and Crop Protection 13, https://doi.org/10.1007/978-3-030-35955-3_8

demand for food, to feed the ever increasing population, is increasing tremendously. The cost of inputs such as seed, fertilizers and pesticides is also increasing worldwide. Crop losses due to pest damage, diseases and weeds are a big challenge for agricultural scientists. In the current scenario, efficient and effective use of resources, farm mechanization and modernization represent a way to overcome these constraints and to reduce the cost of crop production, and maximize outputs.

Applications of nanotechnology have a promising future in agricultural science, and they can be a great source of innovation to improve yields and significantly contribute to precision agriculture farming practices. Nanotechnology has also considerable potential for application in improving health of the environment (Ying 2009). Detection of environmental pollution and treatment devices are developed using nanostructured materials in a preventive approach, by developing different tools (Duebendorf 2008; Nowack 2009). Additionally, these materials are being used in manufacturing of various products available in the market, including antibacterial dressings and nanoparticles-based eco-friendly pesticides (Bhattacharyya et al. 2010; Gha-Young et al. 2008).

NPs have received greater interest in multidisciplinary research fields because of their extraordinary attributes. Physical and chemical properties are influenced by the NPs size, which is a key factor (Wang et al. 2007). Nanomaterials show chemical, biological, and physical properties that are wholly diverse and distinct in their corresponding macro state (Li et al. 2001). Physiochemical facets are associated with changes in size and shape of NPs (Barrak et al. 2016). Physicochemical properties such as optical properties, chemical reactivity, mechanical strength, and surface area pose uniqueness to nanomaterials and enable them for a wide range of applications (Gupta et al. 2013). NPs are grouped into different classes on the basis of their chemical and physical nature, such as: (1) lipid (2) metallic (3) ceramics, (4) semiconductor (5), polymer and (6) carbon NPs.

There are three discrete layers in NPs composition (core region, surface and shell layer) and the core region refers as true NP (Shin et al. 2016; Khan et al. 2017). Structural characterization is used to explore their bonding nature and composition (Ullah et al. 2017). Different techniques are used to explore structural characteristics of nanomaterials (Ingham 2015). Among characterization techniques, the most common are:

- Scanning electron microscopy (SEM)
- X-rays diffraction (XRD)
- Transmission electron microscopy (TEM)
- Infrared spectroscopy (IR)
- Zeta sizing
- Energy dispersive X-rays (EDX)
- X-rays photoelectron spectroscopy (XPS)

Use of ecofriendly approaches, which generate less harmful waste, is a global need. This situation sensitizes scientists to develop and adopt "Green/biosynthesis" strategies and methods. In green synthesis of NPs, nontoxic and ecofriendly resources are used which prevent from production of detrimental waste. Therefore, instead of chemical synthesis, plants and microbe-based synthesis of NPs are catching more attention and popularity.

Actions of NPs depending on their size, e.g., silver NPs have antimicrobial potential, which are lost in their macro state (Sofi et al. 2012). NPs properties and efficiency depend on the ways of synthesis (physical, chemical and biological methods) (Rai and Yadav 2013). Physico-chemical methods are expensive and usually require toxic compounds. These methods have harmful impacts on human and environmental health due to detrimental radiations, synthetic reductants and stabilizing agents (Joerger et al. 2000; Panigrahi et al. 2004; Oliveira et al. 2005; Gan et al. 2012). Whereas biological methods (plants and microbes) having higher efficacy, are ecofriendly and cost effective (Kumar et al. 2012; Bonde et al. 2012). The major constraint of biogenic approaches is difficult to achieve, due to dispersion and lack of control over particles size and shape (Li et al. 2007; Nayak et al. 2011). In fungi-mediated biosynthesis of nanomaterials, fast growing fungi have more potential. Filamentous fungi are especially exploited because of their quick growth on substrate and metabolites production. Polysaccharides of fungal cell wall are the main ingredient for metal ions reduction (Sastry et al. 2003).

8.2 Nanotechnology: Helping Agriculture with Special Emphasis on Plant Protection

Nanobiotechnology is considered an important management technique, among other disease management strategies, due to biosensors (nanomaterial-based disease detection devices), nanofungicides and related delivery systems (Rai and Ingle 2012; Satalkar et al. 2016; Mishra and Singh 2015). This revolutionary science, changed green revolution into green nano-bio-revolution (Khan and Rizvi 2014). NPs synthesis and their application are two main aspects of this technique.

Climate change, environmental health, pests and plant diseases, wastage of natural resources, costly inputs and urbanization are the big challenges which the agriculture sector is facing these days, in developing countries such as Pakistan. tools to sustain agriculture and solve the above mentioned problems are the need of the day. Cost-effective agricultural technologies through innovation has a great potential in improving the farm productivity, and reduce the cost of production. For example, using *nanoscale carriers* might facilitate delivery of efficient nutrients, pesticides, herbicides, and other plant growth promoting inputs. Nanoencapsulation is one of the mechanisms through which controlling and efficiently releasing chemicals to the host plants for disease control. This mechanism also increases the pesticides and other chemicals stability, reducing degradation, and the amounts used for application as well.

Nanofabrication tools have been helpful in improving plant disease management strategies by understanding the mechanism of physico-chemical and biological interactions between host cells and pathogens (Cursino et al. 2009). For example, discovery of *microfabricated xylem vessels* with nano-sized features enabled researchers to study the xylem-inhabiting bacteria at an advanced level. Historically, destructive sampling techniques were used to monitor the changes in populations of these bacteria at different distances from the sites of inoculation. However, scientists were unable to get information on the bacteria colonization, movement and re-colonization at new areas. By the development of *microfabricated xylem vessels* it is now possible to study this mechanism (Zaini et al. 2009).

It is claimed that the development of clay nanotubes as pesticide carriers has the capability to contact plants in a better way and to reduce the amounts of pesticide up to 80%, reducing the cost of treatments (Murphy 2008). In the *photocatalysis* reaction the NPs are exposed to UV light which form, as a result, electron hole pairs (positive holes and negative electrons). These act as efficient oxidizing agents with higher disinfection and degradation rates. Photocatalysis processes can decompose various toxic compounds such as pesticides, which normally take longer degradation time under normal conditions (Malato et al. 2002). It can be used in environmental protection due to the capacity to purify and decontaminate the air. Their application has also being used widely in wastewater treatment (Melemeni et al. 2009).

Bioremediation of resistant pesticides is another favourable property of NPs. These convert harmful compounds having resistance to degradation, or which normally degrade at lower rates, in nontoxic compounds, as for bioremediation of resistant compounds such as pesticides. In this way NPs play an important role in food safety by converting toxic compounds which may enter in the food stuff and cause severe effects in consumers (Lhomme et al. 2008).

The NPs also act as *disinfectants* and have wide applications in food engineering and postharvest technology, due to negative electron hole pairs produced by their excitation, upon contact with bacteria. Due to small size and the a large number of possible combinations, the NPs may be used as a *nanobarcode* in many fields of biotechnology agricultural encoding (Mathew et al. 2009).

The biotechnology advancements in combination with nanotechnology have made it possible for the plant scientists to improve host plant resistance against biotic and abiotic stresses. The nanobarcode allow both kinds of applications, i.e. biological and non-biological. Among the biological application the nanobarcodes are being used as ID tags for analysis of gene expression and histopathological studies (Branton et al. 2008). Nanobarcode with *non-biological nanobarcodes* used as nanoscale tags are applied for tracking agricultural food products. This nanobarcode technology will probably lead to the development of auto-ID technologies, in cases where seem tagging with conventional barcodes is unpractical.

Biosensors are applied for microorganisms which produce various beneficial as well as harmful volatile compounds. Foul odor is one of the characteristic symptoms for identification of bacterial food rot, but human nose cannot detect or distinguish that specific odor, furthermore it may be injurious to human health. Rapid detection biosensors are more appropriate for detection of these odors. They are also used for contaminants in different kinds of materials, such as raw food and food products, as well as in water. These rapid biosensors require less time for the completion of microbial testing. In present situation, proper use of nanomaterials in biosensors (early disease detection), management of plant pathogens and accurate distribution of active compounds is a preferable method (Scott and Chen 2003; Johnston 2010).

Nano-sized particles may be very target-specific in disease management and improvement of crop plants (Ghormade et al. 2011). The NPs surface-to-volume ratio is directly proportional to their reactivity and biochemical activity (Dubchak et al. 2010). The mode of action of these particles is simple, as they bind to the cell wall of phytopathogens causing deformations due to energy transfer, that ultimately cause cell death (Schaller et al. 2004). Targeted and slow delivery of active compounds may be achieved by their encapsulation, which prevent from runoff and leaching (Chen and Yada 2011; Gruère 2012).

8.3 Nanoparticles for Management of Plant Diseases

Excessive use of fungicides and pesticides may affect the environment and also induce the evolution of new resistant populations (Dzhavakhiya et al. 2012; Alghuthaymi et al. 2015; Chen et al. 2015). There is hence a need to find alternative and more appropriate ways for management (Vu et al. 2015). Nanotechnology may play a key role in providing healthier food by promoting precision farming (Gruère 2012), in particular when facing huge losses due to a wide range of phytopathogens (Tournas 2005). The latter are also a cause of malnutrition. Food demand will grow by twofold in coming years, therefore food security will become a grand challenge to feed people (Tilman et al. 2002). Agrochemicals do not always reach target sites, owing to depletion of chemicals at the time of application, off-target deposition and photodegradation. All these processes cause wastage of agrochemicals and increase the cost of production, in particular when new resistant pathogens appear (Schaller et al. 2004; Hettiarachchi and Wickramarachchi 2011; Chowdappa et al. 2013).

Fungi are the most represented (approx. 70%) among the destroyers of crop plants including pulses, cereals, fruits, fiber crops, etc. (Agrios 2005). Fungal diseases cause pre and post-harvest losses which adversely affect economy because a large amount of money must be spent on agrochemicals, annually (González-Fernández et al. 2010). Therefore, for sustainable management, effective and precise strategies would be adopted to minimize losses, such as chemical (fungicides), cultural and biological control (biological agents, bioformulations). Generally, chemical control considers highly efficient method in order to attain rapid control, but as agrochemicals effectively kill pathogens, they also destroy valuable microbes, which are soil inhabitants (Manczinger et al. 2002).

Metallic NPs are gaining massive popularity and proved to be an alternative to agrochemicals. They have outstanding potential to eliminate targeted microbes from plants, soil and hydroponics (Park et al. 2006; Sharma et al. 2012). NPs can be applied in two different ways, directly through application on phytopathogens, and in formulating fungicides (Khan and Rizvi 2014).

NPs may be applied as foliar sprays to destroy pathogens. They also stimulate plants' growth (Agrawal and Rathore 2014). Low concentrations may effectively control pathogens (Nel et al. 2003; Park et al. 2006). Metallic NPs damage fungal cell walls causing hyphal plasmolysis (Min et al. 2009). However, according to different scientists their mechanism of action may vary and rely on different mechanisms (Zeng et al. 2008; Prabhu and Poulose 2012; Lemire et al. 2013). They include:

1. Permeability of plasma membrane, which is disturbed preventing a proper functioning due to attachment of the NPs proteins sulfur groups .
2. DNA damage.
3. Disturbance of electron transport chain and protein oxidation .
4. Reactive oxygen species (ROS) may be generated, which cause cellular damage.
5. Hindrance in nutrients uptake.

All above mentioned mechanisms are interlinked, and exert combined effects against phytopathogens (Alghuthaymi et al. 2015; Abd-Elsalam and Prasad 2018). Bio-reduction reaction of different metals (iron, silver, zinc, gold, copper etc.) has been evaluated for metallic NPs synthesis. Different *in vitro* and *in vivo* assays have been also carried out, in order to investigate antimicrobial potential of metallic NPs. Silver NPs showed higher toxicity to phytopathogens and therefore are considered as useful management tools (Alghuthaymi et al. 2015; Mishra and Singh 2015).

Globally, different scientists have checked the toxicity of NPs to pathogens and safety for beneficial organisms (plants, human beings, animals and some microbes), using minute concentrations of nanofungicides. Therefore it is recommended to use safer management approaches as compared to agrochemicals (Thomas and McCubbin 2003; Zeng et al. 2008). However, Woo et al. (2009) evaluated fungicidal efficacy of NPs on a fungal pathogen causing oak wilt. They recorded inhibition of fungal growth. Similarly, the antifungal potential was tested against *Magnaporthe grisea* and *Bipolaris sorokiniana* of cereals (Jo et al. 2009) with a documented disease severity and progress. Silver NPs are extensively applied in biosystems, as they cause, in fungi, disturbance in normal cellular functioning and ion efflux transport system (Morones et al. 2005).

8.4 Metallic Nanoparticles: Effective Tool for Plant Disease Management

Metallic NPs having remarkable physical and chemical properties appear as effective, eco-friendly and safer alternatives to synthetic fungicides (Kumar and Yadav 2009; Aziz et al. 2016). These metal-based NPs are supposed to supersede synthetic agrochemicals in future (Medici et al. 2015; Ismail et al. 2017; Abd-Elsalam and Prasad 2018).

Several plant species designated as hyper-accumulators, concentrate metals at high levels and then assimilate them as NPs (Dubey et al. 2009). The metallic NPs commonly used as antimicrobial agents are silver, silica, gold, zinc and copper based.

Silver NPs showed an antimicrobial capability, both in ionic and nano forms, and when studied and tested were capable to kill plant pathogens (Sharon et al. 2010). Silver shows a great inhibitory mode of action against a variety of pathogens, including fungal as well as bacterial pathogens (Clement and Jarrett 1994; Kim et al. 2006; Wei et al. 2009). *Copper NPs* have the capability to degrade the cellular material of fungi as well as bacteria by producing an action comparable to fungicides hydroxyl radicals (Esteban Tejeda et al. 2009; Brunel et al. 2013). These nano-sized copper proved to be effective in controlling bacterial blight of rice and mungbean leaf spot disease (Gogoi et al. 2009). *Silica NPs* could sustain plants by enhancing resistance against diseases and stimulated physiological mechanisms (Brecht et al. 2004). *Iron NPs* directly interface with fungal cell surface and affect permeability of membrane, reducing cell growth and ultimately cause death by generating an oxidative stress (Xie et al. 2011). *Zinc NPs* produce hydroxyl and superoxide radicals and act as nanofungicide, deforming fungal cell wall, hyphae, hindering the conidial development and causing cellular death (Borkow and Gabbay 2005; Patra et al. 2012). *Gold NPs* toxicity was determined by Wang et al. (2011) against *Salmonella* spp. documenting higher toxic effects than those of its macro-form. Although silica has no direct antimicrobial activity, it can enhance resistance in host plants against diseases and other stresses (Brecht et al. 2003). Nano silica-silver was proved to be effective against certain fungal and bacterial plant diseases (Park et al. 2006). Mesoporous nano sized silica NPs have pores organized in a regular scheme with increased surface area. It improved site-specific chemicals delivery, efficiency, and effectiveness.

8.5 Nanofungicides: Step Towards Sustainable Control of Fungal Plant Pathogens

Alternative plant pathogens control methods to reduce the use of chemicals is a practical way for sustainable management by keeping the environment clean and healthy as well as producing safe food. In this regards several attempts have been made to develop efficient carrier systems to lower the concentration of applied pesticides (Ghormade et al. 2011). In recent years significant progress has been noticed in nano-scale materials development, having considerably different properties from corresponding bulk materials. Use of nanotechnology offers a promising future and an innovative way for safer delivery of agrochemicals (Ghormade et al. 2011). In spite of a late start to exploit the potential of nanomaterial for pathogens management, there have been exciting success stories in plant disease management by developing and applying NPs based pesticides, are best alternatives for sustainable management (Rai et al. 2009). Nanofungicide formulations enhance the active compounds solubility through a gradual and targeted release at slow rates, increasing the bioavailability of agrochemicals (Lauterwasser 2006: Kah and Hofmann 2014). Several attempts have been made so far for the preparation of different nanofungicides in different ways (Yan et al. 2005). Nanofungicide developed and tested so far proved effective in plant protection strategies (Bordes et al. 2009; Bergeson 2010). *Nanoemulsions (NEs)* with small size, lower viscosity and higher stability are the better option for nanofungicides (Bernardes et al. 2014). *Metallic NPs* previously discussed in detail in this chapter e.g. silver NPs are active and effective antimicrobial agents (Retchkiman-Schabes et al. 2006). *Nanocapsules* are a nanosystem in which the active fungicide ingredient is placed within a core, surrounded by membrane. Nanoencapsulation has also potential in nanofungicides formulations. Polymeric and solid-lipid nanocaspsules, which are loaded with tebuconazole and carbendazim, have been produced as nanofungicides (Bhan et al. 2014; Campos et al. 2015). *Nanogels* of chitosan, evaluated in combination with copper and pheromones, were proved effective against *Fusarium graminearum* (Brunel et al. 2013; Bhagat et al. 2013). *Nanospheres* are comprised of irregular nanoscale spherical particles, in which some "-cidal compounds" or active agents of fungicides are dispersed and/or dissolved in polymeric matrices (Sotthivirat et al. 2007).

Various natural and synthetic polymers and inorganic compounds have been tested to explore their potential in nanofungicide formulations for crop protection (Chuan et al. 2013). There is great need to develop formulations to improve their potency and stability, keeping in consideration the safety of the systems to environment and human beings. In this perspective, nanotechnology has a huge capacity to develop such new formulations and systems (Kanto et al. 2004).

References

Abd-Elsalam, K. A., & Prasad, R. (2018). *Nanobiotechnology applications in plant protection*. Dordrecht: Springer. https://www.springer.com/us/book/9783319911601.

Agrawal, S., & Rathore, P. (2014). Nanotechnology pros and cons to agriculture: A review. *International Journal of Current Microbiology and Applied Sciences, 3*(3), 43–55.

Agrios, G. N. (2005). *Plant pathology* (922 pp 5th edn.). Burlington: Elsevier Academic Press.

Alghuthaymi, M. A., Almoammar, H., Rai, M., Said-Galiev, E., & Abd-Elsalam, K. A. (2015). Myconanoparticles: Synthesis and their role in phytopathogens management. *Biotechnology and Biotechnological Equipment, 29*, 221–236.

Aziz, N., Pandey, R., Barman, I., & Prasad, R. (2016). Leveraging the attributes of *Mucor hiemalis*-derived silver nanoparticles for a synergistic broad-spectrum antimicrobial platform. *Frontiers in Microbiology, 7*, 1984. https://doi.org/10.3389/fmicb.2016.01984.

Barrak, H., Saied, T., Chevallier, P., Laroche, G., M'nif, A., & Hamzaoui, A. H. (2016). Synthesis, characterization, and functionalization of ZnO nanoparticles by N-(trimethoxysilylpropyl) ethylenediamine triacetic acid (TMSEDTA): Investigation of the interactions between Phloroglucinol and ZnO@ TMSEDTA. *Arabian Journal of Chemistry*, 1–8. https://doi.org/10.1016/j.arabjc.2016.04.019.

Bergeson, L. L. (2010). Nanosilver pesticide products: What does the future hold? *Environmental Quality Management, 19*, 73–82.

Bernardes, P. C., Nélio, J. D. A., & Nilda de Fátima, F. S. (2014). Nanotechnology in the food industry. *Bioscience Journal, 30*, 1919–1932.
Bhagat, D., Samanta, S. K., & Bhattacharya, S. (2013). Efficient management of fruit pests by pheromone nanogels. *Scientific Reports, 3*, 1294.
Bhan, S., Lalit, M., & Srivastava, C. N. (2014). Relative larvicidal potentiality of nano-encapsulated temephos and imidacloprid against *Culex quinquefasciatus*. *Journal of Asia-Pacific Entomology, 17*, 787–791.
Bhattacharyya, A., Bhaumik, A., Rani, P. U., Mandal, S., & Epidi, T. T. (2010). Nanoparticles A recent approach to insect pest control. *African Journal of Biotechnology, 9*, 3489–3493.
Bonde, S. R., Rathod, D. P., Ingle, A. P., Ade, R. B., Gade, A. K., & Rai, M. K. (2012). *Murraya koenigii*-mediated synthesis of silver nanoparticles and its activity against three human pathogenic bacteria. *Nanoscience Methods, 1*, 25–36.
Bordes, P., Pollet, E., & Avérous, L. (2009). Nano-biocomposites: Biodegradable polyester/nanoclay systems. *Progress in Polymer Science, 34*, 125–155.
Borkow, G., & Gabbay, J. (2005). Copper as a biocidal tool. *Current Medicinal Chemistry, 12*, 2163–2175.
Branton, D., Deamer, D. W., Marziali, A., & el al, B. H. (2008). The potential and challenges of nanopore sequencing. *Nature Biotechnology, 26*, 1146–1153.
Brecht, M., Datnoff, L., Nagata, R., & Kucharek, T. (2003). *The role of silicon in suppressing tray leaf spot development in St. Augustine grass* (pp. 1–4). Gainesville: Publication in University of Florida.
Brecht, M. O., Datnoff, L. E., Kucharek, T. A., & Nagata, R. T. (2004). Influence of silicon and chlorothalonil on the suppression of gray leaf spot and increase plant growth in St. Augustine grass. *Plant Disease, 88*(4), 338–344.
Brunel, F., El Gueddari, N. E., & Moerschbacher, B. M. (2013). Complexation of copper (II) with chitosan nanogels: Toward control of microbial growth. *Carbohydrate Polymers, 92*, 1348–1356.
Campos, E. V. R., De Oliveira, J. L., Da Silva, C. M. G., Pascoli, M., Pasquoto, T., Lima, R., & Fraceto, L. F. (2015). Polymeric and solid lipid nanoparticles for sustained release of carbendazim and tebuconazole in agricultural applications. *Scientific Reports, 5*, 13809.
Chen, H., & Yada, R. (2011). Nanotechnologies in agriculture: New tools for sustainable development. *Trends in Food Science & Technology, 22*, 585–594.
Chen, L., Song, Y., Tang, B., Song, X., Yang, H., Li, B., et al. (2015). Aquatic risk assessment of a novel strobilurin fungicide: A microcosm study compared with the species sensitivity distribution approach. *Ecotoxicology and Environmental Safety, 120*, 418–427. https://doi.org/10.1016/j.ecoenv.2015.06.027.
Chowdappa, P., Kumar, N. B., Madhura, S., Kumar, M. S., Myers, K. L., Fry, W. E., & Cooke D. E. (2013). Emergence of 13_A 2 blue lineage of *Phytophthora infestans* was responsible for severe outbreaks of late blight on tomato in south-west India. *Journal of Phytopathology, 161*, 49–58.
Chuan, L., He, P., Pampolino, M. F., Johnston, A. M., Jin, J., Xu, X., & Zhou, W. (2013). Establishing a scientific basis for fertilizer recommendations for wheat in China: Yield response and agronomic efficiency. *Field Crops Research, 140*, 1–8.
Clement, J. L., & Jarrett, P. S. (1994). Antibacterial silver. *Metal-Based Drugs, 1*, 467–482.
Cursino, L., Li, Y., Zaini, P. A., De La Fuente, L., Hoch, H. C., & Burr, T. J. (2009). Twitching motility and biofilm formation are associated with tonB1 in Xylella fastidiosa. *FEMS Microbiology Letters, 299*, 193–199.
Dubchak, S., Ogar, A., Mietelski, J. W., & Turnau, K. (2010). Influence of silver and titanium nanoparticles on arbuscular mycorrhiza colonization and accumulation of radiocaesium in *Helianthus annuus*. *Spanish Journal of Agricultural Research, 1*, 103–108.
Dubey, M., Bhadauria, S., & Kushwah, B. S. (2009). Green synthesis of nanosilver particles from extract of Eucalyptus hybrida (safeda) leaf. *Digest Journal of Nanomaterials and Biostructures, 4*, 537–543.

Duebendorf, St. G. T. (2008). How do nanoparticles behave in the environment? NanoEco-Empa organized international meeting on nanoparticles in the environment. In B. Nowack and N. Mueller (Eds.), *Environmental Science and Technology*, http://pubs.acs.org/cgi-bin/asap.cgi/esthag/asap/html/es7029637.html

Dzhavakhiya, V., Shcherbakova, L., Semina, Y., Zhemchuzhina, N., & Campbell, B. (2012). Chemosensitization of plant pathogenic fungi to agricultural fungicides. *Frontiers in Microbiology, 3*, 1–9.

Esteban Tejeda, L., Malpartida, F., Esteban Cubillo, A., Pecharroman, C., & Moya, J. S. (2009). Antibacterial and antifungal activity of a soda-lime glass containing copper nanoparticles. *Nanotechnology, 20*, 505701.

Gan, P. P., Ng, S. H., Huang, Y., & Li, S. F. Y. (2012). Green synthesis of gold nanoparticles using palm oil mill effluent (POME): A low-cost and eco-friendly viable approach. *Bioresource Technology, 113*, 132–135.

Gha-Young, K., Joonmok, S., Min-Su, K., & Seung-Hyeon, M. (2008). Preparation of a highly sensitive enzyme electrode using gold nanoparticles for measurement of pesticides at the ppt level. *Journal of Environmental Monitoring, 10*, 632–637.

Ghormade, V., Deshpande, M. V., & Paknikar, K. M. (2011). Perspectives for nano-biotechnology enabled protection and nutrition of plants. *Biotechnology Advances, 29*, 792–803.

Gogoi, R., Dureja, P., & Singh, P. K. (2009). Nanoformulations – A safer and effective option for agrochemicals. *Indian Farming, 59*, 7–12.

González-Fernández, R., Prats, E., Jorrín-Novo, J. V. (2010). Proteomics of plant pathogenic fungi. *Journal of Biomedicine & Biotechnology*, *2010*, art. n. 932527.

Gruère, G. P. (2012). Implications of nanotechnology growth in food and agriculture in OECD countries. *Food Policy, 37*, 191–198.

Gupta, K., Singh, R. P., Pandey, A., & Pandey, A. (2013). Photocatalytic antibacterial performance of TiO_2 and Ag-doped TiO_2 against *S. aureus, P. aeruginosa* and *E. coli*. *Beilstein Journal of Nanotechnology, 4*, 345.

Hettiarachchi, M. A., & Wickramarachchi, P. A. S. R. (2011). Synthesis of chitosan stabilized silver nanoparticles using gamma ray irradiation and characterization. *Journal of Science, 6*, 65–75.

Ingham, B. (2015). X-ray scattering characterisation of nanoparticles. *Crystallography Reviews, 21*, 229–303.

Ismail, M., Prasad, R., Ibrahim, A. I. M., & Ahmed, I. S. A. (2017). Modern prospects of nanotechnology in plant pathology. In R. Prasad, M. Kumar, & V. Kumar (Eds.), *Nanotechnology* (pp. 305–317). Singapore: Springer Nature.

Jo, Y. K., Kim, B. H., & Jung, G. (2009). Antifungal activity of silver ions and nanoparticles on phytopathogenic fungi. *Plant Disease, 93*, 1037–1043.

Joerger, R., Klaus, T., & Granqvist, C. G. (2000). Biologically produced silver–carbon composite materials for optically functional thin-film coatings. *Advanced Materials, 12*, 407–409.

Johnston, C. T. (2010). Probing the nanoscale architecture of clay minerals. *Clay Minerals, 45*(3), 245–279.

Kah, M., & Hofmann, T. (2014). Nanopesticide research: Current trends and future priorities. *Environment International, 63*, 224–235.

Kanto, T., Miyoshi, A., Ogawa, T., Maekawa, K., & Aino, M. (2004). Suppressive effect of potassium silicate on powdery mildew of strawberry in hydroponics. *Journal of General Plant Pathology, 70*, 207–211.

Khan, M. R., & Rizvi, T. F. (2014). Nanotechnology: Scope and application in plant disease management. *Plant Pathology Journal, 13*, 214–231.

Khan, I., Yamani, Z. H., & Qurashi, A. (2017). Sonochemical-driven ultrafast facile synthesis of SnO_2 nanoparticles: Growth mechanism structural electrical and hydrogen gas sensing properties. *Ultrasonics Sonochemistry, 34*, 484–490.

Kim, Y. H., Lee, D. K., Cha, H. G., Kim, C. W., Kang, Y. C., & Kang, Y. S. (2006). Preparation and characterization of the antibacterial Cu nanoparticle formed on the surface of SiO2 nanoparticles. *The Journal of Physical Chemistry B, 110*, 24923–24928.

Kumar, V., & Yadav, S. K. (2009). Plant-mediated synthesis of silver and gold nanoparticles and their applications. *Journal of Chemical Technology and Biotechnology, 84*, 151–157.
Kumar, R. R., Priyadharsani, K. P., & Thamaraiselvi, K. (2012). Mycogenic synthesis of silver nanoparticles by the Japanese environmental isolate *Aspergillus tamarii. Journal of Nanoparticle Research, 14*, 860.
Lauterwasser, C. (2006). *Small sizes that matter: Opportunities and risks of nanotechnologies. Report in cooperation with the OECD International Futures Programme*. Munchen: Allianz Center for Technology. 45 pp.
Lemire, J. A., Harrison, J. J., & Turner, R. J. (2013). Antimicrobial activity of metals: Mechanisms, molecular targets and applications. *Nature Reviews. Microbiology, 11*, 371.
Lhomme, L., Brosillon, S., & Wolbert, D. (2008). Photocatalytic degradation of pesticides in pure water and a commercial agricultural solution on TiO_2 coated media. *Chemosphere, 70*, 381–386.
Li, L. S., Hu, J., Yang, W., & Alivisatos, A. P. (2001). Band gap variation of size-and shape-controlled colloidal CdSe quantum rods. *Nano Letters, 1*, 349–351.
Li, S., Shen, Y., Xie, A., Yu, X., Qiu, L., Zhang, L., & Zhang, Q. (2007). Green synthesis of silver nanoparticles using Capsicum annuum L. extract. *Green Chemistry, 9*, 852–858.
Malato, S., Blanco, J., Cáceres, J., Fernández-Alba, A. R., Agüera, A., & Rodrıguez, A. (2002). Photocatalytic treatment of water-soluble pesticides by photo-Fenton and TiO_2 using solar energy. *Catalysis Today, 76*, 209–220.
Manczinger, L., Antal, Z., & Kredics, L. (2002). Ecophysiology and breeding of mycoparasitic *Trichoderma* strains. *Acta Microbiologica et Immunologica Hungarica, 49*, 1–14.
Mathew, A. P., Laborie, M. P., & Oksman, K. (2009). Cross-linked chitosan/chitin crystal nanocomposites with improved permeation selectivity and pH stability. *Biomacromolecules, 10*, 1627–1632.
Medici, S., Peana, M., Nurchi, V. M., Lachowicz, J. I., Crisponi, G., & Zoroddu, M. A. (2015). Noble metals in medicine: Latest advances. *Coordination Chemistry Reviews, 284*, 329–350.
Melemeni, M., Stamatakis, D., Xekoukoulotakis, N. P., Mantzavinos, D., & Kalogerakis, N. (2009). Disinfection of municipal wastewater by TiO_2 photocatalysis with UV-A, visible and solar irradiation and BDD electrolysis. *Global NEST Journal, 11*, 357–363.
Min, J. S., Kim, K. S., Kim, S. W., Jung, J. H., Lamsal, K., Kim, S. B., & Lee, Y. S. (2009). Effects of colloidal silver nanoparticles on sclerotium-forming phytopathogenic fungi. *Plant Pathology Journal, 25*, 376–380.
Mishra, S., & Singh, H. B. (2015). Biosynthesized silver nanoparticles as a nanoweapon against phytopathogens: Exploring their scope and potential in agriculture. *Applied Microbiology and Biotechnology, 99*, 1097–1107.
Morones, J. R., Elechiguerra, J. L., Camacho, A., Holt, K., Kouri, J. B., Ramirez, J. T., & Yacaman, M. J. (2005). The bactericidal effect of silver nanoparticles. *Nanotechnology, 16*, 2346.
Murphy, K. (Ed.) (2008). *Nanotechnology: Agriculture's Next"Industrial" Revolution* (Williston, VT: Financial Partner, Yankee Farm Credit, ACA). Spring, pp 3–5.
Nayak, R., Pradhan, N., Behera, D., Pradhan, K., Mishra, S., Sukla, L., & Mishra, B. (2011). Green synthesis of silver nanoparticle by *Penicillium purpurogenum* NPMF: The process and optimization. *Journal of Nanoparticle Research, 13*, 3129–3137.
Nel, A., Xia, T., Madler, L., & Li, N. (2003). Toxic potential of materials at the nano level. *Science, 311*, 622–627.
Nowack, B. (2009). Is anything out there? What life cycle perspectives of nano-products can tell us about nanoparticles in the environment. *Nano Today, 4*, 11–12.
Oliveira, M. M., Ugarte, D., Zanchet, D., & Zarbin, A. J. (2005). Influence of synthetic parameters on the size, structure, and stability of dodecanethiol-stabilized silver nanoparticles. *Journal of Colloid and Interface Science, 292*, 429–435.
Panigrahi, S., Kundu, S., Ghosh, S., Nath, S., & Pal, T. (2004). General method of synthesis for metal nanoparticles. *Journal of Nanoparticle Research, 6*, 411–414.

Park, H. J., Kim, S. H., Kim, H. J., & Choi, S. H. (2006). A new composition of nanosized silica-silver for control of various plant diseases. *The Plant Pathology Journal, 22*, 295–302.

Patra, P., Mitra, S., Debnath, N., & Goswami, A. (2012). Biochemical-, biophysical-, and microarray-based antifungal evaluation of the buffer-mediated synthesized nano zinc oxide: An in vivo and in vitro toxicity study. *Langmuir, 28*, 16966–16978.

Prabhu, S., & Poulose, E. K. (2012). Silver nanoparticles: Mechanism of antimicrobial action, synthesis, medical applications, and toxicity effects. *International Nano Letters, 2*, 32.

Rai, M., & Ingle, A. (2012). Role of nanotechnology in agriculture with special reference to management of insect pests. *Applied Microbiology and Biotechnology, 94*, 287–293.

Rai, M., & Yadav, A. (2013). Plants as potential synthesiser of precious metal nanoparticles: Progress and prospects. *IET Nanobiotechnology, 7*, 117–124.

Rai, M., Yadav, A., & Gade, A. (2009). Silver nanoparticles as a new generation of antimicrobials. *Biotechnology Advances, 27*, 76–83.

Retchkiman-Schabes, P. S., Canizal, G., Becerra-Herrera, C. R., Zorrilla, H. B., & Liu, J. A. (2006). Ascencio biosynthesis and characterization of Ti/Ni bimetallic nanoparticles. *Optical Materials, 29*, 95–99.

Sastry, M., Ahmad, A., Khan, M. I., & Kumar, R. (2003). Biosynthesis of metal nanoparticles using fungi and actinomycete. *Current Science, 85*, 162–170.

Satalkar, P., Elger, B. S., & Shaw, D. M. (2016). Defining nano, nanotechnology and nanomedicine: Why should it matter? *Science and Engineering Ethics, 22*, 1255–1276.

Schaller, M., Laude, J., Bodewaldt, H., Hamm, G., & Korting, H. C. (2004). Toxicity and antimicrobial activity of a hydrocolloid dressing containing silver particles in an ex vivo model of cutaneous infection. *Skin Pharmacology and Physiology, 17*, 31–36.

Scott, N, Chen, H. (2003). *Nanoscale science and engineering for agriculture and food systems*. USDA National Planning Workshop (November 18–19, 2002), Washington, DC, 63 pp.

Sharma, K., Sharma, R., Shit, S., Gupata, S. (2012). *Nanotechnological application on diagnosis of a plant disease*. Int. Conf. on Advances in Biological and Medical Sciences, 1–2.

Sharon, M., Choudhary, A. K., & Kumar, R. (2010). Nanotechnology in agricultural diseases and food safety. *Journal of Phytolpathology, 2*, 83–92.

Shin, W. K., Cho, J., Kannan, A. G., Lee, Y. S., & Kim, D. W. (2016). Cross-linked composite gel polymer electrolyte using mesoporous methacrylate-functionalized SiO_2 nanoparticles for lithium-ion polymer batteries. *Scientific Reports, 6*, 26332.

Sofi, W., Gowri, M., Shruthilaya, M., Rayala, S., & Venkatraman, G. (2012). Silver nanoparticles as an antibacterial agent for endodontic infections. *BMC Infectious Diseases, 12*, 60.

Sotthivirat, S., Lubach, J. W., Haslam, J. L., Munson, E. J., & Stella, V. J. (2007). Characterization of prednisolone in controlled porosity osmotic pump pellets using solid-state NMR spectroscopy. *Journal of Pharmaceutical Science, 96*(5), 1008–1017.

Thomas, S., & McCubbin, P. (2003). A comparison of the antimicrobial effects of four silver-containing dressings on three organisms. *Journal of Wound Care, 12*, 101–107.

Tilman, D., Cassman, K. G., Matson, P. A., Naylor, R., & Polasky, S. (2002). Agricultural sustainability and intensive production practices. *Nature, 418*, 671–677.

Tournas, V. H. (2005). Spoilage of vegetable crops by bacteria and fungi and related health hazards. *Critical Reviews in Microbiology, 31*, 33–44.

Ullah, A. R., Joyce, H. J., Tan, H. H., Jagadish, C., & Micolich, A. P. (2017). The influence of atmosphere on the performance of pure-phase WZ and ZB InAs nanowire transistors. *Nanotechnology, 28*, 454001.

Vu, H. T., Keough, M. J., Long, S. M., & Pettigrove, V. J. (2015). Effects of the boscalid fungicide Filan® on the marine amphipod *Allorchestes compressa* at environmentally relevant concentrations. *Environmental Toxicology and Chemistry, 35*, 1130–1137.

Wang, L., Li, X., Zhang, G., Dong, J., & Eastoe, J. (2007). Oil-in-water nanoemulsions for pesticide formulations. *Journal of Colloid and Interface Science, 314*, 230–235.

Wang, S., Lawson, R., Ray, P. C., & Yu, H. (2011). Toxic effects of gold nanoparticles on *Salmonella typhimurium* bacteria. *Toxicology and Industrial Health, 27*, 547–554.

Wei, D., Sun, W., Qian, W., Ye, Y., & Ma, X. (2009). The synthesis of chitosan-based silver nanoparticles and their antibacterial activity. *Carbohydrate Research, 344*, 2375–2382.

Woo, K. S., Kim, K. S., Lamsal, K., Kim, Y. J., Kim, S. B., Mooyoung, J., et al. (2009). An in vitro study of the antifungal effect of silver nanoparticles on oak wilt pathogen *Raffaelea* sp. *Journal of Microbiology and Biotechnology, 19*, 760–764.

Xie, Y., He, Y., Irwin, P. L., Jin, T., & Shi, X. (2011). Antibacterial activity and mechanism of action of zinc oxide nanoparticles against *Campylobacter jejuni*. *Applied and Environmental Microbiology, 77*, 2325–2331.

Yan, J., Huang, K., Wang, Y., & Liu, S. (2005). Study on anti-pollution nano-preparation of dimethomorph and its performance. *Chinese Science Bulletin, 50*, 108–112.

Ying, J. Y. (2009). Nanobiomaterials. *Nano Today, 4*, 1–2.

Zaini, P. A., De La Fuente, L., Hoch, H. C., & Burr, T. J. (2009). Grapevine xylem sap enhances biofilm development by *Xylella fastidiosa*. *FEMS Microbiology Letters, 295*, 129–134.

Zeng, H., Li, X. F., Zhang, G. Y., & Dong, J. F. (2008). Preparation and characterization of beta cypermethrin nanosuspensions by diluting O/W microemulsions. *Journal of Dispersion Science and Technology, 29*, 358–361.

Chapter 9
Transgenic Approaches in Plants: Strategic Control for Disease Management

Siddra Ijaz, Maria Babar, Hafiza Arooj Razzaq, and Bukhtawer Nasir

Abstract Plants naturally interact with microbial communities in the ecosystem. In this biological interaction, some microbes are in good relationship (symbiotic) to plants, while others may pose threat to them. The phytopathogens cause numerous diseases, that plants counteract by their innate immune responses. Under different instances, however, they surrender against these agents. Hence, disease management strategies, either applying pesticides or by developing breeding programs for qualitative and quantitative disease resistance in plants, have been implemented. Since ancient times. However, actual advancement in science and technology deciphered these strategies into modern approaches. Thereby, development of transgenic plants is a powerful strategy for disease management. In this chapter the development of transgenic plants for disease management using different approaches is elaborated in detail. Development of transgenic plants by nuclear and plastid transformations to introduce and even increase disease resistance through heterologous, homologous, and ectopic expression of transgenesis discussed. Moreover, a comprehensive detail of transgenesis is described, involving insertion of pathogenesis related (PR) genes (antimicrobial), R genes, genes to improve physical barriers (structures). Mechanisms exploited are shown, including the modulation of signaling pathways expression, controlling qualitative and quantitative resistance traits, and overexpression of transcription factors, involved in defense pathways.

Keywords Transgenic plants · R genes · Pathogenesis related (PR) genes · Plant microbe interaction · Nuclear transformation

---

S. Ijaz (✉) · M. Babar · H. A. Razzaq · B. Nasir
Centre of Agricultural Biochemistry and Biotechnology (CABB), University of Agriculture Faisalabad, Faisalabad, Pakistan

© Springer Nature Switzerland AG 2020

I. Ul Haq, S. Ijaz (eds.), *Plant Disease Management Strategies for Sustainable Agriculture through Traditional and Modern Approaches*, Sustainability in Plant and Crop Protection 13, https://doi.org/10.1007/978-3-030-35955-3_9

9.1 Introduction

Substantial increase in global crop production is the current need to fulfill the demand of a rapidly growing population that is exceeding more than 7 billion. Approx. 800 million people are from developing countries, most suffering a severe food deprivation for survival. According to US Census Bureau 2015, world population is expected to exceed 9 billion by 2050. A significant loss in the potential yield of crop plants is due to diseases caused by as fungi, bacteria, viruses etc. Plant diseases contribute 15% loss in the food production worldwide (Oerke and Dehne 2004).

Trade and globalization are becoming the major factors exacerbating plant diseases impact, by spreading phytopathogens which cause emergence of new diseases in previously free areas. Crop yield reductions due to diseases are much higher in developing countries, bringing the food shortage scenario to the least level where it becomes famine and ultimately threatens the survival of people. Plant diseases, however, may be managed by phytosanitory measures and crop husbandry, with the implementation of advanced technologies (Herrera-Estrella et al. 1997).

The use of agrochemicals for disease management has potent limitations. The chemicals (pesticides, fungicides etc.) are undoubtedly expensive as well as have hazardous impacts on the ecosystem. Thus, they come under rigorous regulatory control (Dunwell 2000). The extensive application of chemical control may also results in the co-evolution of resistant phytopathogens and ultimately become ineffective for these new pathotypes (Herrera-Estrella et al. 1997). The detrimental plant microbes pose indeed great challenge to plant scientists, who have responsibility to fight against food production threats, worldwide.

An alternative strategy to chemical control is dwelling in the defensive capabilities of the crop plants, as well as to strengthen their defensive competences naturally active. By opting this way, the use of chemicals to control phytopathogens may diminish (Pinstrup-Andersen and Babinard 2001). Plants produce diverse antimicrobial compounds as an arsenal of their defense mechanisms upon pathogens attack, that secure from succumbing to the invading microbes. Plant-pathogen interactions result in the activation of plant defense mechanisms that culminate in a variety of physicochemical changes in the host. Invasion of pathogens leads to the alteration of plant cell wall, callose deposition and phenolic compounds accumulation such as phytoalexins, protease inhibitors etc. (Broglie and Broglie 1993a, b).

Genetic improvement through conventional plant breeding is a cost effective way of managing diseases in crop husbandry (Evenson and Gollin 2003). The theory of plant breeding is based on the crossing between best individuals and then expecting improvements in one or more traits. This occurs in spite of uncertainties about genetic changes responsible for resultant traits (Tester and Langridge 2010). Plants secondary metabolites (phenolics, tannins, saponins glucosinolates, alkaloids etc.) are the components of defense systems, though their presence has otherwise negative effects on consumption, such as bitter taste, indigestion etc. Hence, efforts of breeders focus on the reduction of these compounds to improve food quality but thereby plants are prone to diseases. Above and beyond this scenario, inadvertent loss of genes conferring resistance may due to breeders' lack of knowledge while selecting desirable germplasm. The loss of genes for resistance from germplasm has been noticed upon encounter devastating outbreak of diseases. Hence, It makes sense to put back the resistance genes into the crop plants from where they were lost, either accidently or deliberately.

The exploration and utilization of innate resistance mechanisms in plants is undoubtedly the best way to handle pathogens and pathogenesis, in an ecofriendly manner. The significant to this approach is to develop and deploy new resistant cultivars in the field in its best-required time, when resistant cultivars overcome by phytopathogens. However, genetic restrictions with conventional plant breeding approaches may impede this process, which becomes a rate-limiting step in mitigating losses attributable to epidemics.

Since last 20 years, tremendous development has been made in understanding the molecular interaction between plants and microbes. Along concomitant development of transgenic technology, knowledge on the plant-pathogen interaction mechanism has been translated into numerous novel approaches, which may speed up the conventional breeding to develop disease resistant plants. Hence, in this chapter, transgenic technology is discussed as a strategic control to manage the plant diseases.

9.2 Transgenic Technology

Transgenic technology involves the genome engineering which is based on Recombinant DNA Technology, which widens the range of alteration in plant genome in a more targeted way, thus saving input time. Various methods for transformation have been developed for monocots and dicots, which include direct insertion through biolistic or through natural genetic material, exchanged with the help of *Agrobacterium*. Gene transfer in plants through vector system was initiated when molecular mechanism behind tumor formation was clarified. It was observed that members of the genus *Agrobacterium*, which has the potential to infect dicot plants, contain an extrachromosomal plasmid which naturally penetrates into the host cell, developing a T-DNA in the host genome. The plasmid of *A. tumefaciens*, which induces tumors in the host, is known as pTi plasmid. Due to the integration of bacterial plasmid genes in the plant genome, an alteration in cell metabolism occurs which results in the tumor formation. The genes involved in tumor formation, as well as those encoding for opines, have been identified through genetic analysis. T-DNA transfer and its integration in the host cell is greatly controlled by vir-genes of *Agrobacterium*, located in the Ti plasmid. The T-DNA segment contains structures of 25 base pairs named as left and right border, on both of its ends. The DNA fragment which is to be transferred is ligated in between these borders. The expression of foreign gene in the host plant is obtained by using modified strains of *Agrobacterium* which integrate the foreign gene as well as its expression.

In addition to *Agrobacterium*-mediated transformation, biological biolistic method is the direct technology of gene transfer. In this method, the plant cell is fired with special gun by metallic micro particles, which are coated with vectors containing the desired gene. This method has been used for stable transformation in rice, wheat, maize and some other crops.

Electroporation is another DNA delivery method based upon the permeability of the cell membrane and application of high voltage. When voltage is applied on the plant protoplasts and transforming DNA mixture, the permeability of the cell membrane alters and allows the penetration of DNA.

In microinjection of DNA delivery, micro needles (2 micron in diameter) are used. The total efficiency of this method is around 10–20%.

9.3 Transgenic Plants and Fungal Disease Management

Fungi are eukaryotes basically categorized into three distinct groups i.e. eumycota, oomycetes and slime molds. They are considered to be ancient with early fossils evidence around 460-455 M years old (Redecker et al. 2000). The exact number of fungal species is not known, but at least 99,000 species have been discovered, and 1200 are identified annually (Blackwell 2011). Fungi associate with plants in quite distinct manners: it can be friendly as seen in symbiotic relationship, but fungi can be worst enemies of plants and cause deadly diseases and destroy them, completely. Fungi are also associated with plants as decomposers and utilize only their remain to fulfil their nutrients (Knogge 1996).

Among fungi, Oomycetes are the most destructive ones and proved to be their worst enemies. A number of diseases are caused due to the fungi from this group. Back in 1845, late blight of potato by *Phytophthora infestans* brought huge losses which resulted in severe starvation and famine. Similarly, sudden oak death due to *Phytophthora ramorum*, damping off caused by *Pythium* spp., downy mildews and white rust are a few examples of fungal diseases (Fry and Grünwald 2010). Fungal infections occur through wounds resulting in lesions and necrosis in a precise area or the whole plant. Fungi are responsible for number of root and crown diseases, for example, some root rots are caused by members of Basidiomycota including *Armillaria* and *Heterobasidion.* Among vascular wilt causing fungi, *Verticilium dahliae*, *Fusarium oxysporum* and *Verticilium albo-atrum* are the most important ones. Panama disease is one important and famous vascular disease caused by *F. oxysporum* f.sp. *cubense*, which destroyed banana production very badly in Latin America in early XX century. Gros Michel was found to be a highly susceptible cultivar towards Panama disease. Later, Cavendish cultivar saved banana industry. Unfortunately, today a new strain of Panama disease, the Tropical race 4 of *F. oxysporum* f.sp. *cubense*, is devastating the susceptible Cavendish cultivar, raising high concerns among scientists and producers (Ploetz 2005).

9.3.1 *Fungal Invasion and Plant Defense Strategies*

The primary step of any pathogen during infection is host-pathogen communication (Boyle and Finlay 2003). Plant lacking the specific recognition signals or factors on their cell surface may not be infected by the specific pathogen. These recognition molecules include different types of polysaccharides and some glycoproteins. Fungi developed very specialized structures namely, haustoria and appressoria, to invade the host plant (Zeilinger et al. 2015). The appressorium plays a crucial role in developing infection as it contains melanin in its cell wall. The appressorium degrades the plant cell membrane by generating a hydrostatic turgor through glycerol accumulation in the melanin layer, while its infection process includes various enzymes. The appressorium matrix contains enzymes, such as cellulase and cutinase, which help fungal hyphae to penetrate the host cell wall, which afterwards invade the epidermal cell layer of the leaf and differentiate into bulbous and lobed infectious hyphae. *Magnaporthe grisea* (rice blast causal agent) produces appressoria which puncture the rice leaf cuticle resulting in its invasion (Mitchell et al. 2004; Zeilinger et al. 2015). There are also some exceptional cases such as *Cladosporium fulvum* (a biotrophic fungus), which does not produce haustoria but grows in apoplast (the space outside the plasma membrane) and relies on that for nutrients.

In response to virulence factors from phytopathogens, plants trigger their immune response (Chisholm et al. 2006; Jones and Dangl 2006). The first line of defense allows plants to hinder the pathogen entry and so-called non-host resistance. This defense line work either inhibiting pathogen penetration, depriving nutrients or other factors essential for proliferation. If pathogen gain access to enter into plant through cell wall, then they encounter by second line of defense (pattern recognition receptors) which is triggered by pathogen associated patterns (PAMPs) e.g. bacterial flagellin. In the battle between plant and pathogen, the strong pathogen (by delivering the effectors to suppress plant defense) only can overcome the basal defense by PAMP- triggered immunity (PTI) response. In this environment, the pathogens are arrested by another effective and rapid line of defense as effector mediated plant response, called effector triggered immunity (ETI) that is regulated by resistance genes expressing PR proteins for recognition of different pathogen effectors e.g. bacterial avr-gene (avirulence gene). As a result of this triggered immunity responses, several genes are switched on to contribute in the inhibition of pathogen growth are referred as *R*-gene mediated resistance (Parker et al. 2001; Dangl and Jones 2001; Chisholm et al. 2006; Jones and Dangl 2006).

(a) **First line of defense**

In any organism, resistance is the capability to overcome the damaging effects of an invading pathogen (Agrios 1988). Plant disease resistance is represented by few symptoms, which actually show the failure of a pathogen to grow and spread in the plant, which often occurs in the form of a hypersensitive reaction.

Plants exhibit first line defense mechanisms in the form of physical barriers i.e. wax, cell wall containing lignin, cutin and bark. These physical barriers protect the plant form pathogen invasion as well as provide rigidity throughout its life (Van Baarlen et al. 2007). Additional to these preventive structures, plants also secrete various chemicals, which inhibit the pathogen activities, acting directly through lytic effects. The chemicals include terpenoids, pyrethrins, diterpenoids, alkaloids, cyanogenic glycosides, atropines, phytoanticipines and phytohormones. Among these chemicals saponins, avanacins (in oats) and tomatine (in tomato) possess anti-fungal membrane degrading potential (Hammond-Kosack and Jones 1997). Those fungi, which do not possess saponinases enzymes, are unable to degrade saponin and are not able to cause any infection in the host plant.

Many plant proteins have potential to inhibit pathogen proteinases and several other degrading enzymes. Plants also contain lectin protein, which get attached to some sugars resulting in lysis and degradation of various fungi. Many hydrolytic enzymes are present on the plant cell surface, which help in degradation of fungal cell wall components (Freeman and Beattie 2008).

Even, if by passing all these first defense line, a pathogen gets entry into the host then some plant cellular and biochemical defense mechanisms get activated. They may be specific or non-specific to the invading pathogen. On the plant cell membrane, receptor proteins are present which are able to sense the pathogens and their secreted factors. The activation of plant defense mechanisms is genetically controlled (Piña-Vázquez et al. 2012). Once the pathogen successfully develops an infection, the plant defense system strives to create hurdles for colonization and further penetration. For example, in lignification, the cell wall (lignified) develops a barrier against hyphal progression into cells. Due to the rigidity and insolubility, lignin is the best barrier against fungal attacks. There are examples of crops which possess lignin and have inhibited fungi invasion, i.e. potato vs *Phytophtora infestans*, wheat vs *Septoria nodorum* and cucumber, vs *Cladosporium cucumerium* (Jones and Dangl 2006).

(b) **Second line of defense**

Induced resistance is quite a common phenomenon in plants in which, upon pathogen inoculation, an increase in resistance has been observed. In host plants, the induced biochemical activities are considered to be the last line of defense against pathogens. The biochemical induced plant defense immunity contains two stages/layers i.e. PTI and ETI. PTI is induced by the detection of the invading pathogen's conserved PAMPs, such as EF-Tu (elongation factor Tu) through PRRs (pattern recognition receptors) present on plant cell surface. PTI plays a role in the activation of primary resistant responses such as closure of stomata, MAPK cascades activation, generation of reactive oxygen species and transcription of genes involved in resistance. ETI has been considered as an the amplified form of PTI which is activated through plant R (resistance) genes after the detection of pathogenic factors. It is linked to hypersensitive response of plant (Hou et al. 2011; Mandadi and Scholthof 2013). Pathogens, on its entry into host plant, secrete Type III secreted effectors (T3Ses) which are virulence proteins and play a role in pathogenesis by disturbing the normal chemical and biochemical processes of the plant (Grant et al. 2006; Block et al. 2008). The T3SEs confer virulence potential by lowering or suppressing plant PTI and ETI responses (Grant et al. 2006; Guo et al. 2009). For example, HopAI1 inactivates the defense mechanism of the cell wall as well as the transcription of PAMPs related genes, through MAPK3 and MAPK6 inactivation, while 10 HopU1 was found to suppress the host immunity by interfering with plant defense related genes (Zhang et al. 2007).

Induced resistance was firstly observed in tobacco infected by tobacco mosaic virus, and was not limited to the area of lesions but spread to other plant parts (Ross 1961). Not only the mature plant parts but also the new leaves manifested induced resistance, showing that some kind of signal was produced on primary infection. It was later amplified and triggered the plant response more efficiently against infection (Bozarth and Ross 1964). Various studies revealed that induced resistance is non-specific (Hammerschmidt and Kuc 1995), as shown by the primary infection of *Colletotrichum lagenarium* in cucumber that enhanced resistance vs other pathogens as well (Kuć 1982). Induced resistance subsequently lowers the expression of disease symptoms, which results in no infection appearance.

9.3.2 Transgenic Technology: A Strategic Control for Fungal Diseases in Plants

Nucleic acids regulate plant defense and immunity. Resistance against pathogens increases by epigenetic changes which occur through response proteins, and by lowering the response inhibitors. Chromatin modifications result in changes of the expression of different plant defense-related components (Jaskiewicz et al. 2011). Plant defense is greatly affected by processes such as DNA and histone methylation, and RNA interference (Holeski et al. 2012). Epigenetic regulations increase the host response towards the specific threats so epigenetic factors are considered to be another way for regulation of resistance in plants. In addition, PDR (pathogen derived resistance) is conferred through inoculation of plant with a less virulent strain of the pathogen, which makes the plant more resistant towards severe virulent strains, as shown by the progeny of barley plants that exhibited higher resistance against *Rhynchosporium commune* due to the chemical elicitors present in their parental plants (Burketova et al. 2015).

Through genetic engineering, transgenic crops were developed which exhibit resistance against fungi by expressing genes encoding exogenous phytoalexins. For example, Casben synthetase, which is involved in the biosynthesis of the phytoalexin casben, is induced in plant tissues upon exposure to fungi or fungi-related factors. By transforming the gene, which encodes for casben synthetase in a transgenic development, enhanced plant resistance was observed. The resistance against fungi can be increased by engineering the plant genes which encode for such products, which alter and modify phytoalexin structures making them more toxic to fungal diseases such as early blight and Fusarium wilt (Punja 2006).

Various strategies have been applied to develop transgenic plants by transforming the genes which encode for glucanases and chitinases, which in turn develop disease resistance. For example wheat plants, engineered with chitinase and b-1,3-glucanase genes, showed resistances against powdery mildew (Bliffeld et al. 1999). Similarly, *Brassica* sp. lines engineered with barley chitinase and β-1,3-glucanase genes exhibited enhanced resistance against *Leptosphaeria maculans* (Melander et al. 2006).

Antimicrobial peptides (AMPs) can be generated in crops for resistance against fungal pathogens. These compounds may be expressed in a stable way in plants through biolistic or *Agrobacterium* mediated. Moreover, several plants have been engineered with AMP genes to boost their resistance against various fungi. AMPs introduced in potato showed complete resistance against *Verticilium dahlia*. Similarly, rice and barley were transformed with these genes, which showed resistance against fungal pathogens. The most commonly used AMP gene is Wj-AMP1, isolated from *Wasabia japonica*. This gene was expressed in plants of *Nicotiana benthamiana* which exhibited higher fungal resistance. Moreover, it also conferred resistance against *Magnaporthe grisea* and *Botrytis cinerea*. Similar results were observed for potato, which showed resistance against *B. cinerea*. Example of Wj-AMP1 transgenes also include its expression in the orchid *Phalaenopsis* against *Erwinia carotovora* and in rice for resistance against *Magnaporthe grisea*. For resistance enhancement, suitable promoters play crucial roles as well. In most transgenic plants, CaMV 35S or UBQ (constitutive promoters) are used for AMP gene expression. Increased antimicrobial potential have been observed in fused AMPs, for example, in *Arabidopsis*, CWP2 was fused to RsAFP, resulting in an increased resistance towards *F. oxysporum* (Ribeiro et al. 2013).

PRs (pathogenesis related proteins) are plant proteins whose expression is triggered during a pathogen attack. These proteins have not been identified due to their anti-pathogen activity but because of their occurrence in attacked/infected plants. Eleven PRs families have been discovered and nine plant families have been identified containing PRs proteins (Van Loon et al. 1994). In plants, the PR families-4, PR families-5, thionins and ribosomes inactivating proteins exhibit antifungal properties (Vigers et al. 1991; Linthorst and Van Loon 1991). These proteins are mostly found in storage tissues and organs such as seeds as well as tubers (Broekaert et al. 1995).

In the past few years, it was shown that plants produce specific proteins in response to different environmental changes and stress, among which the most popular ones are the heat shock proteins, induced at the time of critical temperature rise from optimal level (Vierling 1991). Various types of proteins are induced during different environmental conditions in order to overcome deleterious effects on plant survival. In cold environments, alfalfa produces a set of proteins which help in decreasing the harsh effect of low temperatures on the plant membrane.

Abscisic acid, a plant hormone, enhances resistance to freezing conditions by inducing a similar set of proteins (Mohapatra et al. 1988; Heino et al. 1990). The pathogenesis-related proteins may also be categorized as stress proteins, as they are induced in response to some kind of stress during pathogen attack, playing critical role in acquired resistance (Van Loon 1989). PRs exhibit anti pathogenic activities as well (Linthorst and Van Loon 1991; Van Loon et al. 1994). As for fungal cell wall degradation, chitinases can be combined with glucanases. PR-9 peroxidases could be involved in cell wall strengthening through lignin accumulatin. PR 1 and PR 5 are strongly induced and it seems that they have some kind of interaction with membranes. Their specific function, however, has not been clarified yet.

The PRs induced in plants are mostly acidic. The basic type of PRs is present in a vacuole in smaller amounts, while at the time of a pathogen's attack and infection, they are induced as well expressed in specific tissues, in controlled amounts (Eyal and Fluhr 1991; Linthorst and Van Loon 1991). Intercellular PRs proteins make up the first line of defense against infection. In case of failure in plant protection the vacuole accumulated proteins are released as a second line of defense. This results in the engulfment of invading pathogen cells through lytic enzymes (Mauch and Staehelin 1989). In older foliar and floral parts, the constitutive expression of such proteins may be regarded as a defense mechanism against challenging pathogens. Tissue specific expression of PR genes reveals that these defense proteins might also play a crucial role in normal plant growth, metabolic and developmental processes.

In transgenic plants, the constitutive expression of PRs can reduce growth of pathogen and the appearance of symptoms (Ryals et al. 1994). It might be also possible that pathogens evolved such kind of mechanisms capable to lower the PRs effect on them. Due to those modified mechanisms, fungi, which contain chitin in their cell wall, do not get damaged from chitinases when their cell wall chitin is protected by a layer (Schlumbaum et al. 1986). Symptoms due to *Rhizoctonia solani*, a soil borne fungus, decreased in canola or tobacco plants containing vacuolar chitinase (class I) from bean, cucumber or tobacco PR-8 (Lawton et al. 1993). However, increased chitinase showed no protection against *Cercospora nicotianae* (Neuhaus et al. 1991). The *in vitro* as well as *in vivo* antifungal activity of chitinases greatly increase in presence of 1,3-glucanases, as observed in tobacco in which co-expression of chitinase and glucanases boosted its resistance against *C. nicotianae* (Zhu and Weir 1994; Jach et al. 1995). Similarly, the combined expression of chitinase PR-3d and glucanases PR-2e enhanced resistance of tomato towards *F. oxysporum* while the transgenic plant expressing only one of these genes failed to resist against this fungus (Jongedijk et al. 1995).

However, the proteins targeted in the apoplast were not found to be very effective, showing that tonoplast leakage is crucial to inhibit pathogen growth and progress. Moreover, the acidic tobacco chitinase (PR-3a) in combination with glucanases (PR-2b) was also not effective. A similar scenario was observed for PR-1 and PR-5 which exhibited antifungal potential against oomycetes lacking cell wall chitin. PR-1a constitutively expressed in tobacco lowered *Peronospora tabacina* as well as *Phytophthora parasitica* symptoms, but was not effective against *C. nicotianae* which is not an oomycete (Alexander et al. 1993).

Through genetic engineering, various classes of PR protein genes have been transferred to a variety of crops which, as a result, became resistant against different fungal pathogens. To enhance resistance in sweet orange against *Ph. citrophthora*, a

PR5 gene from tomato has been expressed in transgenic plants (Fagoaga et al. 2001). The famous Fuji apple cultivar is greatly threatened by powdery mildew disease caused by *Podosphaera leucotricha*. To make this cultivar resistant the NPR1 gene was introduced. The NPR1 (MhNPR1) gene interacts with powdery mildew which in turn enhances the resistance of the Fuji apple (Chen et al. 2012). In transgenic tomatoes and carrots, AtNPR1 gene was overexpressed in order to make them resistant against some fungal phytopathogens (Lin et al. 2004). Similarly, in grapevine, VvNPR1 gene was overexpressed to confer resistance against a number of fungal pathogens.

In addition to PR gene, also the chitinases confer resistance against fungal pathogens. Chitinases are involved in the degradation of chitin, resulting in resistance to fungi. Genes encoding for chitinases have been introduced in multiple crops for the control of fungal diseases. The CHiC gene of tobacco has been introduced in carrot against *B. cinerea* (Punja and Raharjo 1996). A rice chitinase gene, namely RCC2, was isolated and transferred to strawberry which showed resistance against *Sphaerotheca humuli* (Asao et al. 1997). Transgenic litchi plants were developed through the introduction of a rice chitinase, which showed high resistant responses, as compared to non-transformed (normal) plants (Das et al. 2012). To combat guava wilt disease caused by *F. oxysporum*, a *Trichoderma* endochitinase gene was transferred to guava (Mishra et al. 2016). Glucanases are an enzymes degrading the glucan component of the fungal cell wall. Transgenic potato plants showed boosted resistance against *Ph. infestans* after expressing soybean glucanases (Borkowska et al. 1998). Over expression of the soybean β-1,3-glucanase gene in kiwi fruit showed increased enzyme activity against *B. cinerea* (Nakamura et al. 1999). The introduction of alfalfa glucanases in brinjal made them resistant to *Verticillium dahliae* (Singh et al. 2014).

Defensin is an AMP which plays crucial role in defense mechanism against fungi. A transgenic tomato transformed with the defensin gene of *Capsicum annuum*, showed increased resistance vs various fungal pathogens (Zainal et al. 2009). RIP (plant ribosome inactivating proteins) interfere with elongation factor and inactivate it by binding to ribosomes. Some RIPs have been observed to cause toxicity towards fungi (Stirpe et al. 1992). A RIP gene introduced in tobacco increased its resistance against *R. solani* (Logemann et al. 1992). A transgenic tobacco developed through the combination of a RIP and a chitinase exhibited enhanced resistance to *R. solani* (Jach et al. 1995).

The economic losses due to oomycetes are high and various strategies have been applied for their control. The best ones are based on resistance ® genes and RNA silencing of pathogen transcripts. Plants R genes are complementary to avirulence (Avr) genes of pathogens, and such kind of interaction between host and pathogen is called incompatible For a compatible interaction there exist no complementation between R and Avr genes which result in disease appearance. After the introduction of a gene to gene model (Flor 1954), 11 different R genes in potato were generated through introgression, named as MaR1-MaR11 (Black et al. 1953). Among them, R1, R3a and R10 were successfully used in European breeding programs (Ballvora et al. 2002). Potato plants possessing R3a genes show resistance against AVR3a of *Ph. infestans* (Voegele et al. 2001).

Unfortunately, the resistance acquired through R genes in breeding programs is quickly overcome by invading pathogens as these possess a wide range of effector receptors. Screening of specific R genes through marker assisted selection is quite time consuming and hectic. The potential of any R gene and its specificity for a specific effector can be verified through co-expression in plants deprived of resistance genes. After identification of specific R genes, these must be applied in the field through R gene stacking and mix varieties.

Among R genes, NLRs (NOD-like receptor) genes form a complex gene families. Their products change their state from a stable, inactive ADP-bound form to an ATP releasing state, upon detection and interaction with pathogens' effectors, triggering the subsequent NLR activation (Borrelli et al. 2018). NLRs are commonly present in cytoplasm and a contain nucleotide-binding (NB-ARC) and a leucine-rich repeat (LRRs) domain, located on the C-terminal. LRRs have the potential to detect various effector structures. Coiled coil (CC)(NBS-LRR, CNL) genes are found in monocots as well as dicots, but TOLL/interleukin-1 receptor (TIR) TNL genes (TIR-NBS-LRR genes) are only found in dicots (Jacob et al. 2013). The translocation of NLR genes to an unlinked locus enhances its functional range (Wu et al. 2017). On interaction with pathogenic effectors, NLRs become activated (Dodds et al. 2006).

Along with PR and R genes, RNA silencing is a technique which generates such hairpin RNA transcripts, which specifically target pathogen virulence genes. This technique has been used in control of fungi such as *Phytophthora* spp. (Vega-Arreguín et al. 2014).

9.4 Transgenic Technology for Bacterial Diseases in Plants

Among the huge diversity of bacteria, only a few of these prokaryotic microscopic organisms are known to cause diseases in plants. *Erwinia amylovora* was the first plant bacterial pathogen identified as the causal agent of apple fire blight (Burrill 1878). Bacterial diseases have an economic impact as they affect several agricultural crops including cereals, vegetables, fruits etc. In some cases, the protective agrochemical technologies are enough to control their damage but not so effectively. On the other hand, chemical control agents are the sources of potential damage to the environment. Occurrence of plant germplasm resistant to several bacterial diseases aids in the development of resistant cultivars. Antimicrobial agents, active against a huge spectrum of bacterial strains threatening plants, have been also explored (Al Akeel et al. 2014).

Use of resistant germplasm is a promising approach with potential to control diseases more effectively. Resistant germplasm sources are often been found in wild relatives or landraces of crop plants (Hegde and Mishra 2009). Classical hybridization strategies are limited to intra-species crosses, though for transferring genes conferring disease resistances from wild type to commercially available cultivars, a complex, laborious and time consuming process is needed. Current advancement in plant transformation strategies is due to unequivocal understanding of the plant-pathogen biological interaction. By the use of transgenic technology, a number of strategies for resistant plant development against various bacterial diseases have been proposed, including inhibition of bacterial pathogenicity, virulence factors inhibition, boosting up plant defenses and artificially induced programmed cell death (Dzhavakhiya and Shcherbakova 2007).

9.4.1 Mechanism of Plant Bacterial Interaction

Phyllosphere and rhizosphere are both a microbial habitat (Lindow and Brandl 2003). A number of bacteria may survive and even bloom on the plant exterior as epiphytes. In plant pathogenesis, bacterial pathogens must first enter the host plant tissues. Differing from fungal pathogens, bacteria lack the ability for plant epidermal penetration and depend on accidental wounds or natural openings to penetrate the host internal tissues. However, the molecular mechanism of bacterial entry via natural openings is not identified yet. The microscopic surface openings, i.e. stomata, are assumed to be a passive port for bacterial entrance, thus playing a vital role in defense mechanism. The stomatal closure serves as a barrier against invasion. In *Arabidopsis*, the guard cells of stomata start producing NO_2 and OST1 kinase (guard cell specific) upon interaction with bacterial surface molecules (requiring FLS2 receptors). Hence, to overcome this scenario, bacterial pathogens have continuously evolved to produce several specific virulence factors to affectively re-open the stomatal openings for pathogenesis (Melotto et al. 2006).

Presently, virulence studies on several pathogenic bacteria have been mainly motivated by host-pathogen interactions (once inside host tissues). *Pseudomonas syringae* is used as a model to unravel various fundamental mechanisms required for host-bacterial infection (Ausubel 2005; Chisholm et al. 2006). Several strains of *P. syringae* cumulatively have been studied on a number of host plant species. Phytopathogenic bacteria rapidly evolved their virulence factors by modulating associated genomic regions such as the *hrp* gene encoding type III secretion system, to undermine host defense (Greenberg and Vinatzer 2003; Nomura and He 2005). This is an effectively system used by bacterial pathogens to inject virulence effectors in the host cell. However, it is not the only factor causing diseases in plants. Several phytotoxins produced by pathogenic bacteria are necessary for maximum virulence, such as coronatine (COR), a polyketide toxin produced by lethal *P. syringae* DC3000, causing infection in tomato and *Arabidopsis* (Brooks et al. 2004; Cui et al. 2005).

9.4.2 Engineering Plant Genomes for Bacterial Resistance

Due to the remarkable development in the field of plant transformation, various transgenic crop plants have been released for commercial production. Transgenic crops presently marketed have resistant traits to herbicides, insect pests, and pathogenic diseases, as well as improved oil quality and pollination control (Mitten et al. 1999). Genes conferring agronomically valuable traits have been targeted and introduced into several crops. Banana (*Musa acuminata*) is being cultivated in more than 120 countries, and among them India is one of the most important producers. However, banana crop production has been severely affected by the Banana Xanthomonas wilt (BXW) (Tripathi et al. 2009). Two banana cultivars of, "Nakinyika" and "Sukali Ndiizi", have been genetically engineered with a pepper gene (plant ferredoxin-like protein, Pflp), for resistance against *Xanthomonas campestris* pv. *musacearum*, a causal organism of BXW (Namukwaya et al. 2012). In the bioassay of transgenic bananas, 67% of them showed resistance to BXW with no disease symptom, in comparison to wild-type plants, which showed severe disease symptoms. The Pflp are important photosynthetic proteins that elicit the defense response in several crops against bacterial challenges. These Pflp had been introduced in tomato, tobacco, *Oncidium* orchids, calla lily and rice against *Pseudomonas*, *Erwinia*, *Xanthomonas* and *Ralstonia* (Tang et al. 2001; Huang et al. 2007; Yip et al. 2007).

Heterologous transgene expression of sweet pepper *pflp* gene in *Arabidopsis* to underpin resistance against bacterial pathogens such as *Ralstonia solanacearum* and *Pectobacterium carotovorum* has been studied (Lin et al. 2010; Gir et al. 2014). A ferredoxin-I protein (pflp) of sweet pepper is associated to a HR (hypersensitive reaction) response which ultimately enhanced production of AOS (active oxygen species) (Dayakar et al. 2003). Galambos et al. (2013) transformed the hybrid of grapevine namely "Richter 110" (V. berlandieri × V. rupestris) with oncogene silencing of *Agrobacterium* to develop crown gall resistant lines. Besides, transgenic grape plant have been developed against Pierce's disease caused by *Xylella fastidiosa*, by overexpressing a rpfF gene (Lindow et al. 2014). Transgenic chili (*Capsicum annuum*) cv. "Nockwang" was successfully developed by introducing a gene, Tsi1 (tobacco stress-induced 1) into hopocotyl and cotyledon, as explant (Shin et al. 2002). The Tsi1 gene product plays a role in the regulation of several stress related genes. These transgenic chili showed high level of resistance against *X. campestris* pv. *vesicatoria*.

9.4.2.1 Engineering Systemic Acquired Resistance (SAR) Pathways Vs Bacterial Diseases

SAR is induced in plants in response to pathogen attacks, through a number of related genes that have a primary importance for their association to local or systemic immunity. Heterologous expression of salicylic acid responsive genes and several PR genes exhibit long term immune responses (Dong et al. 2004; An and Mou 2011). NPR (non-expressor of pathogenesis related) genes are associated to pathogenic attacks and play a central role in SAR responses. da Silva et al. (2018) suggested that NPR1 genes and its orthologs are key regulators of plant defense, thus representing a suitable choice in genome engineering for resistance, in several crops. Overexpression of *Arabidopsis thaliana* NPR1::35S (CaMV) gene cassette in *Solanum tuberosum*, *Lycopersicum esculentum*, *Oryza sativa*, *Brassica napus*, *Daucus carota* and *Fragaria ananassa* showed broad spectrum resistance against various phytopathogens, particularly bacteria (Table 9.1). Transgenic apple lines had been developed for overexpression of MpNPR1-1 (homologue of NPR1) gene in two important cultivars, M26 and Galaxy, under a constitutive promotor (CaMV 35S). This gene exhibited broad-spectrum resistance against several pathogens, including *E. amylovora*, the causal agent of the apple fire blight (Malnoy et al. 2007).

Huanglongbing (citrus greening) is a serious disease of citrus worldwide, caused by *Candidatus* Liberibacter spp. (Duan et al. 2009). To control citrus greening, the AtNPR gene was overexpressed in sweet orange cv. "Valencia" and "Hamlin", under constitutive promoters CaMV 35S and phloem-specific *Arabidopsis* SUC2 (AtSUC2), (Dutt et al. 2015). These transgenics showed reduced disease severity, even a few of them were observed as disease-free after 3 years of planting in a greening infested field. Overexpression of NH1 (NPR1 homologue) in rice showed high levels of resistance against *X. oryzae* pv. *oryzae* (Chern et al. 2005). Other orthologs of AtNPR1 (BnNPR1 and LhSorNPR1) have been overexpressed in several *Bassica napus* and Lilium longiflorum against bacterial attacks (Potlakayala et al. 2007; Wang et al. 2017).The NPR have significant role in plant SAR responses. However, SAR is a complex process controlled by a substantial set of genes and induced upon several physiological and biochemical processes (da Silva et al. 2018).

R- and PR genes (part of SAR responses), are the key components of the plant innate immune system (Ali et al. 2018). During infection the plant defense responses, especially PTI and ETI, are triggered by several array of genes activation and repression. For example, NBS–LRR type of *R* genes i.e. *RPS2* and *RPM1,* triggered disease resistance responses among two thousand genes that are differentially expressed (Tao et al. 2003). Overexpression of several R-genes has been observed among several transgenic crops. For example, the *pto* gene with a serine-threonine protein kinase identity, was introduced in tomato against *P. syringae* pv. *tomato*. RPS2, with a leucine-rich repeat protein and, RPM1 with a leucine zip-like protein identity, were introduced in Arabidopsis against *P. syringae* pv. *tomato*. The Xa7 gene was introduced in rice against *X. oryzae* pv. *oryzae* (Martin et al. 1993; Bent et al. 1994; Grant et al. 1995; Yang et al. 2000). A R gene, Bs4C-R, encoding a structurally unique protein that localizes in the endoplasmic reticulum membrane, and the AvrBs4 encoding for proteins for a hypersensitive response, were introduced in pepper and tomato against *Xanthomonas* spp. (Minsavage et al. 1999). The transgenesis by introducing Bs4C-R, from pepper species into rice, made the recipient plant resistant against bacterial blight (Wang et al. 2018).

Transgenesis has gained importance due to its remarkable defense capability. Among plant defense proteins, pathogenesis-related proteins (PR proteins) are

Table 9.1 Some transgenic plants developed against various bacterial pathogens

Genes	Bacterial pathogens	Transgenic plants	References
Ttr	*Pseudomonas syringae*	Tobacco	Anzai et al. (1989) and Batchvarova et al. (1998)
PTO	*P. syringae*	Tomato	Martin et al. (1993)
RPS2	*P. syringae*	Arabidopsis	Bent et al. (1994)
RPM1	*P. syringae*	Arabidopsis	Grant et al. (1995)
Lysozyme	*P. syringae*	Tobacco	Nakajima et al. (1997)
Attacin E	*Erwinia amylovora*	Apple	Reynoird et al. (1999)
Msr A1	*E. carotovora*	Potato	Osusky et al. (2000)
SP-cec B	*Xanthomonas oryzae*	Rice	Sharma et al. (2012)
Xa7	*X. oryzae*	Rice	Yang et al. (2000)
Myp30	General bacteria	Tobacco	Li et al. (2001)
Lactoferrin	*Ralstonia solanacearum*	Tomato	Lee et al. (2002)
PPO Cdna	*Pseudomonas syringae*	Potato	Li and Steffens (2002)
Tsi1	*X. campestris*	Chilli pepper	Shin et al. (2002)
MSI-99	Bacterial speck disease	Tomato	DeGray et al. (2001) and Alan and Earle (2002)
Lactoferrin	*E. amylovora*	Pear	Malnoy et al. (2003)
NH1	*X. oryzae*	Rice	Chern et al. (2005)
AtNPR1, BnNPR1	*P. syringae*	Canola	Potlakayala et al. (2007)
Caffeine alkaloid gene	*P. syringae*	Tobacco	Ashihara et al. (2008)
PG1, RC101	*E. carotovora*	Tobacco	Lee et al. (2011a, b)
Lysozyme	*E. chrysanthemi*	Potato	Rivero et al. (2012)
Pflp	*Xanthomonas*	Rice	Tang et al. (2001)
	Pectobacterium chrysanthemi	Oncidium orchids	Liau et al. (2003)
	Pseudomonas, Ralstonia	Tomato	Huang et al. (2007)
	E. carotovora	Calla lily	Yip et al. (2007)
	R. solanacearum, Pec. carotovorum	Arabidopsis	Lin et al. (2010) and Gir et al. (2014)
	X. campestris	Banana	Namukwaya et al. (2012)
rpfF	*Xylella fastidiosa*	Grape cv. Freedom	Lindow et al. (2014)
NPR1	*Candidatus* Liberibacter asiaticus	Sweet orange	Dutt et al. (2015)

(continued)

Table 9.1 (continued)

Genes	Bacterial pathogens	Transgenic plants	References
AtNPR1	*R. solanacearum*, *X. campestris*	Tomato	Lin et al. (2004)
	X. oryzae, *E. chrysanthemi*	Rice	Fitzgerald et al. (2004) and Quilis et al. (2008)
	X. hortorum	Carrot	Wally et al. (2009)
	X. citri	Citrus	Zhang et al. (2010), Dutt et al. (2015) and Boscariol-Camargo et al. (2016)
	X fragariae	Strawberry	Silva et al. (2015)
Thionin	Citrus canker	Citrus	Hao et al. (2016)
AtPDF1.1	*Pec. carotovorum*	Arabidopsis thaliana	Hsiao et al. (2017)
LhSorNPR1	*P. syringae*	Lily, tomato, Arabidopsis	Wang et al. (2017)
Bs4C-R	*X. oryzae*	Rice	Wang et al. (2018)
WRKY40	*P. syringae*	Arabidopsis	Chakraborty et al. (2018)
OsTGA2	*X. oryzae*	Rice	Moon et al. (2018)

important contributors for defensive processes against a variety of pathogens (Breen et al. 2017; Gamir et al. 2017). These proteins play a role in signaling of systemic acquired resistance (SAR) response (van Loon et al. 2006). Among 14 PR-1 proteins of cacao, TcPR-1f and TcPR-1 g mimic receptor-like kinase (RLK) proteins (Motamayor et al. 2013). Therefore, the kinase domains of TcPR-1f and TcPR-1 g were successfully cloned, expressed and purified from *Escherichia coli* BL21 (DE3)-R3 cells (Tosarini et al. 2018).

Another secretory PR protein, AtPDF1.1 (*A. thaliana* plant defensing type 1.1) controls iron homeostasis by chelating apoplastic iron. Iron deficiency induces the ethylene signaling pathway resulting into induction of plant defense mechanism. The overexpression of AtPDF1.1 in *Arabidopsis thaliana* showed increased resistance against *Pectobacterium carotovorum* (Hsiao et al. 2017). Over expression of thionin (PR protein) in citrus rootstock (Citrange Carrizo *Citrus triptera* × *Citrus sinensis*) enhanced resistance against citrus canker (Hao et al. 2016) because of the cysteine residues that induce characteristic opening of the pathogen cell membrane pores thereby resulting into the leakage of calcium and potassium ions from the cell (Pelegrini and Franco 2005; Oard 2011).

Antimicrobial peptides (AMPs), also called defense peptides, are ubiquitous and characterized by an α-helical structure, found in many organisms (DeGray et al. 2001). They are cysteine rich PR proteins that have versatile defense functions (Ali et al. 2018). Cercosporin as AMP interacts with bacterial membrane and produces pores. Previously, Huang et al. (1997) introduced SB-37 and MB-39 genes (natural cercosporin with its synthetic analogues) in tobacco plants that became resistant to several pathogens. In another study, Reynoird et al. (1999) introduced genes to confer bacterial resistance in apple against fire blight. However, in the case of tobacco the observed resistance was not effective against *P. syringae* pv. *tabaci* and *R. solanacearum* due to a lower expression of the transgene or host protease degradation. For that intracellular expression is needed, that was accomplished by Sharma et al. (2012), by introducing a SP-cecB gene construct into rice against *X. oryzae*, which showed intracellular secretion.

Cercosporin-based, smaller lytic peptides namely attacins, show also considerable resistance toward bacterial attacks. To overcome the attack of *E. amylovora* on apple transgenes expressing attacins gene showed resistance to bacterial fire blight (Reynoird et al. 1999). Similarly, apple cv. Galaxy, Royal Gala and M26 were transformed with introduction of attacin LP under a constitutive control, conferring resistance to fire blight (Aldwinckle et al. 2003). Moreover, the gene of attacin E has also been expressed in pear (*Pyrus communis*) to develop resistance to *E. amylovora* as well (Reynoird et al. 1999).

A further antimicrobial peptide with broad spectrum resistance capability, melittin, was introduced into potato against *E. carotovora* (Osusky et al. 2000). Megainins (MII) and their analogs are another example of defense peptides reported as effective against different pathogens (Jacob and Zasloff 1994). One of the magainin analog, namely Myp30, have been expressed in transgenic tobacco against bacterial and fungal pathogens (Li et al. 2001). MSI99, a synthetic analog of megainin, has been engineered in tomato to make them resistant to bacterial speck disease (Alan and Earle 2002; DeGray et al. 2001). The expression profiling of antimicrobial disulphide-bonded peptides, namely Protegrin-1 (PG1) and Retrocyclin-101 (RC101), have been investigated targeting the chloroplasts of tobacco (Lee et al. 2011a, b). To facilitate expression, GFP-fused peptides were expressed in transgenic plastids. The antibacterial activity of PG1 and RC101 expressed in tobacco was confirmed by *in-planta* bioassay with *E. carotovora*. Therefore, the engineering of these unique peptides in relation to several molecular forms, provides a powerful means for testing peptide chimerae (Fox 2013).

The engineering strategies against several bacterial virulence factors i.e. pectin enzymes, toxins, hormones and exo-polysaccharides, led to the decreased susceptibility or increased resistance to several diseases in plants. To enhance resistances against the tabotoxin produced by *P. syringae* pv. *tabaci* the ttr gene (tabotoxin resistance gene) was introduced in tobacco plants (Anzai et al. 1989; Batchvarova et al. 1998).

9.4.2.2 Expression of Transcription Factors and Small RNAs in Transgenic Plants Vs Bacterial Diseases

The promotor regions of plant defense genes have a prime importance in gene regulation. The numerous W-box elements (cis-regulatory), provide site for binding of *WRKY* transcription factors (DNA binding proteins). As pathogen attack stimulates a considerable alteration in the plant defense gene expression, in turn *WRKY* transcription factors binding produces signaling for both PTI and ETI defense lines (Rushton et al. 2010). It had been explored that *WRKYs* expressively contribute to

RBOH (respiratory burst oxidase homolog) transactivation as well as in ETI-induced ROS (Reactive Oxygen Species) bursts (Yoshioka et al. 2003). Chickpea *WRKY40* positively regulates several defense responses against pathogens. Therefore, over-expression of chickpea *WRKY40* (heterologous) in *Arabidopsis* under the control of *35S* promotor and *nos* terminator, enhances callose deposition and ROS production, while reduces *in-planta* multiplications of lethal *P. syringae* DC3000 (Chakraborty et al. 2018). In PR gene promotors a GCC-box have also been observed for binding of EREBPs (ethylene responsive element binding proteins) transcriptional factors. The ectopic expression of these factor gene enhances resistance against biotic and abiotic stresses. The *Tsi1* gene (*Tobacco-stress-induced gene 1*) encodes for EREBP and has sequence or domains homology that efficiently bind to the GCC box of the promotor region. The overexpression of *Tsi1* gene in tobacco lead to develop bacterial resistance against *P. syringae* pv. *tabaci* (Park et al. 2001). Another example of gene regulation, the TGA family (TGACGTCA cis-element binding proteins) associated with basic leucine zipper (bZIP) transcription factors, is mainly regulated by NPR proteins, which in turn boost PR gene expression ultimately resulting in disease resistance (Fan and Dong 2002; Fu and Dong 2013; Pirnia 2016). The members of TGA family are transcriptional activators of PR genes such as TGA2, TGA5 and TGA6. Overexpression of *OsTGA2* in rice show increased resistance against bacterial leaf blight of rice, caused by *X. oryzae* pv. *oryzae* (Moon et al. 2018).

To date, endogenous small RNAs are known to be successive controllers of gene expression among several cellular processes. Some current studies depicted small RNA-based gene silencing that might aid to reprogramming gene expression for plant immunity. In *Arabidopsis*, there the first recognized miRNA was miR393, part of the antibacterial PTI response, modulating the auxin signaling process (Navarro et al. 2006). The bacterial PAMP peptide (flg22) was induced by miR393 inoculation. miR393 targets the auxin receptor *TIR1* (*Transport-Inhibitor-Response 1*) and two of its functional paralogs (*AFB2* and *AFB3*), that declined upon treatment with flg22 . The over-expression of another paralog, namely *TIR*, increased disease sensitivity toward *Pst DC3000*. Overexpressing lines with miR393 instead inhibited bacterial growth.

RNAi technology is a key tool for regulation of the plant defense response against several pathogen. It works as switching off certain endogenous genes expression. The development of transgenic tomato, walnut, tobacco and apple plants producing hairpin RNA constructs against oncogenes, *ipt* and *iaaM*, of *Agrobacterium* led to resistant to crown gall pathogen (Escobar et al. 2001; Viss et al. 2003; Alburquerque et al. 2012).

9.4.2.3 Mammalian Proteins and Bacterial Resistance in Plants

The reduction of iron availability in transgenic plants is one of the key regulator to confer resistance against bacterial attacks. The well-known iron-chelating agent, lactoferrin, is one of the main glycoproteins involved in transgenic plant resistance.

Lactoferrin is naturally found in mammalian milk and shows a remarkable bactericidal action (Bortner et al. 1989). For this, transgenic tobacco plants were developed by introducing the lactoferrin gene that effectively show high resistance against *R. solanacearum* (Zhang et al. 1998). Similarly, partial resistance was observed in transgenic tomato against bacterial wilt (Lee et al. 2002). In another example, a transgenic pear was developed with resistance to fire blight by introducing bovine lactoferrin cDNA (Malnoy et al. 2003). Therefore, engineering of the lactoferrin gene into several cereal, vegetables, and fruit crops showed resistance to numerous bacterial pathogens (García-Montoya et al. 2012).

Another protein, the lysozyme enzyme, shows an antimicrobial activity by attacking the bacterial cell wall to breakdown the peptidoglycan component. The lysozymes of chicken, human and T4 bacteriophages have been reported for this bactericidal effect. However, upon engineering, very little expression of the lysozyme transgenes was observed, that made, however, plants less susceptible to pathogens (During et al. 1993). The susceptibility of potato toward *E. carotovora* has been lowered by introducing lysozyme gene of the T4 bacteriophage (Rivero et al. 2012). A human lysozyme gene was introduced in tobacco plants that exhibited fewer symptoms of *P. syringae* pv. *tabaci* (Nakajima et al. 1997). Similarly, a lower susceptibility to *E. chrysanthemi* (black leg disease) was developed in transgenic potato by introducing the chicken lysozyme gene (Hirai et al. 2004). However, significant levels of resistance in several crop plants has still to be achieved, against many bacterial diseases.

Other molecules involved in plant defense signaling show great importance toward disease resistance. For example, the enzyme polyphenol oxidases (PPOs) are abundant among flowering plants. PPOs catalyze the oxidation reaction of phenols to quinones. They are involved in plant defense processes from several pathogenic shocks. Furthermore, the PPO role in resistance was investigated in tomato overexpressing a potato PPO cDNA introduced via *Agrobacterium*-mediated plant transformation, under constitutive control of the CaMV 35S promoter (Li and Steffens 2002). An enhanced (around 30 fold) PPO transcript production and activity (around five to tenfold) was observed in this transgenic tomato. The tomato plants with PPO-overexpression exhibited higher resistance to *P. syringae* pv. *tomato*.

Another example concerns caffeine alkaloids, which have potential in chemical defense of plants against several pathogens (Uefuji et al. 2005; Kim and Sano 2008). Biochemical properties of transgenic lines (with increased production of caffeine) were investigated, showing that transcripts of proteinase inhibitor (PI) and PR proteins readily accumulated (Kim and Sano 2008). As PR proteins are usually related to disease opposition, therefore transgenic tobacco plants were developed for resistance to *P. syringae*. They showed stimulation of the expression of defense-related genes that was triggered by caffeine based, endogenous defense mechanisms(Ashihara et al. 2008).

9.5 Transgenic Plants: A Strategic Control of Viral Diseases

In *Arabidopsis*, induction of mutation by knockout of the DBP1 (DNA binding proteins phosphatase 1) genic region showed no effect on plants growth, though it yielded resistance against *Turnip mosaic virus* and plum pox virus (Castelló et al. 2010). The conformation of DBP1 domain is related to its involvement in transcriptional regulation and signal transduction processes in the cell (Carrasco et al. 2005). Other proteins, such as Tobamovirus multiplication 1 and Tobamovirus multiplication 3, have been identified as associated proteins with strong relation to the accumulation of *Youcai mosaic virus* in the plants (Kumar et al. 2012). The mutation in these genes has been found to be related to resistance to this virus (Yamanaka et al. 2002; Chen et al. 2007).

For expression of viral antigens in plants, plastid transformation has been employed successfully. In *Nicotiana tabacum*, the VP6 capsid protein (from Rotavirus) and the β subunit of the cholera toxin have introduced for developing transgenic plants expressing viral antigens.

References

Agrios, G. N. (1988). *Plant pathology* (3rd ed.). San Diego: Academic.

Al Akeel, R., Al-Sheikh, Y., Mateen, A., Syed, R., Janardhan, K., & Gupta, V. C. (2014). Evaluation of antibacterial activity of crude protein extracts from seeds of six different medical plants against standard bacterial strains. *Saudi Journal of Biological Sciences, 21*, 147–151.

Alan, A. R., & Earle, E. D. (2002). Sensitivity of bacterial and fungal plant pathogens to the lytic peptides, MSI-99, Magainin II, and Cecropin B. *Molecular Plant-Microbe Interactions, 15*, 701–708.

Alburquerque, N., Petri, C., Faize, L., & Burgos, L. (2012). A short-length single chimeric transgene induces simultaneous silencing of *Agrobacterium tumefaciens* oncogenes and resistance to crown gall. *Plant Pathology, 61*(6), 1073–1081.

Aldwinckle, H. S., Borejsza-Wysocka, E. E., Malnoy, M., et al. (2003). Development of fire blight resistant apple cultivars by genetic engineering. *Acta Horticulturae, 622*, 105–111.

Alexander, D., Goodman, R. M., Gut-Rella, M., Glascock, C., Weymann, K., Friedrich, L., et al. (1993). Increased tolerance to two oomycete pathogens in transgenic tobacco expressing pathogenesis-related protein 1a. *Proceedings of the National Academy of Sciences of the United States of America, 90*, 7327–7331.

Ali, S., Ganai, B. A., Kamili, A. N., Bhat, A. A., Mir, Z. A., Bhat, J. A., et al. (2018). Pathogenesis-related proteins and peptides as promising tools for engineering plants with multiple stress tolerance. *Microbiological Research, 212*, 29–37.

An, C., & Mou, Z. (2011). Salicylic acid and its function in plant immunityF. *Journal of Integrative Plant Biology, 53*, 412–428.

Anzai, H., Yoneyama, K., & Yamaguchi, I. (1989). Transgenic tobacco resistant to a bacterial disease by the detoxification of a pathogenic toxin. *Molecular and General Genetics, 219*, 492–494.

Asao, H., Nishizawa, Y., Arai, S., Sato, T., Hirai, M., Yoshida, K., et al. (1997). Enhanced resistance against a fungal pathogen *Sphaerotheca humuli* in transgenic strawberry expressing a rice chitinase gene. *Plant Biotechnology, 14*, 145–149.

Ashihara, H., Sano, H., & Crozier, A. (2008). Caffeine and related purine alkaloids: Biosynthesis, catabolism, function and genetic engineering. *Phytochemistry, 69*, 841–856.

Ausubel, F. M. (2005). Are innate immune signaling pathways in plants and animals conserved? *Nature Immunology, 6*(10), 973.

Ballvora, A., Ercolano, M. R., Weiß, J., Meksem, K., Bormann, C. A., Oberhagemann, P., et al. (2002). The R1 gene for potato resistance to late blight (*Phytophthora infestans*) belongs to the leucine zipper/NBS/LRR class of plant resistance genes. *The Plant Journal, 30*, 361–371.

Batchvarova, R., Nikolaeva, V., Slavov, S., et al. (1998). Transgenic tobacco cultivars resistance to *Pseudomonas syringae* pv. *tabaci*. *Theoretical and Applied Genetics, 97*, 986–989.

Bent, A. F., Kunkel, B. N., Dahlbeck, D., Brown, K. L., Schmidt, R., Giraudat, J., et al. (1994). RPS2 of *Arabidopsis thaliana*: A leucine-rich repeat class of plant disease resistance genes. *Science, 265*, 1856–1860.

Black, W., Mastenbroek, C., Mills, W. R., & Peterson, L. C. (1953). A proposal for an international nomenclature of races of *Phytophthora infestans* and of genes controlling immunity in Solanum demissum derivatives. *Euphytica, 2*, 173–179.

Blackwell, M. (2011). The fungi: 1, 2, 3… 5.1 million species? *American Journal of Botany, 98*(3), 426–438.

Bliffeld, M., Mundy, J., Potrykus, I., & Fütterer, J. (1999). Genetic engineering of wheat for increased resistance to powdery mildew disease. *Theoretical and Applied Genetics, 98*, 1079–1086.

Block, A., Li, G., Fu, Z. Q., & Alfano, J. R. (2008). Phytopathogen type III effector weaponry and their plant targets. *Current Opinion in Plant Biology, 11*, 396–403.

Borkowska, M., Krzymowska, M., Talarczyk, A., Awan, M. F., Yakovleva, L., Kleczkowski, K., & Wielgat, B. (1998). Transgenic potato plants expressing soybean ß-1, 3-endoglucanase gene exhibit an increased resistance to *Phytophthora infestans*. *Zeitschrift fur Naturforschung C: Journal of Biosciences, 1*, 1012–1016.

Borrelli, G. M., Mazzucotelli, E., Marone, D., Crosatti, C., Michelotti, V., Valè, G., & Mastrangelo, A. M. (2018). Regulation and evolution of NLR genes: A close interconnection for plant immunity. *International Journal of Molecular Science, 19*, 1662. https://doi.org/10.3390/ijms19061662.

Bortner, C. A., Arnold, R. R., & Miller, R. D. (1989). Bactericidal effect of lactoferrin on Legionella pneumophila: Effect of the physiological state of the organism. *Canadian Journal of Microbiology, 35*, 1048–1051.

Boscariol-Camargo, R. L., Takita, M. A., & Machado, M. A. (2016). Bacterial resistance in AtNPR1 transgenic sweet orange is mediated by priming and involves EDS1 and PR2. *Tropical Plant Pathology, 41*, 341–349.

Boyle, E. C., & Finlay, B. B. (2003). Bacterial pathogenesis: Exploiting cellular adherence. *Current Opinion in Cell Biology, 15*, 633–639.

Bozarth, R. F., & Ross, A. F. (1964). Systemic resistance induced by localized virus infections: Extent of changes in uninfected plant parts. *Virology, 24*, 446–455.

Breen, S., Williams, S. J., Outram, M., Kobe, B., & Solomon, P. S. (2017). Emerging insights into the functions of pathogenesis-related protein 1. *Trends in Plant Science, 22*, 871–879.

Broekaert, W. F., Terras, F. R., Cammue, B. P., & Osborn, R. W. (1995). Plant defensins: Novel antimicrobial peptides as components of the host defense system. *Plant Physiology, 108*, 1353.

Broglie, R., & Broglie, K. (1993a). Chitinase gene expression in transgenic plants: A molecular approach to understanding plant defence responses. *Philosophical Transactions of the Royal Society of London. Series B: Biological Sciences, 342*, 265–270.

Broglie, R., & Broglie, K. (1993b). Production of disease-resistant transgenic plants. *Current Opinion in Biotechnology, 4*, 148–151.

Brooks, D. M., Hernández-Guzmán, G., Kloek, A. P., Alarcón-Chaidez, F., Sreedharan, A., Rangaswamy, V., et al. (2004). Identification and characterization of a well-defined series of coronatine biosynthetic mutants of *Pseudomonas syringae* pv. tomato DC3000. *Molecular Plant-Microbe Interactions, 17*, 162–174.

Burketova, L., Trda, L., Ott, P. G., & Valentova, O. (2015). Bio-based resistance inducers for sustainable plant protection against pathogens. *Biotechnology Advances, 33*, 994–1004.

Burrill, T. J. (1878). Pear blight. *Transactions of the Illinois State Horticulture Science, 11*, 114–116.

Carrasco, J. L., Ancillo, G., Castelló, M. J., & Vera, P. (2005). A novel DNA-binding motif, hallmark of a new family of plant transcription factors. *Plant Physiology, 137*, 602–606.

Castelló, M. J., Carrasco, J. L., & Vera, P. (2010). DNA-binding protein phosphatase AtDBP1 mediates susceptibility to two potyviruses in *Arabidopsis*. *Plant Physiology, 153*, 1521–1525.

Chakraborty, J., Ghosh, P., Sen, S., & Das, S. (2018). Epigenetic and transcriptional control of chickpea WRKY40 promoter activity under *Fusarium* stress and its heterologous expression in *Arabidopsis* leads to enhanced resistance against bacterial pathogen. *Plant Science, 276*, 250–267.

Chen, C. H., Huang, H., Ward, C. M., Su, J. T., Schaeffer, L. V., Guo, M., & Hay, B. A. (2007). A synthetic maternal-effect selfish genetic element drives population replacement in *Drosophila*. *Science, 316*, 597–600.

Chen, X. K., Zhang, J. Y., Zhang, Z., Du, X. L., Du, B. B., & Qu, S. C. (2012). Overexpressing MhNPR1 in transgenic Fuji apples enhances resistance to apple powdery mildew. *Molecular Biology Reports, 39*, 8083–8089.

Chern, M., Fitzgerald, H. A., Canlas, P. E., Navarre, D. A., & Ronald, P. C. (2005). Overexpression of a rice NPR1 homolog leads to constitutive activation of defense response and hypersensitivity to light. *Molecular Plant-Microbe Interactions, 18*, 511–520.

Chisholm, S. T., Coaker, G., Day, B., & Staskawicz, B. J. (2006). Host-microbe interactions: Shaping the evolution of the plant immune response. *Cell, 124*, 803–814.

Cui, J., Bahrami, A. K., Pringle, E. G., Hernandez-Guzman, G., Bender, C. L., Pierce, N. E., & Ausubel, F. M. (2005). *Pseudomonas syringae* manipulates systemic plant defenses against pathogens and herbivores. *Proceedings of the National Academy of Sciences, 102*, 1791–1796.

da Silva, A. C., de Oliveira Silva, F. M., Milagre, J. C., Omena-Garcia, R. P., Abreu, M. C., Mafia, R. G., Nunes-Nesi, A., & Alfenas, A. C. (2018). Eucalypt plants are physiologically and metabolically affected by infection with *Ceratocystis fimbriata*. *Plant Physiology and Biochemistry, 123*, 170–179.

Dangl, J. L., & Jones, J. D. (2001). Plant pathogens and integrated defence responses to infection. *Nature, 411*, 826.

Das, M. P., Rebecca, L. J., Sharmila, S., Ankita, B., & Dhiraj, K. (2012). Identification and optimization of cultural conditions for chitinase production by *Bacillus amyloliquefaciens* SM3. *Journal of Chemical and Pharmaceutical Research, 4*, 4969–4974.

Dayakar, B. V., Lin, H. J., Chen, C. H., Ger, M. J., Lee, B. H., Pai, C. H., et al. (2003). Ferredoxin from sweet pepper (*Capsicum annuum* L.) intensifying harpin pss-mediated hypersensitive response shows an enhanced production of active oxygen species (AOS). *Plant Molecular Biology, 51*, 913–924.

DeGray, G., Rajasekaran, K., Smith, F., Sanford, J., & Daniell, H. (2001). Expression of an antimicrobial peptide via the chloroplast genome to control phytopathogenic bacteria and fungi. *Plant Physiology, 127*, 852–862.

Dodds, P. N., Lawrence, G. J., Catanzariti, A. M., Teh, T., Wang, C. I., Ayliffe, M. A., et al. (2006). Direct protein interaction underlies gene-for-gene specificity and coevolution of the flax resistance genes and flax rust avirulence genes. *Proceedings of the National Academy of Sciences, 103*, 8888–8893.

Dong, H. P., Peng, J., Bao, Z., Meng, X., Bonasera, J. M., Chen, G., et al. (2004). Downstream divergence of the ethylene signaling pathway for harpin-stimulated Arabidopsis growth and insect defense. *Plant Physiology, 136*, 3628–3638.

Duan, M., Zhang, T., Hu, W., Sundström, L. F., Wang, Y., Li, Z., & Zhu, Z. (2009). Elevated ability to compete for limited food resources by 'all-fish' growth hormone transgenic common carp *Cyprinus carpio*. *Journal of Fish Biology, 75*, 1459–1472.

Dunwell, J. M. (2000). Transgenic approaches to crop improvement. *Journal of Experimental Botany, 51*(Suppl_1), 487–495.
During, K., Porsch, P., Fladung, M., & Lorz, H. (1993). Transgenic potato plants resistant to the phytopathogenic bacterium *Erwinia carotovora*. *The Plant Journal, 3*, 587–598.
Dutt, M., Barthe, G., Irey, M., & Grosser, J. (2015). Transgenic citrus expressing an Arabidopsis NPR1 gene exhibit enhanced resistance against Huanglongbing (HLB; Citrus greening). *PLoS One, 10*, e0137134.
Dzhavakhiya, V. G., Shcherbakova, L. A. (2007). Creation of disease-resistant plants by gene engineering. In *Comprehensive and molecular phytopathology* (pp 439–466). Amsterdam: Elsevier.
Escobar, M. A., Civerolo, E. L., Summerfelt, K. R., & Dandekar, A. M. (2001). RNAi-mediated oncogene silencing confers resistance to crown gall tumorigenesis. *Proceedings of the National Academy of Sciences of the United States of America, 98*, 13437–13442.
Evenson, R. E., & Gollin, D. (2003). Assessing the impact of the green revolution, 1960–2000. *Science, 300*, 758–762.
Eyal, Y., & Fluhr, R. (1991). Cellular and molecular biology of pathogenesis related proteins. *Oxford Surveys of Plant Molecular and Cell Biology, 7*, 223–254.
Fagoaga, C., Rodrigo, I., Conejero, V., Hinarejos, C., Tuset, J. J., Arnau, J., et al. (2001). Increased tolerance to *Phytophthora citrophthora* in transgenic orange plants constitutively expressing a tomato pathogenesis related protein PR-5. *Molecular Breeding, 7*, 175–185.
Fan, W., & Dong, X. (2002). In vivo interaction between NPR1 and transcription factor TGA2 leads to salicylic acid–mediated gene activation in Arabidopsis. *The Plant Cell, 14*, 1377–1389.
Fitzgerald, H. A., Chern, M. S., Navarre, R., & Ronald, P. C. (2004). Overexpression of (At) NPR1 in rice leads to a BTH-and environment-induced lesion-mimic/cell death phenotype. *Molecular Plant-Microbe Interactions, 17*, 140–151.
Flor, H. H. (1954). Seed-flax improvement. III. Flax rust. *Advances in Agronomy, 6*, 152–161.
Fox, J. L. (2013). Antimicrobial peptides stage a comeback. *Nature Biotechnology, 31*, 379–382.
Freeman, B. C., & Beattie, G. A. (2008). An overview of plant defenses against pathogens and herbivores. *The Plant Health Instructor*. https://doi.org/10.1094/PHI-I-2008-0226-01.
Fry, W. E., & Grünwald, N. J. (2010). Introduction to oomycetes. *Plant Health Instructor*, 1207–1210.
Fu, Z. Q., & Dong, X. (2013). Systemic acquired resistance: Turning local infection into global defense. *Annual Review of Plant Biology, 64*, 839–863.
Galambos, A., Zok, A., Kuczmog, A., Oláh, R., Putnoky, P., Ream, W., & Szegedi, E. (2013). Silencing *Agrobacterium* oncogenes in transgenic grapevine results in strain-specific crown gall resistance. *Plant Cell Reports, 32*, 1751–1757.
Gamir, J., Darwiche, R., van't Hof, P., Choudhary, V., Stumpe, M., Schneiter, R., & Mauch, F. (2017). The sterol-binding activity of pathogenesis-related protein 1 reveals the mode of action of an antimicrobial protein. *The Plant Journal, 89*, 502–509.
García-Montoya, I. A., Cendón, T. S., Arévalo-Gallegos, S., & Rascón-Cruz, Q. (2012). Lactoferrin a multiple bioactive protein: An overview. *Biochimica et Biophysica Acta, 1820*, 226–236.
Gir, S. K., Goba, A., Andersen, K. G., Sealfon, R. S., Park, D. J., Kanneh, L., et al. (2014). Genomic surveillance elucidates Ebola virus origin and transmission during the 2014 outbreak. *Science, 345*, 1369–1372.
Grant, G., More, L. J., McKenzie, N. H., Dorward, P. M., Buchan, W. C., Telek, L., & Pusztai, A. (1995). Nutritional and haemagglutination properties of several tropical seeds. *The Journal of Agricultural Science, 124*(3), 437–445.
Grant, S. R., Fisher, E. J., Chang, J. H., Mole, B. M., & Dangl, J. L. (2006). Subterfuge and manipulation: Type III effector proteins of phytopathogenic bacteria. *Annual Review of Microbiology 60*, 425–449.
Greenberg, J. T., & Vinatzer, B. A. (2003). Identifying type III effectors of plant pathogens and analyzing their interaction with plant cells. *Current Opinion in Microbiology, 6*, 20–28.

Guo, M., Tian, F., Wamboldt, Y., & Alfano, J. R. (2009). The majority of the type III effector inventory of *Pseudomonas syringae* pv. *tomato* DC3000 can suppress plant immunity. *Molecular Plant-Microbe Interactions, 22*, 1069–1080.

Hammerschmidt, R., & Kuc, J. (1995). *Induced resistance to disease in plants*. Dordrecht: Kluwer Academic Publishers. 182 pp.

Hammond-Kosack, K. E., & Jones, J. D. (1997). Resistance gene-dependent plant defense responses. *The Plant Cell, 8*, 1773.

Hao, G., Stover, E., & Gupta, G. (2016). Overexpression of a modified plant thionin enhances disease resistance to citrus canker and huanglongbing (HLB). *Frontiers in Plant Science, 7*, 1078.

Hegde, V. S., & Mishra, S. K. (2009). Landraces of cowpea, *Vigna unguiculata* (L.) Walp., as potential sources of genes for unique characters in breeding. *Genetic Resources and Crop Evolution, 56*, 615–627.

Heino, P., Sandman, G., Lång, V., Nordin, K., & Palva, E. T. (1990). Abscisic acid deficiency prevents development of freezing tolerance in *Arabidopsis thaliana* (L.) Heynh. *Theoretical and Applied Genetics, 79*, 801–806.

Herrera-Estrella, L., De la Fuente, J. M., Ramírez-Rodríguez, V., & Cabrera-Ponce, J. L. (1997). Aluminum tolerance in transgenic plants by alteration of citrate synthesis. *Science, 276*, 1566–1568.

Hirai, D., Suzuki, T., Yanagida, D., et al. (2004). An evaluation of disease resistance of *Agrobacterium*-mediated transgenic potato (*Solanum tuberosum* L.) containing the chicken lysozyme gene or the wild spinach chitinase gene. *Bull Hokkaido Prefect Agricultural Experiment Station, 86*, 19–26.

Holeski, L. M., Jander, G., & Agrawal, A. A. (2012). Transgenerational defense induction and epigenetic inheritance in plants. *Trends in Ecology & Evolution, 27*, 618–626.

Hou, S., Yang, Y., Wu, D., & Zhang, C. (2011). Plant immunity: Evolutionary insights from PBS1, Pto, and RIN4. *Plant Signaling & Behavior, 6*, 794–799.

Hsiao, P. Y., Cheng, C. P., Koh, K. W., & Chan, M. T. (2017). The *Arabidopsis* defensin gene, AtPDF1. 1, mediates defence against *Pectobacterium carotovorum* subsp. *carotovorum* via an iron-withholding defence system. *Scientific Reports, 23*, 9175.

Huang, M. E., Cadieu, E., Souciet, J. L., & Galibert, F. (1997). Disruption of six novel yeast genes reveals three genes essential for vegetative growth and one required for growth at low temperature. *Yeast, 13*, 1181–1194.

Huang, H. E., Ger, M. J., Chen, C. Y., Pandey, A. K., Yip, M. K., Chou, H. W., & Feng, T. Y. (2007). Disease resistance to bacterial pathogens affected by the amount of ferredoxin-I protein in plants. *Molecular Plant Pathology, 8*, 129–137.

Jach, G., Görnhardt, B., Mundy, J., Logemann, J., Pinsdorf, E., Leah, R., et al. (1995). Enhanced quantitative resistance against fungal disease by combinatorial expression of different barley antifungal proteins in transgenic tobacco. *The Plant Journal, 8*, 97–109.

Jacob, L., & Zasloff, M. (1994). Potential therapeutic applications of magainins and other antimicrobial agents of animal origin: Antimicrobial peptides. *Ciba Foundation Symposium, 186*, 197–223.

Jacob, F., Vernaldi, S., & Maekawa, T. (2013). Evolution and conservation of plant NLR functions. *Frontiers in Immunology, 4*, 297.

Jaskiewicz, M., Conrath, U., & Peterhänsel, C. (2011). Chromatin modification acts as a memory for systemic acquired resistance in the plant stress response. *EMBO Reports, 12*, 50–55.

Jones, J. D., & Dangl, J. L. (2006). The plant immune system. *Nature, 444*, 323.

Jongedijk, E., Tigelaar, H., Van Roekel, J. S., Bres-Vloemans, S. A., Dekker, I., van den Elzen, P. J., et al. (1995). Synergistic activity of chitinases and β-1, 3-glucanases enhances fungal resistance in transgenic tomato plants. *Euphytica, 85*, 173–180.

Kim, Y. S., & Sano, H. (2008). Pathogen resistance of transgenic tobacco plants producing caffeine. *Phytochemistry, 69*, 882–888.

Knogge, W. (1996). Fungal infection of plants. *The Plant Cell, 8*, 1711.

Kuć, J. (1982). Induced immunity to plant disease. *Bioscience, 32*, 854–860.

Kumar, S., Jin, M., & Weemhoff, J. L. (2012). Cytochrome P450-mediated phytoremediation using transgenic plants: A need for engineered cytochrome P450 enzymes. *Journal of Petroleum & Environmental Biotechnology, 3*, 1000127.

Lawton, K., Uknes, S., Friedrich, L., Gaffney, T., Alexander, D., Goodman, R., et al. (1993). The molecular biology of systemic acquired resistance. In B. Fritig & M. Legrand (Eds.), *Mechanisms of defense responses in plants* (pp. 410–420). Dordrecht: Kluwer Academic Publishers.

Lee, Y. H., Yoon, I. S., Suh, S. C., & Kim, H. I. (2002). Enhanced disease resistance in transgenic cabbage and tobacco expressing a glucose oxidase gene from *Aspergillus niger*. *Plant Cell Replication, 20*, 857–863.

Lee, W. S., Fu, S. F., Verchot-Lubicz, J., & Carr, J. P. (2011a). Genetic modification of alternative respiration in *Nicotiana benthamiana* affects basal and salicylic acid-induced resistance to potato virus X. *BMC Plant Biology, 11*, 41.

Lee, S. B., Li, B., Jin, S., & Daniell, H. (2011b). Expression and characterization of antimicrobial peptides Retrocyclin-101 and Protegrin-1 in chloroplasts to control viral and bacterial infections. *Plant Biotechnology Journal, 9*, 100–115.

Li, L., & Steffens, J. C. (2002). Overexpression of polyphenol oxidase in transgenic tomato plants results in enhanced bacterial disease resistance. *Planta, 215*, 239–247.

Li, Q., Lawrence, C. B., Xing, H. Y., Babbitt, R. A., Bass, W. T., Maiti, I. B., & Everett, N. P. (2001). Enhanced disease resistance conferred by expression of an antimicrobial magainin analogue in transgenic tobacco. *Planta, 212*, 635–639.

Liau, C. H., Lu, J. C., Prasad, V., Hsiao, H. H., You, S. J., Lee, J. T., et al. (2003). The sweet pepper ferredoxin-like protein (pflp) conferred resistance against soft rot disease in *Oncidium* orchid. *Transgenic Research, 12*, 329–335.

Lin, W. C., Lu, C. F., Wu, J. W., Cheng, M. L., Lin, Y. M., Yang, N. S., et al. (2004). Transgenic tomato plants expressing the Arabidopsis NPR1 gene display enhanced resistance to a spectrum of fungal and bacterial diseases. *Transgenic Research, 13*, 567–581.

Lin, Y. H., Huang, H. E., Wu, F. S. Ger, M. J., Liao, P. L., Chen, Y. R., et al. (2010). Plant ferredoxin-like protein (PFLP) outside chloroplast in *Arabidopsis* enhances disease resistance against bacterial pathogens. *Plant Science, 179*, 450–458.

Lindow, S. E., & Brandl, M. T. (2003). Microbiology of the phyllosphere. *Applied and Environmental Microbiology, 69*(4), 1875–1883.

Lindow, S., Newman, K., Chatterjee, S., Baccari, C., Iavarone, A. T., & Ionescu, M. (2014). Production of *Xylella fastidiosa* diffusible signal factor in transgenic grape causes pathogen confusion and reduction in severity of Pierce's disease. *Molecular Plant-Microbe Interactions, 27*, 244–254.

Linthorst, H. J., & Van Loon, L. C. (1991). Pathogenesis-related proteins of plants. *Critical Reviews in Plant Sciences, 10*(2), 123–150.

Logemann, J., Jach, G., Tommerup, H., Mundy, J., & Schell, J. (1992). Expression of a barley ribosome-inactivating protein leads to increased fungal protection in transgenic tobacco plants. *Nature Biotechnology, 10* 305.

Malnoy, M., Venisse, J. S., Brisset, M. N., & Chevreau, E. (2003). Expression of bovine lactoferrin cDNA confers resistance to *Erwinia amylovora* in transgenic pear. *Molecular Breeding, 12*, 231–244.

Malnoy, M., Jin, Q., Borejsza-Wysocka, E. E., He, S. Y., & Aldwinckle, H. S. (2007). Overexpression of the apple MpNPR1 gene confers increased disease resistance in *Malus domestica*. *Molecular Plant-Microbe Interactions, 20*, 1568–1580.

Mandadi, K. K., & Scholthof, K. B. (2013). Plant immune responses against viruses: How does a virus cause disease? *The Plant Cell, 25*, 1489–1505.

Martin, G. B., Brommonschenkel, S. H., Chunwongse, J., Frary, A., Ganal, M. W., Spivey, R., et al. (1993). Map-based cloning of a protein kinase gene conferring disease resistance in tomato. *Science, 262*, 1432–1436.

Mauch, F., & Staehelin, L. A. (1989). Functional implications of the subcellular localization of ethylene-induced chitinase and [beta]-1, 3-glucanase in bean leaves. *Plant Cell, 1*, 447–457.

Melander, M., Kamnert, I., Happstadius, I., Liljeroth, E., & Bryngelsson, T. (2006). Stability of transgene integration and expression in subsequent generations of doubled haploid oilseed rape transformed with chitinase and β-1, 3-glucanase genes in a double-gene construct. *Plant Cell Reports, 25*, 942–952.

Melotto, M., Underwood, W., Koczan, J., Nomura, K., & He, S. Y. (2006). Plant stomata function in innate immunity against bacterial invasion. *Cell, 126*, 969–980.

Minsavage, G., Jones, J., Stall, R., Miller, S., & Ritchie, D. (1999). Hypersensitive resistance in Capsicum pubescens PI 235047 to *Xanthomonas campestris* pv. *vesicatoria* (Xcv) is elicited by avrBs3-2. *Phytopathology, 89*(6), S53.

Mishra, M., Jalil, S. U., Mishra, R. K., Kumari, S., & Pandey, B. K. (2016). In vitro screening of guava plantlets transformed with endochitinase gene against *Fusarium oxysporum* f. sp. *psidii*. *Czech Journal of Genetics and Plant Breeding, 52*, 6–13.

Mitchell, R. S., Beitzel, B. F., Schroder, A. R., Shinn, P., Chen, H., Berry, C. C., et al. (2004). Retroviral DNA integration: ASLV, HIV, and MLV show distinct target site preferences. *PLoS Biology, 2*, e234.

Mitten, D. H., MacDonald, R., & Klonus, D. (1999). Regulation of foods derived from genetically engineered crops. *Current Opinion in Biotechnology, 10*, 298–302.

Mohapatra, S. S., Poole, R. J., & Dhindsa, R. S. (1988). Abscisic acid-regulated gene expression in relation to freezing tolerance in alfalfa. *Plant Physiology, 87*, 468–473.

Moon, S. J., Park, H. J., Kim, T. H., Kang, J. W., Lee, J. Y., Cho, J. H., et al. (2018). OsTGA2 confers disease resistance to rice against leaf blight by regulating expression levels of disease related genes via interaction with NH1. *PLoS One, 13*(11), e0206910.

Motamayor, J. C., Mockaitis, K., Schmutz, J., Haiminen, N., Livingstone, D., Cornejo, O., et al. (2013). The genome sequence of the most widely cultivated cacao type and its use to identify candidate genes regulating pod color. *Genome Biology, 14*, r53.

Nakajima, H., Muranaka, T., Ishige, F., et al. (1997). Fungal and bacterial resistance in transgenic plants expressing human lysozyme. *Plant Cell Replication, 16*, 674–679.

Nakamura, Y., Sawada, H., Kobayashi, S., Nakajima, I., & Yoshikawa, M. (1999). Expression of soybean b-1, 3-endoglucanase cDNA and effect on disease tolerance in kiwifruit plants. *Plant Cell Reports, 18*, 527–532.

Namukwaya, B., Tripathi, L., Tripathi, J. N., Arinaitwe, G., Mukasa, S. B., & Tushemereirwe, W. K. (2012). Transgenic banana expressing Pflp gene confers enhanced resistance to *Xanthomonas* wilt disease. *Transgenic Research, 21*, 855–865.

Navarro, L., Dunoyer, P., Jay, F., Arnold, B., Dharmasiri, N., Estelle, M., et al. (2006). A plant miRNA contributes to antibacterial resistance by repressing auxin signaling. *Science, 312*, 436–439.

Neuhaus, J. M., Ahl-Goy, P., Hinz, U., Flores, S., & Meins, F. (1991). High-level expression of a tobacco chitinase gene in *Nicotiana sylvestris*. Susceptibility of transgenic plants to Cercospora nicotianae infection. *Plant Molecular Biology, 16*, 141–151.

Nomura, K., & He, S. Y. (2005). Powerful screens for bacterial virulence proteins. *Proceedings of the National Academy of Sciences of the United States of America, 102*(10), 3527–3528.

Oard, S. V. (2011). Deciphering a mechanism of membrane permeabilization by α-hordothionin peptide. *Biochimica et Biophysica Acta (BBA) - Biomembranes, 1808*(6), 1737–1745.

Oerke, E. C., & Dehne, H. W. (2004). Safeguarding production – Losses in major crops and the role of crop protection. *Crop Protection, 23*, 275–285.

Osusky, M., Zhou, G., Osuska, L., Hancock, R. E., Kay, W. W., & Misra, S. (2000). Transgenic plants expressing cationic peptide chimeras exhibit broad-spectrum resistance to phytopathogens. *Nature Biotechnology, 18*, 1162–1166.

Park, J. M., Park, C. J., Lee, S. B., Ham, B. K., Shin, R., & Paek, K. H. (2001). Overexpression of the tobacco Tsi1 gene encoding an EREBP/AP2–type transcription factor enhances resistance against pathogen attack and osmotic stress in tobacco. *The Plant Cell, 13*, 1035–1046.

Parker, J. E., Aarts, N., Austin, M. A., Feys, B. J., Moisan, L. J., Muskett, P., & Rusterucci, C. (2001). Genetic analysis of plant disease resistance pathways. In *Rice biotechnology: Improving yield, stress tolerance and grain quality, Novartis found symposium 236* (pp. 153–161). Chichester: Wiley.

Pelegrini, P. B., & Franco, O. L. (2005). Plant c-thionins: Novel insights on the mechanism of action of a multi-functional class of defense proteins. *International Journal of Biochemistry and Cell Biology, 37*, 2239–2253.

Piña-Vázquez, C., Reyes-López, M., Ortíz-Estrada, G., De la Garza, M., & Serrano-Luna, J. (2012). Host-parasite interaction: Parasite-derived and-induced proteases that degrade human extracellular matrix. *Journal of Parasitology Research, 2012*, 748206. https://doi.org/10.1155/2012/748206.

Pinstrup-Andersen, P., & Babinard, J. (2001). Globalization and human nutrition: Opportunities and risks for the poor in developing countries. *African Journal of Food, Agriculture, Nutrition and Development, 1*, 9–18.

Pirnia, S. (2016). *Novel plant DNA binding protein: Non-expresser of pathogenesis related 1 gene (NPR1) involved in disease resistance*. PhD thesis, Brock University, St. Catharines, ON.

Ploetz, R. C. (2005). Panama disease: An old nemesis rears its ugly head: Part 1. The beginnings of the banana export trades. *Plant Health Progress, 6*, 18.

Potlakayala, S. D., DeLong, C., Sharpe, A., & Fobert, P. R. (2007). Conservation of non-expressor of pathogenesis-related genes1 function between *Arabidopsis thaliana* and *Brassica napus*. *Physiological and Molecular Plant Pathology, 71*, 174–183.

Punja, Z. K. (2006). Recent developments toward achieving fungal disease resistance in transgenic plants. *Canadian Journal of Plant Pathology, 28*(S1), S298–S308.

Punja, Z. K., & Raharjo, S. H. (1996). Response of transgenic cucumber and carrot plants expressing different chitinase enzymes to inoculation with fungal pathogens. *Plant Disease, 80*, 999–1005.

Quilis, J., Peñas, G., Messeguer, J., Brugidou, C., & Segundo, B. S. (2008). The Arabidopsis AtNPR1 inversely modulates defense responses against fungal, bacterial, or viral pathogens while conferring hypersensitivity to abiotic stresses in transgenic rice. *Molecular Plant-Microbe Interactions, 21*, 1215–1231.

Redecker, D., Kodner, R., & Graham, L. E. (2000). Glomalean fungi from the Ordovician. *Science, 289*, 1920–1921.

Reynoird, J. P., Mourgues, F., Norelli, J. L., Aldwinckle, H. S., Brisset, M. N., & Chevreau, E. (1999). First evidence for differences in fire blight resistance among transgenic pear clones expressing attacin gene. *Plant Science, 149*, 23–31.

Ribeiro, S. M., Porto, W. F., Silva, O. N., de Oliveira Santos, M., Dias, S. C., & Franco, O. L. (2013). Plant antifungal peptides. In *Handbook of biologically active peptides* (2nd ed., pp. 169–179). Ann Arbor: ProQuest.

Rivero, M., Furman, N., Mencacci, N., Picca, P., Toum, L., Lentz, E., et al. (2012). Stacking of antimicrobial genes in potato transgenic plants confers increased resistance to bacterial and fungal pathogens. *Journal of Biotechnology, 157*, 334–343.

Ross, A. F. (1961). Localized acquired resistance to plant virus infection in hypersensitive hosts. *Virology, 14*, 329–339.

Rushton, P. J., Somssich, I. E., Ringler, P., & Shen, Q. J. (2010). WRKY transcription factors. *Trends in Plant Science, 15*, 247–258.

Ryals, J., Uknes, S., & Ward, E. (1994). Systemic acquired resistance. *Plant Physiology, 104*, 1109–1111.

Schlumbaum, A., Mauch, F., Vögeli, U., & Boller, T. (1986). Plant chitinases are potent inhibitors of fungal growth. *Nature, 324*, 365.

Sharma, T. R., Rai, A. K., Gupta, S. K., Vijayan, J., Devanna, B. N., & Ray, S. (2012). Rice blast management through host-plant resistance: Retrospect and prospects. *Agricultural Research, 1*, 37–52.

Shin, R., Park, J. M., An, J. M., & Paek, K. H. (2002). Ectopic expression of Tsi1 in transgenic hot pepper plants enhances host resistance to viral, bacterial, and oomycete pathogens. *Molecular Plant-Microbe Interactions, 15*, 983–989.

Silva, K. J., Brunings, A., Peres, N. A., Mou, Z., & Folta, K. M. (2015). The *Arabidopsis* NPR1 gene confers broad spectrum disease resistance in strawberry. *Transgenic Research, 24*, 693–704.

Singh, D., Ambroise, A., Haicour, R., Sihachakr, D., & Rajam, M. V. (2014). Increased resistance to fungal wilts in transgenic eggplant expressing alfalfa glucanase gene. *Physiology and Molecular Biology of Plants, 20*, 143–150.

Stirpe, F., Barbieri, L., Battelli, M. G., Soris, M., & Lappi, D. A. (1992). Ribosome-inactivating proteins from plants. *Biotechnology, 10*, 405–412.

Tang, K., Sun, X., Hu, Q., Wu, A., Lin, C. H., Lin, H. J., et al. (2001). Transgenic rice plants expressing the ferredoxin-like protein (AP1) from sweet pepper show enhanced resistance to *Xanthomonas oryzae* pv. *oryzae*. *Plant Science, 160*, 1035–1042.

Tao, Y., Zeng, Z. B., Li, J., Hartl, D. L., & Laurie, C. C. (2003). Genetic dissection of hybrid incompatibilities between *Drosophila simulans* and *D. mauritiana*. II. Mapping hybrid male sterility loci on the third chromosome. *Genetics, 164*, 1399–1418.

Tester, M., & Langridge, P. (2010). Breeding technologies to increase crop production in a changing world. *Science, 327*(5967), 818–822.

Tosarini, T. R., Ramos, P. Z., Profeta, G. S., Baroni, R. M., Massirer, K. B., Couñago, R. M., & Mondego, J. M. (2018). Cloning, expression and purification of kinase domains of cacao PR-1 receptor-like kinases. *Protein Expression and Purification, 146*, 78–84.

Tripathi, L., Mwangi, M., Abele, S., Aritua, V., Tushemereirwe, W. K., & Bandyopadhyay, R. (2009). Xanthomonas wilt: A threat to banana production in East and Central Africa. *Plant Disease, 93*, 440–451.

Uefuji, H., Tatsumi, Y., Morimoto, M., Kaothien-Nakayama, P., Ogita, S., & Sano, H. (2005). Caffeine production in tobacco plants by simultaneous expression of three coffee N-methyltrasferases and its potential as a pest repellant. *Plant Molecular Biology, 59*, 221–227.

Van Baarlen, P., Van Belkum, A., Summerbell, R. C., Crous, P. W., & Thomma, B. P. (2007). Molecular mechanisms of pathogenicity: How do pathogenic microorganisms develop cross-kingdom host jumps? *FEMS Microbiology Reviews, 31*, 239–277.

Van Loon, L. C. (1989). Stress proteins in infected plants. Plant-microbe interactions, molecular and genetic. *Perspectives, 3*, 198–237.

Van Loon, L. C., Pierpoint, W. S., Boller, T. H., & Conejero, V. (1994). Recommendations for naming plant pathogenesis-related proteins. *Plant Molecular Biology Reporter, 12*, 245–264.

van Loon, L. C., Rep, M., & Pieterse, C. M. (2006). Significance of inducible defense-related proteins in infected plants. *Annual Review of Phytopathology, 44*, 135–162.

Vega-Arreguín, J. C., Jalloh, A., Bos, J. I., & Moffett, P. (2014). Recognition of an Avr3a homologue plays a major role in mediating nonhost resistance to *Phytophthora capsici* in *Nicotiana* species. *Molecular Plant-Microbe Interactions, 27*, 770–780.

Vierling, E. (1991). The roles of heat shock proteins in plants. *Annual Review of Plant Biology, 42*(1), 579–620.

Vigers, A. J., Roberts, W. K., & Selitrennikoff, C. P. (1991). A new family of plant antifungal proteins. *Molecular Plant-Microbe Interactions, 4*, 315–323.

Viss, W. J., Pitrak, J., Humann, J., Cook, M., Driver, J., & Ream, W. (2003). Crown-gall-resistant transgenic apple trees that silence *Agrobacterium tumefaciens* oncogenes. *Molecular Breeding, 12*, 283–295.

Voegele, R. T., Struck, C., Hahn, M., & Mendgen, K. (2001). The role of haustoria in sugar supply during infection of broad bean by the rust fungus *Uromyces fabae*. *Proceedings of the National Academy of Sciences of the United States of America, 98*, 8133–8138.

Wally, O., Jayaraj, J., & Punja, Z. K. (2009). Broad-spectrum disease resistance to necrotrophic and biotrophic pathogens in transgenic carrots (*Daucus carota* L.) expressing an Arabidopsis NPR1 gene. *Planta, 231*, 131–141.

Wang, W. Z., Yang, B. P., Feng, X. Y., Cao, Z. Y., Feng, C. L., Wang, J. G., et al. (2017). Development and characterization of transgenic sugarcane with insect resistance and herbicide tolerance. *Frontiers in Plant Science, 8*, 1535.

Wang, M., Nishihama, R., Onishi, M., & Pringle, J. R. (2018). Role of the Hof1–Cyk3 interaction in cleavage-furrow ingression and primary-septum formation during yeast cytokinesis. *Molecular Biology of the Cell, 29*, 597–609.

Wu, C. H., Abd-El-Haliem, A., Bozkurt, T. O., Belhaj, K., Terauchi, R., Vossen, J. H., & Kamoun, S. (2017). NLR network mediates immunity to diverse plant pathogens. *Proceedings of the National Academy of Sciences, 114*, 8113–8118.

Yamanaka, T., Imai, T., Satoh, R., Kawashima, A., Takahashi, M., Tomita, K., et al. (2002). Complete inhibition of tobamovirus multiplication by simultaneous mutations in two homologous host genes. *Journal of Virology, 76*, 2491–2497.

Yang, B., Zhu, W., Johnson, L. B., & White, F. F. (2000). The virulence factor AvrXa7 of *Xanthomonas oryzae* pv. *oryzae* is a type III secretion pathway-dependent nuclear-localized double-stranded DNA-binding protein. *Proceedings of the National Academy of Sciences of the United States of America, 97*, 9807–9812.

Yip, M. K., Huang, H. E., Ger, M. J., Chiu, S. H., Tsai, Y. C., Lin, C. I., & Feng, T. Y. (2007). Production of soft rot resistant calla lily by expressing a ferredoxin-like protein gene (pflp) in transgenic plants. *Plant Cell Reports, 26*, 449–457.

Yoshioka, H., Numata, N., Nakajima, K., Katou, S., Kawakita, K., Rowland, O., et al. (2003). *Nicotiana benthamiana* gp91phox homologs NbrbohA and NbrbohB participate in H2O2 accumulation and resistance to Phytophthora infestans. *The Plant Cell, 15*, 706–718.

Zainal, Z., Marouf, E., Ismail, I., & Fei, C. K. (2009). Expression of the *Capsicuum annum* (chili) defensin gene in transgenic tomatoes confers enhanced resistance to fungal pathogens. *American Journal of Plant Physiology, 4*, 70–79.

Zeilinger, S., Gupta, V. K., Dahms, T. E., Silva, R. N., Singh, H. B., Upadhyay, R. S., et al. (2015). Friends or foes? Emerging insights from fungal interactions with plants. *FEMS Microbiology Reviews, 40*, 182–207.

Zhang, Z., Coyne, D. P., Vidaver, A. K., & Mitra, A. (1998). Expression of human lactoferrin cDNA confers resistance to *Ralstonia solanacearum* in transgenic tobacco plants. *Phytopathology, 88*, 730–734.

Zhang, J., Shao, F., Li, Y., Cui, H., Chen, L., Li, H., et al. (2007). A *Pseudomonas syringae* effector inactivates MAPKs to suppress PAMP-induced immunity in plants. *Cell Host & Microbe, 1*, 175–185.

Zhang, X., Francis, M. I., Dawson, W. O., Graham, J. H., Orbović, V., Triplett, E. W., & Mou, Z. (2010). Over-expression of the *Arabidopsis* NPR1 gene in citrus increases resistance to citrus canker. *European Journal of Plant Pathology, 128*, 91–100.

Zhu, J., & Weir, B. S. (1994). Analysis of cytoplasmic and maternal effects. II. Genetic models for triploid endosperms. *Theoretical and Applied Genetics, 89*, 160–166.

Chapter 10
Exploiting RNA Interference Mechanism in Plants for Disease Resistance

Anita Puyam and Kiranjot Kaur

Abstract RNA interference (RNAi) is a novel technique in the field of functional genomics. It has an immense potential for managing plant diseases by down regulating expression of phytopathogens' genes (invader's gene) and other negative regulators of resistance pathways. This technique has become a breakthrough in the field of managing plant diseases rather than implementing biological and chemical control measures. RNAi mechanism involves the silencing of specific genes responsible for infection in the host plant, in a homology-dependent manner, before their translation. Incorporation of RNAi over the time has become one of the most promising technology, which reduces the risks incurred in the production of transgenic plants. The idea of gene silencing has been successful under laboratory conditions, and it is now gaining importance for field applications as well. However, problem presently to solve include delivering RNAi gene silencing in the field, in a convenient way for managing fungal, bacterial and viral plant diseases, on host-pathogen related targeted sites. This chapter will give an insight on the strategies of delivering RNAi mediated gene silencing and managing plant diseases in a most practical way for the farmers.

Keywords RNA interference · Dicer · Functional genomics · Gene silencing · Gene knock down

A. Puyam (✉)
Rani Lakshmi Bai Central Agricultural University, Jhansi, Uttar Pradesh, India

K. Kaur
Khalsa College, Amritsar, India

© Springer Nature Switzerland AG 2020

I. Ul Haq, S. Ijaz (eds.), *Plant Disease Management Strategies for Sustainable Agriculture through Traditional and Modern Approaches*, Sustainability in Plant and Crop Protection 13, https://doi.org/10.1007/978-3-030-35955-3_10

10.1 Introduction

The present world may face a global food crisis in the near future if no proper measures are taken to deal with problems associated to crop protection and production. This is due to an increase in human population, biotic and abiotic stresses and host-resistance breakdown. These factors contribute to lower agricultural yields decreasing crop production throughout the world (Oerke 2005) with losses estimated around USD 100 billion per year (Fletcher et al. 2006). A potential loss of USD 540 billion per year has been predicted if the no proper control measures are taken against destructive pests and pathogens (Kew Royal Botanical Gardens 2017). Biotic stresses have been one important cause for yield losses, worldwide. In our approach to protect crops from biotic stresses and to boost up production, chemicals use due to non-judicial application of fungicides are contaminating and harming the environment inadvertently. Subsequently, this will pose risks to human health, beneficial micro-macro fauna of soil and the environment (Nicolopoulou-Stamati et al. 2016).

From time to time, different strategies have been reported and established. Yet, there is a need for new and powerful strategies to manage the diseases. Thus, there is a constant research in different parts of the world to develop technologies which will help to protect crops and not hampering the environmental conditions. This calls for novel tools and technology to manage plant diseases which will serve as an alternative method to fungicides and chemicals.

A new promising strategy which has a high potential for plant protection is the delivery of double stranded (ds)RNA, an initiating molecule for the gene silencing process, also termed as RNA interference (RNAi). It is a natural and highly conserved mechanism found in eukaryotic organisms, which plays a critical role in growth and development. It may impart host-resistance against viruses and inactivation of transposons, across different lineages such as fungi, plants and animals (Niehl et al. 2018).

RNAi, also known as post-transcriptionally gene silencing, is a conserved regulatory mechanism of gene expression in which a short nucleotide sequence yields the degradation of the targeted, complementary homologous mRNA (Duan et al. 2012; Fagwalawd et al. 2013). It is an epigenetic change based on RNA-directed DNA methylation (RdDM in plants) at transcriptional level, inhibiting translation at the post-transcriptional level (Senthil-Kumar and Mysore 2010).

RNAi rapidly gained popularity from decades now, and has been studied to knock down targeted genes and their expression in plants, micro-organisms and lower animals. For many years, it has been explored for crop improvements related to production and productivity, nutrition, or enhancement of vegetables and fruits shelf life (International service for acquisition of agri-biotech applications, Personal communication 2012).

RNAi is a key regulator of plant growth, development and response against different type of stresses (Singh 2005). And now, the technology has been successfully explored for developing resistance against pathogens like fungi, bacteria, viruses, insects and nematodes (Singh 2005) as well as for abiotic stresses (cold, salinity, drought etc.) (Khraiwesh et al. 2012). This technology has already been explored as a potential method for gene introgression, pyramiding, and development of transgenics as an eco-friendly approach for management of plant disease (Sanghera et al. 2011). Due to some sociological ethics, trangenics are not completely adopted as a management strategy. So, the major concern in recent times is to explore this mechanism for delivering in the field, other than its application in transgenics. Presently, even though this disease managing technique has been becoming popular among researchers, crop application *i.e.*, delivering RNAi gene silencing in the field, is not yet established or standardized on a large scale. This chapter will give an insight on RNAi-mediated gene silencing and management of plant diseases by this mechanism, in a most practical way for the farmers.

10.2 Mechanism of RNAi in Plant Disease Management

In a normal interaction, pathogen-derived small RNAs (sRNAs) enter the host cells and suppress the their resistance against the pathogen (Weiberg et al. 2013). In *Botrytis cinerea* and *Arabidopsis* interaction, Bc-siR37 suppresses at least 15 Arabidopsis genes, including cell wall-modifying enzymes, receptor-like kinases, and WRKY transcription factors responsible for host resistance (Wang et al. 2017a). In another case, wheat rust pathogen *Puccinia striiformis* (Ps) suppresses wheat Pathogenesis-related protein 2 (PR-2) responsible for host resistance by delivering microRNA-like RNA1 (milR1) into host cells (Wang et al. 2017b). As a consequence, if such genes enhancing pathogenicity are silenced/mutated, resistance may be achieved. For example, in wheat, PsmilR1 precursor silencing leads to increased resistance against a virulent Ps isolate (Wang et al. 2017b). Gene silencing is a result of certain biochemicals co-ordination. These include small RNAs, dicers and argonaute proteins which form the core of the RNAi mechanism. Diverse forms of RNA molecules such as siRNAs, nat-RNAs, micro-RNAs (miRNAs) based on complementary base pairing, result in degradation of target homologous mRNAs.

The whole process is based on the Dicer- Argonaute core that generates siRNAs in higher plants, essential for biological functions (Brodersen and Voinnet 2006). The mechanism can be triggered by double-stranded RNA (dsRNA), introduced as a transgene, a hairpin construct or a viral intruder. In this pathway, dsRNA is processed to 20–25 nt small dsRNAs having staggered ends, known as small interfering RNA (siRNA). These molecules recruit Dicer (host ribonuclease –III like enzyme) having distinctive dsRNA binding site, PAZ (Piwi/Argonaute/Zwille) domains, RNA helicase, RNase III and (Brodersen and Voinnet 2006). Dicer component acts on long dsRNA (which can be sprayed, a construct or a viral intruder) which cleaves into 21–25 nucleotides (nt) si RNAs (Zrachya et al. 2007). With the help of argonautes, the siRNAs are directed towards the multi-component RNAse called RNA-induced silencing complex (RISCs) (Pandolfini et al. 2003). RISC then targets the sense homologous mRNA, based on complementary base pairing for subsequent cleavage and degradation (Baulcombe 2004). RISC along with antisense strands silence the target mRNA based on homologous based pairing, inhibiting the protein synthesis. Thus, the triggered RISC component continues to silenced and degrade its target mRNA (Brodersen and Voinnet 2006).

In fungi, RNAi was first reported in *Neurospora crassa*. Romano and Macino (1992) termed the mechanism as "quelling". The biochemical core of RNAi pathway in fungi consists of highly conserved dicer proteins, argonaute proteins and RNA-dependent-RNA polymerases (RdRps) (Li et al. 2010). Important differences in RdRps in local and systemic silencing was reported in *Arabidopsis thaliana* (Garcia-Ruiz et al. 2010). The role of these type of proteins is to regulate sexual perithecia development (*F. graminearum*), fungal growth, production of conidia, reaction to abiotic or biotic stress (Chen et al. 2015; Son et al. 2017). However, some fungi completely lack RNAi machinery or some of its key components, such as *Saccharomyces cerevisiae* and *Ustilago maydis* (Trieu et al. 2015).

Successful fungal-host interaction establishment is defined by the development of specialized structures called haustoria, responsible for nutrient absorption. The hustorium is a structure in which both the host and the pathogen membrane enclosed extra-haustorial matrix. Their main function is absorbing nutrients and facilitating signal exchange related to pathogenesis. This site serves as an exchange barrier between fungi and silencing signals generated in the host cell. If the generated signals and siRNAs can overcome this barrier and then, gene silencing may be triggered in the haustorial cells, this will interfere with pathogenesis or with any related essential metabolic process (Yin et al. 2011). Duan et al. (2012) suggested the pos-

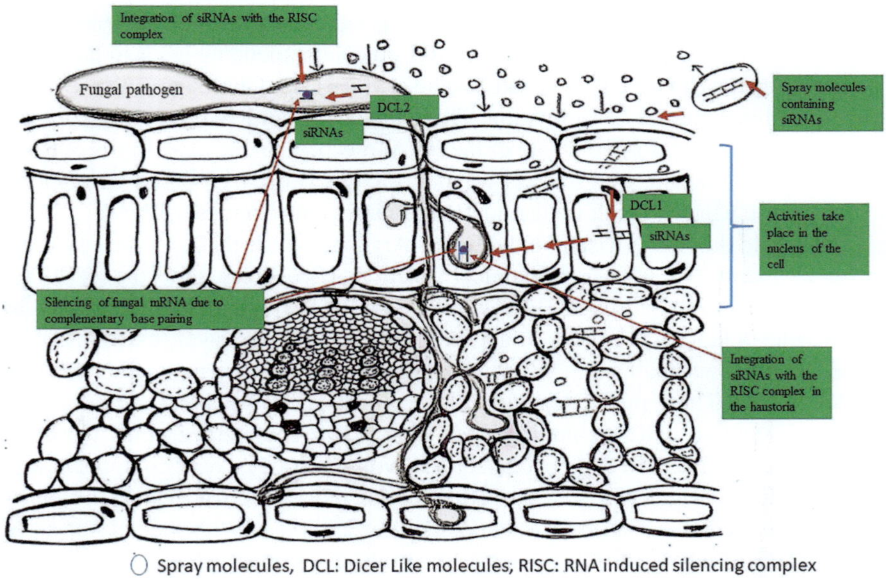

Fig. 10.1 RNAi based gene silencing in fungal pathogens

sibility of trafficking siRNA or dsRNA from the host cell into the fungal pathogen during their interaction, triggering host resistance mediated by the RNA silencing mechanism. The whole concept is referred to as Host-induced gene silencing (HIGS). Once this mechanism of RNAi is activated, mobile signaling molecules are produced and ultimately, the whole plant system is activated resulting in a gene silencing to achieve the desirable traits (Broglie et al. 1991). The locally initiated gene silencing spread systematically, or the signaling molecules were transported cell to cell through the phloem for long transport and through plasmodesmata symplastically. In artificial gene silencing, genes required for pathogenicity or virulence, or for pathogen survival, are targeted (Ozden and Nuh 2017) (Fig. 10.1).

In case of the bacterial disease crown gall, oncogenes (ipt and iaaM) have been identified as a pre-requisite for tumorogenesis (gall formation). If the expression of these bacterial oncogenes is silenced, the tumorogenesis process is restricted or reduced as shown in the case of transgenic plants (*Lycopersicon esculentum* and *Arabidopsis thaliana*) (Escobar et al. 2001).

Nowadays, the mechanism is considered as a promising strategy for disease management. There have been many attempts to exploit this mechanism in different ways. One of the most popular concerns its use in the transgenics, where dsRNAs incorporated in the host are processed into sRNAs subsequently silencing essential and/or pathogenicity genes present in fungi, oomycetes, bacteria, nematodes, viruses, viroids and insects (Cai et al. 2018).

10.3 RNAi-Based Gene Silencing in Diseases Control

RNAi gene silencing has been reported to provide resistance against bacteria, nematodes, parasitic plants, and fungi (Niblett and Bailey 2012; Papolu et al. 2013). It has been shown to provide resistance against insects (*Helicoverpa, Diabrotica*), bacteria (*Staphylococcus*, *Agrobacterium*), nematodes (*Meloidogyne, Heterodera*) and parasitic plants (*Striga*, *Orobanche*, *Triphysaria*) (Niblett and Bailey 2012). It also provided durable and effective resistance to Bayoud disease (*Fusarium oxysporum* f. sp. *albedinis*), red palm weevil (*Rhynchophorus ferrugineus*) and other serious pests of date palm (Niblett and Bailey 2012). Many researchers from different regions of the world have reported the effectiveness of RNAi-induced gene silencing against a number of plant pathogenic fungi *viz. Neurospora crassa* (Romano and Macino 1992), *Cladosporium fulvum* (Segers et al. 1999; Schweizer et al. 2000), *Magnaporthe oryzae* (Kadotani et al. 2003), *Venturia inaequalis* (Fitzgerald et al. 2004), powdery mildew *Blumeria graminis* (Nowara et al. 2010; Tinoco et al. 2010; Koch and Kogel 2014) and *P. graminis* f. sp. *tritici* or *P. striiformis* f. sp. *tritici* (Yin et al. 2011), *Fusarium verticilloides* in tobacco (Tinoco et al. 2010), *Cercospora beticola* (Starkel 2011), *S. sclerotiorum* (Andradeab et al. 2015), *Bremia lactucae* (Govindarajulu et al. 2015), *P. infestans* (Jahan et al. 2015), *F. oxysporum* f. sp. *conglutinans* (Zongli et al. 2015), *S. sclerotiorum* (Andradeab et al. 2015), *Phytophthora infestans* (Jahan et al. 2015), *V. dahliae* in tomato and *A. thaliana*

(Song and Thomma 2016). Activity against bacteria included bacterial pathogens causing crown gall disease (Escobar et al. 2001; Dunoyer et al. 2007), *Xanthomonas oryzae*, the leaf blight bacterium (Jiang et al. 2009). Viral diseases include *Potato virus Y* (Waterhouse et al. 1998), *Vigna mungo yellow mosaic virus* (VMYMV) in black gram (*Vigna mungo*) leaves (Pooggin et al. 2003), *Plum pox virus* infection (Pandolfini et al. 2003), *Bean golden mosaic virus* (BGMV) (Aragao and Faria 2009). The silencing is more common in RNA viruses as compared to DNA viruses (Seemanpillai et al. 2003).

In fungi and host interactions, control of the pathogen resulting from RNAi is attributed to the silencing of the target gene responsible for a particular function related to the pathogen growth, development, including formation of structures related to pathogenicity. For instance, inhibition of mycelium formation or growth was shown by silencing the 14α-demethylase and Chs3b in *Fusarium graminearum* (Koch et al. 2013; Cheng et al. 2015). In some cases, silencing involved an impaired formation of appressoria as reported in for *Phytophthora infestans* when silencing PiGPB1 (Jahan et al. 2015), reducing the pathogen load and the disease progression (Sanju et al. 2015).

In virus-host pathogen system, the target of silencing is mostly related to replication of the viral nucleic acids. Silencing may be achieved by antisense strategies or using the coat protein. It can also be directly achieved by silencing the genes responsible for pathogenicity. In some cases, due to RNAi silencing of certain genes, unwinding of the duplex viral RNA was inhibited. Other approaches are based on harboring vectors for simultaneous expression of both sense and anti-sense transcripts of the helper-component gene. The first report of effectiveness of RNAi technology against a virus was reported in gene silencing of the *Potato Virus Y* and HC-Pro gene (the helper component *proteinase*) (Waterhouse et al. 1998). Inhibition of the viral RNA unwinding was reported as due to *P1/HC* suppressors from the *Potyvirus* (Chapman et al. 2004).

Similarly, in bacteria-host interactions, resistance or control depends on silencing genes responsible for pathogenicity or by silencing those that are negative regulators of the host defense, so that the bacterial pathogen cannot establish the infection. For instance, in *Agrobacterium tumefaciens* causing crown gall, iaaH, iaaM and ipt genes are required for tumorogenesis. If these genes are silenced, no tumors are produced and the disease is controlled (Escobar et al. 2002). In some cases some genes which are negative regulators of defense genes in the hosts are silenced, then the bacterial pathogen is controlled, as demonstrated in *Pseudomonas syringae* (Katiyar-Aggarwal et al. 2006; Katiyar-Aggarwal and Jin 2007).

Nematodes have also been considered for RNAi, as they are important phytoparasites, causing abnormalities in plants. Losses due to nematodes result from histological changes in roots, with galls formation, low nitrogen fixation and poor growth, leading in severe conditions to total crop failure. However, less work has been done on RNAi gene silencing of plant parasitic nematodes. Silencing of genes required by the nematode to establish a feeding site in the host helps in reducing infection, i.e. silencing of 16D10 dsRNA responsible for *Meloidogyne* spp. -host integration, that also reduced the number of eggs laid by the nematode per g of roots. Other targets are the modulation of muscle and nerve activity, such as silencing of FMR Famide-like peptide genes (flp-14 and flp-18) (Papolu et al. 2013).

Despite many researches, complete exploration and practical application of RNAi gene silencing has not been fully applicable due to many challenges in transferring the technology in the field. Different strategies for practical application of the technology are promulgated in different parts of the world, transgenic crops being one of the most common and popular one adopted by researchers. However, due to many social ethics and debates all over the world in mass adoption of transgenic crops, the technology is restricted to a number of regions only.

10.4 Strategies of Delivering RNAi Mediated Gene Silencing

Different methods have been suggested for practical application of RNAi gene silencing in plant disease management such as agro-infiltration, micro-bombardment, Virus induced gene-silencing and spraying.

10.4.1 Agro-Infiltration

It is an RNAi gene silencing system where *Agrobacterium* is used as a carrier to deliver RNA silencing molecules into the intracellular space of leaves or plant tissues (Hily and Liu 2007; Duan et al. 2012; Singh 2005; Persengiev et al. 2004; Yuan et al. 2004; Moritoh et al. 2005; Bertazzon et al. 2012; Dunoyer et al. 2007; Wroblewski et al. 2007). The process is termed as agro-inoculation or agro-infiltration (Hily and Liu 2007). RNAi gene silencing is induced based on the similar mechanism used by *Agrobacterium tumefaciens* to deliver T-DNA to the host. It leads to the transformation of transient plants with *A. tumefaciens* genes harboring the gene of interest (Zrachya et al. 2007). The mechanism has been used to transform tomato plants cv. Micro-Tom using plasmid pMon RNAi CP from *A. tumefaciens* Gv3101/PMP90kk (Singh 2005). In *Nicotiana benthaniana*, this same activity results in down regulating of expression of green fluorescent protein (GFP) gene (Fagwalawd et al. 2013). Besides, agro-infiltration was successfully used in rice for silencing the gene encoded with the OSGEN-1-green fluorescent fusion protein (Moritoh et al. 2005), for silencing of ipt and iaaM genes inducing resistance to apple crown gall disease (Dunoyer et al. 2007), and in different transgenic lettuce lines by silencing hpGUS (Wroblewski et al. 2007). Agro-infiltration has been reported by many researchers in different host systems (Bertazzon et al. 2012). This approach induces a transient RNA silencing system by delivering the RNA silencing molecules directly into the plant tissues (Duan et al. 2012). This will overcome emergent bio-safety concerns due to transgenic plants produced for RNA silencing (Zrachya et al. 2007).

10.4.2 Micro-Bombardment

The mechanism is based on the delivery of hair-pin construct of dsRNA /siRNA or DNA into the plants using a ballistic pressure. GFP expression have been silenced based on micro-bombardment (Wani et al. 2010). Mohanpuria et al. (2008) also reported glutathione synthetase (GSHS) gene silencing in somatic embryos of *Camellia sinensis* L.

10.4.3 Virus Induced Gene Silencing

A virus vector is used as a medium for production of dsRNA (Fig. 10.2). Many reports were made using phytopathogenic viruses, in order to trigger RNAi in plants (Wuriyanghan and Falk 2013; Khan et al. 2013; Nandety et al. 2015; Armas-Tizapantz and Mozntiel-Gonzalez 2016). In TRV-HIGS, *Tobacco rattle virus* acted as a vector to trigger RNAi gene silencing. siRNAs were generated by the virus vector, during replication after inoculating in the host. The siRNAs were then taken up by the fungal pathogen triggering the silencing of genes responsible for some important function of that particular organism (Liu et al. 2002). Since silencing is due to viral siRNAs it is defined as "Virus induced gene silencing" (VIGS). The produced siRNAs will silence the pathogenicity related genes of the pathogen, which may be related to the production of infection structures, normal development of mycelium or sporulation.

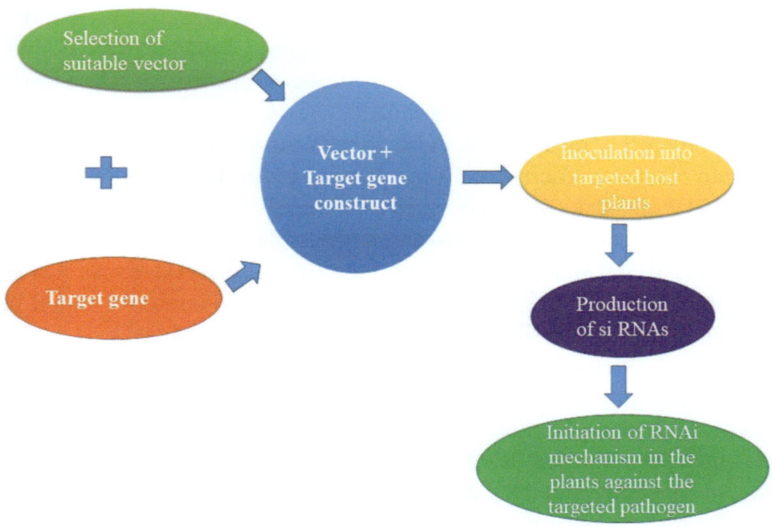

Fig. 10.2 A scheme showing virus induced gene silencing (VIGs) in plants against a targeted pathogen

Either RNA/DNA viruses may be used as vectors to transfer the gene of interest and silence endogenous plant genes (Wani et al. 2010). It was first demonstrated in RNA viruses by using TMV as vector (Brunt et al. 1996). However all RNA virus-derived expression vectors cannot be consider as silencing vectors because some RNA viruses possess anti-silencing proteins, directly interfering with the host silencing machinery (Palmer and Rybicki 2001; Kumagai et al. 1995). Therefore, selecting suitable vectors for proper execution is necessary. On the other hand, DNA viruses are rarely used as expression vectors due to size constraints for movement (Kjemtrup et al. 1998; Scofield et al. 2005). A VIGS-BSMV based system was reported in monocots related to leaf rust resistance for Lr21 mediated response (Chuang and Meyerowtiz 2000). Among all the vectors for silencing gene, *Maize streak virus* (MSV) was reported to be the most promising (Kumagai et al. 1995). This VIGS strategy ensures high resistance, the type of virus, however, determines the efficacy of silencing, whether it may induce a low resistance or a delayed infection with a subsequent recovery from infection (Chuang and Meyerowtiz 2000; Liu et al. 2002). To take advantage of this technology, gene pyramiding with multiple hpRNA constructs may be assayed in transgenic plants with a single hpRNA construct, multiple viral sources, or different sequences (Wani et al. 2010).

10.4.4 Host-Induced Gene Silencing

It is known as HIGS, this is a RNAi-based gene silencing process in which the host-synthesized small RNAs trigger silencing of pests or pathogens' genes when these attack the plant (Yin and Hulbert 2015). Successful application of HIGS *in vitro* for management of different diseases caused by fungi, viruses and nematodes have been reported by many researchers. Initial reports of HIGS-mediated resistance to fungal pathogen was reported for *Fusarium verticillioides* in tobacco plants (Tinoco et al. 2010). Subsequent reports include the silencing of: 14α-demethylase in *F. graminearum* (Koch et al. 2013), Dicer-like protein 1 (Bc-DCL1) in *Botrytis cinerea* (Wang et al. 2016), Avr3 in *Phytophthora infestans* (Sanju et al. 2015). HIGS was applied for management of viral diseases such as *Bean golden mosaic virus* by the AC1 gene (Bonfim et al. 2007), *Rice Dwarf Virus* (RDV) by silencing PNS12 (Shimizu et al. 2009), *Papaya ringspot virus* by targeting the coat protein (Papolu et al. 2013), and *Plum pox virus* (Dolgov et al. 2010; Scorza et al. 2013). Control of phytoparasitic nematodes such as *Pratylenchus*, *Meloidogyne* was reported by silencing Pv010, and flp-14/flp-18, respectively (Walawage et al. 2013; Papolu et al. 2013). Targets for HIGS involved genes, related to inhibition of the disease development in the host, include those essential for primary or secondary metabolism, required for synthesis of structural components such as ergosterol and chitin, and other required for developmental regulation (Yin and Hulbert 2015). For instance, in cereal-*Puccinia* interaction, RNAi silencing acquired resistance involved the silencing of gene responsible for biosynthesis of IAA (the plant auxin hormone), an essential component during pathogenicity (Yin et al. 2015).

Unfortunately, all these strategies were not readily applicable and easily adaptable to the field. The methods were more or less functional, i.e. the transgenic plants. The most emerging technology based on RNAi readily available and adaptable in the field is known as “Spray induced gene silencing”.

10.4.5 Spray Induced Gene Silencing

During the past 4–5 years, scientists have been trying to transfer the technology to the farmers in such a way that it will be easy to adopt and practically applicable to the field, in a way similar to the traditional chemical control. The most recent breakthrough in RNAi research for plant disease control is silencing genes by spraying without altering the DNA of the plant, known as Spray Induced gene silencing (SIGS). This is the most promising strategy for crop protection, acting as a direct disease control agent and as resistance repressors. SIGS refers to exogenous application of long siRNAs and dsRNA that will trigger silencing of one or more genes. The mechanism has been proved effective for controlling both *B. cinerea* and *F. graminearum* (Koch et al. 2016; Wang et al. 2016). SIGS has been emerging as a promising and potential technology against multiple crops. It will be one of the most powerful approach to combat the resistance breakdown, and overcome the problems associated with transgenic acceptance.

Main SIGS principle is to exogenously apply long dsRNA and siRNAs to exploit the RNAi mechanism for disease or pest control. The strategy may provide selective pathogen control without off target effects, thus saving ecosystem and biodiversity. However, the present problem is how to deliver the RNAi molecules, in system that needs to be standardized to be consistent. Mitter et al. (2017) promulgated an efficient RNAi molecule delivery system based on double-layered hydroxide clay nanoparticles made up of stacked of sheets of minerals i.e. magnesium chloride coating, containing the dsRNA molecule of importance. Using this technology, the persistence of RNA molecules on sprayed leaves increases up to 30 days (Mitter et al. 2017). However, the nanoparticle-based technique, first described in 2006, have been tested primarily for human therapeutics (Ladewig et al. 2009). Its principle is that the positively charged clay nanoparticles will bind the negatively charged RNAs and protect from adverse conditions, thereby maintaining their stability. Over time, breaking down of clay particles will occur due to the reaction with atmospheric carbon dioxide forming carbonic acid in presence of humidity, thereby facilitating the gradual release of the active dsRNAs (Mitter et al. 2017; Reddy et al. 2006). The latter are then absorbed by the plants to trigger the interference mechanism without modifying the plant actual DNA. Their effect will result in a built-in defense system used by plants cell to fight off the pathogen infection, by temporarily knocking down some genes expression.

In other cases, Koch et al. (2013, 2016) demonstrated that in *F. graminearum* the RNAi molecules were imbibed by the fungal cell. A 791-nt long CYP3-dsRNAs complementary sequence suppressed the three target CYP51 genes (CYP51A,

CYP51B and CYP51C), encoding cytochrome p450 lanosterol C14-demethylase. Due to this activity of the sprayed nucleotides, fungal membrane integrity was disrupted prior to infection, as shown by the reduced CYP51 genes amounts as measured by reverse transcription quantitative PCR (Koch et al. 2013, 2016).

SIGS showed control of *F. graminearum* in barley (Koch et al. 2016) and *Botrytis cinerea* (Wang et al. 2016). Effectiveness of SIGS in controlling *Botrytis cinerea* silencing Dicerlike 1 (*DCL1*) and (*DCL2)* genes was predicted due to mycelial uptake of exogenously applied siRNAs and long dsRNA *in vitro*, subsequently leading to a low disease incidence. Such effective results due to exogenous spray of siRNAs and long dsRNAs targeting fungal *DCL1* and *DCL2* was found against grey mould disease caused by *B. cinerea*. Treatment was found to be effective against *B. cinerea* for rose petals, strawberry, fruits of tomato, lettuce leaves and grape using either siRNAs or long dsRNAs (Wang et al. 2016). This demonstrates the possibility and applicability of SIGS strategy as an effective and innovative technology to protect multiple crop plants. Silencing of target fungal genes, essential for pathogenesis, will represent an enormous potential technology for crop protection. However, the exploitation of this strategy, as one of the most successful measure on large scale, needs to attend several challenges.

An alternative method may be root absorption and trunk injection which have already been proved in management of insect pests. This was achieved by delivering dsRNA into the plant vascular system through root absorption or injection, as these insects naturally took up dsRNA through chewing or sucking (Andrade and Hunter 2016). The same strategy can also be applied to manage plant pathogens.

10.5 Advantages of RNAi Application in the Field

The expected outcome of exploring RNAi in disease management is to promote an environmental friendly strategy by maintaining biodiversity, without harming the population of beneficial organisms such as insects, fungi, bacteria etc. As the tactic is based on short nucleotides sequence, there will be no danger for toxic residues in soil and the environment, nor will it induce resistance or tolerance on pathogen strains, a major problem in chemical control of plant disease (Wang et al. 2017). dsRNAs are ubiquitously present in the system of the organisms both endogenously and exogenously and essential for their growth and development. They are target-specific, bio-compatible and bio-degradable compounds with low stability in water or soil (Dubelman et al. 2014; Niehl et al. 2018). In natural RNA silencing pathways of plants and other organisms, exogenously entered dsRNAs are naturally subjected to natural degradation mechanisms (Cerutti and Ibrahim 2010).There is less chance for retention or release of any novel residues in food products due to this activity (Niehl et al. 2018). This technique can be considered as a novel and innovative strategy to manage plant diseases.

Moreover, this tactic will also help to overcome the time constraints related to the development of a GMO crop. The strategy will be an efficient means to control viral diseases as till date there is no anti-viral chemicals exploitable for its control. In the coming days, RNAi will likely replace many current plant disease management method, including transgenic crops, and may serve as an alternative strategy. These may create enormous scope and development to the modern concept of sustainable and eco-friendly cropping, and a positive advancement in modern agriculture.

10.6 Challenges for RNAi-based Spray Molecules in Plant Disease Management

10.6.1 Longevity Issues

Unprotected sprayed RNA molecules last only for few days and soon break down, due to environmental adversities. Single gene silencing sprays can protect tobacco plants from viruses for as long as 20 days (Mitter et al. 2017). Lack of persistence of dsRNAs within the pedosphere has been also reported. Degradation of dsRNA within 24 h in all types of soil have been examined, regardless of the length of the molecule (Dubelman et al. 2014). Albright et al. (2017) also reported the degradation of dsRNAs within 96 hours upon entering natural water systems. RNAi application in the field then requires a standardized method to protect its active ingredient, and a way to release it slowly in the environment, in order to affect the target gene for a longer period of time. The longevity of dsRNA and its degradation in the pedosphere versus phyllosphere may show differences due to distinct microbial communities (Bodenhausen et al. 2013). Most likely, the bacterial nucleases, particularly RNase III enzymes, degrade dsRNA molecules in aquatic and the soil environments (Urich et al. 2008; Cho 2017). In a way this lack of persistence may also be considered as advantageous in nature, as the residual dsRNA molecules from foliar sprays may be hardly disseminated from the site of application through the soil. Efficiency of a single application and its persistence in the field to control any disease that emerges at different times in the same field, cannot be considered as realistic nor justified.

10.6.2 Costly Sprays

Synthesizing RNA is a cumbersome and expensive activity and may cost a huge amount of money to treat a small field. This factor may hamper in broad applications of the technology, either in greenhouses and fields. The producing industry cannot assure that the product will be cheaper than other commercially available fungicides. The challenges hence include discovering efficient and economical ways to design dsRNA production, purification and large-scale release (Niehl et al.

2018). However, scientist have developed new alternative technologies to assure cost-efficient large scale production of RNA for agricultural use.

10.6.3 Lack of Feasible Methods to Synthesis dsRNA

Basically, synthetically produced dsRNA are governed by physical annealing of two enzymatically synthesized ssRNA (Niehl et al. 2018). Annealing can be either *in vitro* (Tenllado and Diaz–Ruiz 2001; Carbonell et al. 2008; Konakalla et al. 2016; Koch et al. 2016; Wang et al. 2016) or *in vivo* in RNase III-deficient bacterial cells, following ssRNA synthesis (Gan et al. 2010; Tenllado et al. 2003; Yin et al. 2009). However, these technologies yield relatively low yields of high quality dsRNA. The practice of using bacterial production systems, contains homologous DNA molecules which affect the quality of the RNA preparation and its applicability. Another more evolved method is the utilization of enzymes encoded by specialized dsRNA viruses that synthesized dsRNA by converting ssRNA templates into dsRNA with high processivity using a de novo, primer-independent initiation mechanism (Makeyev and Bamford 2000; Laurila et al. 2002).

Regardless of the innovative strategy applied, most of the methods proved to be unstable for the efficient production of dsRNA. Another more recent strategy is to produce more stable and high-quality long dsRNA molecules by using RNA-dependent RNA polymerase components of the bacteriophage phi6, *in vivo* engineered in *Pseudomonas syringae* bacteria Niehl et al. (2018). This technique may serve an established dsRNA production system, which can be exploited for broader application, that promises for efficient, non-transgenic and eco-friendly approach for protecting crops.

10.6.4 Health and Safety Issues

Release of transgenic crops following social and governmental approvals is still a difficult task. Similarly, use of RNAi sprays in the field must be regulated, considering the possibility of the intrusion of siRNA in food and the food chain system, through human and animal consumption, thereby affecting their gene expression. The product establishment must be guided by certain guidelines and must be approved by Health organizations of the country before its release and vast applications. There is high chance of siRNA delivery into mammalian systems via the digestive tracts (Zhang et al. 2012a; b). However, contradictory reports indicated that ingested plant siRNA could not be detected in mammalian gut (Witwer et al. 2013; Witwer and Hirschi 2014).

10.6.5 Selective Targets

RNAi is target specific, based on complementary base pairing. So, a single spray will be specific to a particular pathogen and will not have a broad spectrum effect. To develop an RNAi-based spray technique, it is needed to identify the broad spectrum targeting sequences which may be effective against a number of pathogens (bacteria, fungi, virus, nematodes, plant-parasitic phaenerogams, phytoplasmas). One of such gene is HMSPG which was demonstrated to be effective against all groups of fungi causing disease in date palm. HMSPG is also reported to be effective against insect, bacteria and nematode pests (Niblett and Bailey 2012).

10.6.6 Non-target Effects

Releasing nucleotides in the environment must consider its effects on non-target organisms. If the released short sequence RNA molecule targeted for a particular pathogenic organism holds complementary regions with some other non-target genes present in that particular eco-system, then it may result in unwanted consequences. In other cases, target organisms may also change their DNA sequences to overcome suppression and develop resistance. However, due to its sequence-specific nature, RNAi-based disease management approaches have less chances of adverse effects in the environment or other non-target species. So, proper investigations must be made before complete application of the technology.

10.7 Conclusion

Decades of researches showed that RNAi gene silencing as a promising tool for the study of functional genomic and also a modern strategy to manage plant diseases, with effectual results. The particular strategy will offer a broader future endeavour to deal with the biotic stresses during crop production. A proper way of delivering the technology will be acceptable as a more feasible method to face changing environmental scenarios, due to actual abundant dependence on chemicals for crop protection. This will also help to overcome the limitations of bio-control agents which require a particular environment and time to establish for effective results. The only problem to commercialize and adopt any RNAi based strategy is to explore most feasible ways of delivering the nucleic acids to the field and also the eventual occurrence of any ill-effect on the crop ecology.

References

Albright, V. C., Wong, C. R., Hellmich, R. L., & Coats, J. R. (2017). Dissipation of double-stranded RNA in aquatic microcosms. *Environmental Toxicology and Chemistry, 36*, 1249–1253.

Andrade, C. E., & Hunter, W. B. (2016). In I. Y. Abdurakhmonov (Ed.), *RNA interference – Natural gene-based technology for highly specific pest control (HiSPeC) in RNA interference* (pp. 391–409). Croatia: InTech.

Andradeab, C. M., Tinocoa, M. L. P., Rietha, A. F., Maiaa, F. C. O., & Aragao, F. J. L. (2015). Host-induced gene silencing in the necrotrophic fungal pathogen *Sclerotinia sclerotiorum. Plant Pathology, 65*, 626–632.

Aragao, F. J., & Faria, J. C. (2009). First transgenic gemini-virus-resistant plant in the field. *Nature Biotechnology, 27*, 1086–1088.

Armas-Tizapantz, A., & Mozntiel-Gonzalez, A. M. (2016). RNAi silencing: A tool for functional genomics research on fungi. *Fungal Biology Reviews, 30*, 91–100.

Baulcombe, D. (2004). RNA silencing in plants. *Nature, 431*, 356–363.

Bertazzon, N., Raiola, A., Castiglioni, C., Gardiman, M., Angelini, E., Borgo, M., & Ferrari, S. (2012). Transient silencing of the grapevine gene VvPGIP1 by agroinfiltration with a construct for RNA interfer-ence. *Plant Cell Reports, 31*, 133–143.

Bodenhausen, N., Horton, M. W., & Bergelson, J. (2013). Bacterial communities associated with the leaves and the roots of *Arabidopsis thaliana*. *PLoS One, 8*, e56329.

Bonfim, K., Faria, J. C., Nogueira, E. O., Mendes, E. A., & Aragao, F. J. (2007). RNAi-mediated resistance to bean golden mosaic virus in genetically engineered common bean (*Phaseolus vulgaris*). *Molecular Plant-Microbe Interactions, 20*, 717–726.

Brodersen, P., & Voinnet, O. (2006). The diversity of RNA silencing pathways in plants. *Trends in Genetics, 22*, 268–280.

Broglie, K. I., Chet, M., & Holliday, M. N. (1991). Transgenic plants with enhanced resistance to fungal pathogen *Rhizoctonia solani*. *Science, 254*, 1194–1197.

Brunt, A. A., Crabtree, K., Dallwitz, M. J., Gibbs, A. J., Watson, L., & Zurcher, E. J. E. (1996). *Plant viruses online: Descriptions and lists from the VIDE database* (Version: 20th August 1996).

Cai, Q., He, B., Kogel, K. H., & Jin, H. (2018). Cross-kingdom RNA trafficking and environmental RNAi — nature's blueprint for modern crop protection strategies. *Current Opinion in Microbiology, 46*, 58–64.

Carbonell, A., Martinez de Alba, A. E., Flores, R., & Gago, S. (2008). Double stranded RNA interferes in a sequence-specific manner with the infection of representative members of the two viroid families. *Virology, 371*, 44–53.

Cerutti, H., & Ibrahim, F. (2010). Turnover of mature miRNAs and siRNAs in plants and algae. *Advances in Experimental Medicine and Biology, 700*, 124–139.

Chapman, E. J., Prokhnevsky, A. I., Gopinath, K., Dolja, V. V., & Carrington, J. C. (2004). Viral RNA silencing suppressors inhibit the micro-RNA pathway at an interphase step. *Genes and Development, 18*, 1179–1186.

Chen, Y., Gao, Q. X., Haung, M. M., Liu, Y., Liu, Z. Y., & Liu, X. (2015). Characterization of RNA silencing components in the plant pathogenic fungus *Fusarium graminearum. Science Reports-UK, 5*, 12500.

Cheng, W., Song, X. S., Li, H. P., Cao, L. H., Sun, K., & Qiu. (2015). Host-induced gene silencing of an essential chitin synthase gene confers durable resistance to Fusarium head blight and seedling blight in wheat. *Plant Biotechnology Journal, 13*, 1335–1345.

Cho, K. H. (2017). The structure and function of the gram-positive bacterial RNA degradosome. *Frontiers in Microbiology, 8*, 1–10.

Chuang, C. F., & Meyerowtiz, E. M. (2000). Specific and heritable genetic interference by double–stranded RNA in *Arabidopsis thaliana*. *PNAS, 97*, 4985–4990.

Dolgov, S., Mikhaylov, R., Serova, T., Shulga, O., & Firsov, A. (2010). Pathogen–derived methods for improving resistance of transgenic plums (*Prunus domestica* L.) for plum pox virus infection. *Julius–Kuhn–Arch, 427*, 133–140.

Duan, C. G., Wang, C. H., & Guo, H. S. (2012). Application of RNA silencing to plant disease resistance. *Silence, 3*, 1–8.

Dubelman, S., Fischer, J., Zapata, F., Huizinga, K., Jiang, C., Uffman, J., & Levine, S. (2014). Environmental fate of double–stranded RNA in agricultural soils. *PLoS One, 9*, e93155.

Dunoyer, P., Himber, C., Ruiz-Ferrer, V., Alioua, A., & Voinnet, O. (2007). Intra– and Intercellular RNA interference in Arabidopsis thaliana requires components of the microRNA and heterochromatic silencing path–ways. *Nature Genetics, 39*, 848–856.

Escobar, M. A., Civerolo, E. L., Summerfelt, K. R., & Dandekar, A. M. (2001). RNAi–mediated oncogene silencing confers resistance to crown gall tumorigenesis. *Proceedings of the National Academy of Sciences of the United States of America, 98*, 13437–13442.

Escobar, M. A., Leslie, C. A., Mcgranahan, G. H., & Dandekar, A. M. (2002). Silencing crown gall disease in walnut (*Julgans regia* L.). *Plant Science, 163*, 591–597.

Fagwalawd, I. D., Kutama, A. S., & Yakasai, M. T. (2013). Current issues in plant disease control: Biotechnology and plant disease. *Bayero Journal of Pure and Applied Sciences (BAJOPAS), 6*, 121–126.

Fitzgerald, A., Van, K. J. A., & Plummer, K. M. (2004). Simultaneous silencing of multiple genes in the apple scab fungus *Venturia inaequalis*, by expression of RNA with chimeric inverted repeats. *Fungal Genetics and Biology, 41*, 963–971.

Fletcher, J., Bender, C., Budowle, B., Cobb, W. T., Gold, S. E., Ishimaru, C. A., & Luster, D. (2006). Plant pathogen forensics: Capabilities, needs, and recommendations. *Microbiology and Molecular Biology Reviews, 70*, 450–471.

Gan, D., Zhang, J., Jiang, H., Jiang, T., Zhu, S., & Cheng, B. (2010). Bacterially expressed dsRNA protects maize against SCMV infection. *Plant Cell Reports, 29*, 1261–1268.

Garcia-Ruiz, H., Takeda, A., Chapman, E. J., Sullivan, C. M., Fahlgren, N., & Brempelis, K. J. (2010). Arabidopsis RNA–dependent RNA polymerases and dicer–like proteins in antiviral defense and small interfering RNA biogenesis during turnip mosaic virus infection. *The Plant Cell, 22*, 481–496.

Govindarajulu, M., Epstein, L., Wroblewski, T., & Michelmore, R. W. (2015). Host–induced gene silencing inhibits the biotrophic pathogen causing downy mildew of lettuce. *Plant Biotechnology Journal, 13*, 875–883.

Hily, J. M., & Liu, Z. (2007). An overview of small RNAs. In C. L. Bassett (Ed.), *Regulation of gene expression in plants* (pp. 123–147). Berlin: Springer–Verlag.

International service for acquisition of agri–biotech applications. (2012). *Pocket K No. 34: RNAi for crop improvement*. Available Source: http://isaaa.org/resources/publications/pocketk/34/default.asp

Jahan, S. N., Asman, A. K. M., Corcoran, P., Fogelqvist, J., Vetukuri, R. R., & Dixelius, C. (2015). Plant mediated gene silencing restricts growth of the potato late blight pathogen *Phytophthora infestans*. *Journal of Experimental Botany, 66*, 2785–2794.

Jiang, C. J., Shimono, M., Maeda, S., Inoue, H., Mori, M., Hasegawa, M., et al. (2009). Suppression of the rice fatty–acid desaturase gene Os- SSI2 enhances resistance to blast and leaf blight diseases in rice. *Molecular Plant-Microbe Interactions, 22*, 820–829.

Kadotani, N., Nakayashiki, H., Tosa, Y., & Mayama, S. (2003). RNA silencing in the pathogenic fungus *Magnaporthe oryzae*. *Molecular Plant-Microbe Interactions, 16*, 769–776.

Katiyar-Aggarwal, S., & Jin, H. (2007). Discovery of pathogen–regulated small RNAs in plants. *Methods in Enzymology, 427*, 215–227.

Katiyar-Aggarwal, S., Rebekah, M., Douglas, D., Omar, B., Andy Villegas, J., Jian-Kang, Z., et al. (2006). Pathogen–inducible endogenous siRNA in plant immunity. *Pest Management Science, 74*, 790–799.

Kew Royal Botanical Gardens. (2017). *State of the world's plants* (pp. 66–71). Kew: Royal Botanic Gardens.

Khan, A. M., Ashfaq, M., Kiss, Z., Khan, A. A., Mansoor, S., & Falk, B. W. (2013). Use of recombinant tobacco mosaic virus to achieve RNAinterference in plants against the Citrus Mealybug, *Planococcus citri* (Hemiptera: Pseudococcidae). *PLoS ONE, 8*, e73657.

Khraiwesh, B., Zhu, J. K., & Zhu, J. (2012). siRNAs in biotic and abiotic stress responses of plants. *Biochimica et Biophysica Acta, 1819*, 137–148.

Kjemtrup, S., Sampson, K. S., Peele, C. G., Nguyen, L. V., & Conkling, M. A. (1998). Gene silencing from plant DNA carried by a geminivirus. *The Plant Journal, 14*, 91–100.

Koch, A., & Kogel, K. H. (2014). New wind in the sails: Improving the agronomic value of crop plants through RNAi–mediated gene silencing. *Plant Biotechnology Journal, 12*, 821–831.

Koch, A., Kumar, N., Weber, L., Keller, H., Imani, J., & Kogel, K. (2013). Host–induced gene silencing of cytochrome P450 lanosterol C14α–demethylase–encoding genes confers strong resistance to Fusarium species. *Proceedings of the National Academy of Sciences of the United States of America, 110*, 19324–19329.

Koch, A., Biedenkopf, D., Furch, A., Weber, L., Rossbach, O., Abdellatef, E., & Linicus, L. (2016). An RNAi–based control of *Fusarium graminearum* infections through spraying of long dsRNAs involves a plant passage and is controlled by the fungal silencing machinery. *PLoS Pathogens, 12*, e1005901.

Konakalla, N. C., Kaldis, A., Berbati, M., Masarapu, H., & Voloudakis, A. E. (2016). Exogenous application of double–stranded RNA molecules from TMV p126 and CP genes confers resistance against TMV in tobacco. *Planta, 244*, 961–969.

Kumagai, M. H., Donson, J., della–Cioppa, G., Harvey, D., Hanley, K., & Grill, L. K. (1995). Cytoplasmic inhibition of carotenoid biosynthesis with virus–derived RNA. *Proceedings of the National Academy of Sciences of the USA, 92*, 1679–1683.

Ladewig, K., Xu, Z. P., & Lu, G. Q. (2009). Layered double hydroxide nanoparticles in gene and drug delivery. *Expert Opinion on Drug Delivery, 6*, 907–922.

Laurila, M. R., Makeyev, E. V., & Bamford, D. H. (2002). Bacteriophage phi6 RNA–dependent RNA polymerase: Molecular details of initiating nucleic acid synthesis without primer. *The Journal of Biological Chemistry, 277*, 17117–17124.

Li, L. D., Chang, S. S., & Liu, Y. (2010). RNA interference pathways in filamentous fungi. *Cellular and Molecular Life Sciences, 67*, 3849–3863.

Liu, Y. L., Schiff, M., & Dinesh Kumar, S. P. (2002). Virus-induced gene silencing in tomato. *The Plant Journal, 31*, 777–786.

Makeyev, E. V., & Bamford, D. H. (2000). Replicase activity of purified recombinant protein P2 of double–stranded RNA bacteriophage phi6. *The EMBO Journal, 19*, 124–133.

Mitter, N., Worrall, E. A., Robinson, K. E., Li, P., Jain, R. G., Taochy, C., & Fletcher, S. J. (2017). Clay nanosheets for topical delivery of RNAi for sustained protection against plant viruses. *Nature Plants, 3*, 1620.

Mohanpuria, P., Rana, N., & Yadav, S. (2008). Transient RNAi based gene silencing of glutathione synthetase reduces glutathione content in *Camellia sinensis* (L.) O. Kuntze somatic embryos. *Biologia Plantarum, 52*, 381–384.

Moritoh, S., Miki, D., Akiyama, M., Kawahara, M., Izawa, T., Maki, H., & Shimamoto, K. (2005). RNAi–mediated silencing of OsGEN–L (OsGEN–like), a new member of the RAD2/XPG nuclease family, causes male sterility by defect of microspore development in rice. *Plant & Cell Physiology, 46*, 699–715.

Nandety, R. S., Kuoy, Y. W., Nouriy, S., & Falk, B. W. (2015). Emerging strategies for RNA interference (RNAi) applications in insects. *Bioengineered, 6*, 8–19.

Niblett, C. L., & Bailey, A. M. (2012). Potential applications of gene silencing or RNA interference (RNAi) to control disease and insect pests of date palm. *Emirates Journal of Food and Agriculture, 24*, 462–469.

Nicolopoulou-Stamati, P., Maipas, S., Kotampasi, C., Stamatis, P., & Hens, L. (2016). Chemical pesticides and human health: The urgent need for a new concept in agriculture. *Frontiers in Public Health, 4*, 148.

Niehl, A., Marjukka, S., Poranen, M. M., & Manfred, H. (2018). Synthetic biology approach for plant protection using dsRNA. *Plant Biotechnology Journal, 16*, 1679–1687.

Nowara, D., Gay, A., Lacomme, C., Shaw, J., Ridout, C., Douchkov, D., Hensel, G., Kumlehn, J., & Schweizer, P. (2010). HIGS: Host induced gene silencing in the obligate biotrophic fungal pathogen Blumeria graminis. *Plant Cell, 22*, 3130–3141.

Oerke, E. C. (2005). Crop losses to pests. *The Journal of Agricultural Science, 144*, 31–43.

Ozden, S., & Nuh, B. (2017). New approach in management against plant fungal disease: Host induced gene silencing. *International Journal of Molecular Science, 1*(1), 20–29.

Palmer, K. E. & Rybicki, E. (2001). Investigation of the potential of maize streak virus to act as an infectious gene vector in maize plants. *Archives of Virology, 146*(6), 1089–1104.

Pandolfini, T., Molesini, B., Avesani, L., Spena, A., & Polverari, A. (2003). Expression of self–complementary hairpin RNA under the control of the rolC promoter confers systemic disease resistance to plum pox virus without preventing local infection. *BMC Biotechnology, 3*, 7.

Papolu, P. K., Gantasala, N. P., Kamaraju, D., Banakar, P., Sreevathsa, R., & Rao, U. (2013). Utility of host delivered RNAi of two FMRF amide like peptides, flp–14 and flp–18, for the management of root knot nematode, *Meloidogyne incognita. PLoS One., 6;8*(11), e80603.

Persengiev, S. P., Zhu, X., & Green, M. R. (2004). Non–specific, concentration–dependent stimulation and repression of mammalian gene expression by small interfering RNAs (siRNAs). *RNA, 10*, 12–18.

Pooggin, M., Shivaprasad, P. V., Veluthambi, K., & Hohn, T. (2003). RNAi targeting of DNA virus in plants. *Nature Biotechnology, 21*, 131–132.

Reddy, M. K. R., Xu, Z. P., Lu, G., & da Costa, J. C. D. (2006). Layered double hydroxides for CO2 capture: Structure evolution and regeneration. *Industrial and Engineering Chemistry Research, 45*, 7504–7509.

Romano, N., & Macino, G. (1992). Quelling: Transient inactivation of gene expression in Neurospora crassa by transformation with homologous sequences. *Molecular Microbiology, 6*, 3343–3353.

Sanghera, G. S., Kashyap, P. L., Singh, G., & Teixeira da Silva, J. A. (2011). Transgenics: Fast track to plant stress amelioration. *Transgenic Plant Journal, 5*, 1–26.

Sanju, S., Siddappa, S., Thakur, A., Shukla, P. K., Srivastava, N., Pattanayak, D., et al. (2015). Host mediated gene silencing of a single effector gene from the potato pathogen *Phytophthora infestans* imparts partial resistance to late blight disease. *Functional & Integrative Genomics, 15*, 697–706.

Schweizer, P., Pokorny, P., Schulze-Lefert, P., & Dudler, R. (2000). Double stranded RNA interference with gene functions at the single cell in cereals. *The Plant Journal, 24*, 895–903.

Scofield, S. R., Huang, L., Brandt, A. S., & Gill, B. S. (2005). Development of a virus–induced gene–silencing system for hexaploid wheat and its use in functional analysis of the Lr21–mediated leaf rust resistance pathway. *Plant Physiology, 138*, 2165–2173.

Scorza, R., Callahan, A., Dardick, C., Ravelonandro, M., Polak, J., Malinowski, T., Zagrai, I., & Cambra, M. (2013). Genetic engineering of plum pox virus resistance: 'HoneySweet' plum—From concept to product. *Plant Cell, Tissue and Organ Culture, 115*, 1–12.

Seemanpillai, M., Dry, I., Randles, J., & Rezaian, A. (2003). Transcriptional silencing of geminiviral promoter–driven transgenes following homologous virus infection. *Molecular Plant-Microbe Interactions, 16*, 429–438.

Segers, G. C., Hamada, W., Oliver, R. P., & Pspanu, P. D. (1999). Isolation and characterisation of five different hydrophobin–encoding cDNA from the fungal tomato pathogen *Cladosporium fulvum. Molecular & General Genetics, 261*, 644–652.

Senthil-Kumar, M., & Mysore, K. S. (2010). RNAi in plants: Recent developments and applications in agriculture. In *Gene silencing: Theory, techniques and applications* (pp. 83–199). New York: NOVA Science Publishers, Inc.

Shimizu, T., Yoshii, M., Wei, T., Hirochika, H., & Omura, T. (2009). Silencing by RNAi of the gene for Pns12, a viroplasm matrix protein of Rice dwarf virus, results in strong resistance of transgenic rice plants to the virus. *Plant Biotechnology Journal, 7*, 24–32.

Singh, R. S. (2005). *Introduction to principles of plant pathology* (pp. 178–189). New Delhi: Oxford and IBH Publishing PVT. LTD.

Son, H., Park, A. R., Lim, J. Y., Shin, C., & Lee, Y. W. (2017). Genome–wide exonic small interference RNA–mediated gene silencing regulates sexual reproduction in the homothallic fungus Fusarium graminearum. *PLoS Genetics, 13*, e1006595.

Song, Y., & Thomma, B. P. (2016). Host–induced gene silencing compromises Verticillium wilt in tomato and Arabidopsis. *Molecular Plant Pathology, 19*, 77–89. https://doi.org/10.1111/mpp.12500.

Starkel, C. (2011). *Host induced gene silencing – Strategies for the improvement of resistance against Cercospora beticola in sugar beet (B. vulgaris L.) and against Fusarium graminearum in wheat (T. aestivum L.) and maize (Z. mays L.)* (PhD Thesis, Berlin, Germany).

Tenllado, F., & Diaz–Ruiz, J. R. (2001). Double–stranded RNA–mediated interference with plant virus infection. *Journal of Virology, 75*, 12288–12297.

Tenllado, F., Barajas, D., Vargas, M., Atencio, F. A., Gonzalez–Jara, P., & Diaz–Ruiz, J. R. (2003). Transient expression of homologous hairpin RNA causes interference with plant virus infection and is overcome by a virus encoded suppressor of gene silencing. *Molecular Plant-Microbe Interactions, 16*, 149–158.

Tinoco, M. L., Dias, B. B., Dall'Astta, R. C., Pamphile, J. A., & Aragao, F. J. (2010). In vivo trans–Specific gene silencing in fungal cells by in planta expression of a double–stranded RNA. *BMC Biology, 31*, 27.

Trieu, T. A., Calo, S., Nicolas, F. E., Vila, A., Moxon, S., & Dalmay, T. (2015). A non–canonical RNA silencing pathway promotes mRNA degradation in basal fungi. *PLoS Genetics, 11*, e1005168.

Turnage, M. A., Muangsan, N., Peele, C. G., & Robertson, D. (2002). Geminivirus–based vectors for gene silencing in arabidopsis. *The Plant Journal, 30*, 107–117.

Urich, T., Lanzén, A., Qi, J., Huson, D. H., Schleper, C., & Schuster, S. C. (2008). Simultaneous assessment of soil microbial community structure and function through analysis of the meta–transcriptome. *PLoS One, 3*, e2527.

Walawage, S. L., Britton, M. T., Leslie, C. A., Uratsu, S. L., Li, Y., & Dandekar, A. M. (2013). Stacking resistance to crown gall and nematodes in walnut rootstocks. *BMC Genomics, 14*, 668.

Wang, M., Weiberg, A., Lin, F. M., Thomma, B. P. H. J., Huang, H. D., & Jin, H. (2016) Bidirectional cross–kingdom RNAi and fungal uptake of external RNAs confer plant protection. *Nature Plants, 2*, 16151.

Wang, M., Thomas, N., & Jin, H. (2017). Cross-kingdom RNA trafficking and environmental RNAi for powerful innovative pre- and post-harvest plant protection. *Current Opinion in Plant Biology, 38*, 133–141.

Wang, M., Weiberg, A., Dellota, E., Jr., Yamane, D., & Jin, H. (2017a). Botrytis small RNA Bc–siR37 suppresses plant defense genes by cross–kingdom RNAi. *RNA Biology, 14*, 421–428.

Wang, B., Sun, Y. F., Song, N., Zhao, M. X., Liu, R., & Feng, H. (2017b). *Puccinia striiformis* f. sp *tritici* microRNA–like RNA 1 (Pst–milR1), an important pathogenicity factor of Pst, impairs wheat resistance to Pst by suppressing the wheat pathogenesis–related 2 gene. *The New Phytologist, 215*, 338–350.

Wani, S. H., Sanghera, G. S., & Singh, N. B. (2010). Biotechnology and plant disease control–role of RNA interference. *American Journal of Plant Sciences, 1*, 55–68.

Waterhouse, P. M., Graham, M. W., & Wang, M. B. (1998). Virus resistance and gene silencing in plants can be induced by simultaneous expression of sense and antisense RNA. *Proceedings of the National Academy of Sciences, 95*, 13959–13964.

Weiberg, A., Wang, M., Lin, F. M., Zhao, H., Zhang, Z., Kaloshian, I., et al. (2013). Fungal small RNAs suppress plant immunity by hijacking host RNA interference pathways. *Science, 342*, 118–123.

Witwer, K. W., & Hirschi, K. D. (2014). Transfer and functional consequences of dietary microRNAs in vertebrates: Concepts in search of corroboration: Negative results challenge the hypothesis that dietary xenomiRs cross the gut and regulate genes in ingesting vertebrates, but important questions persist. *Bioessays, 36*, 394–406.

Witwer, K. W., McAlexander, M. A., Queen, S. E., & Adams, R. J. (2013). Real–time quantitative PCR and droplet digital PCR for plant miRNAs in mammalian blood provide little evidence for general uptake of dietary miRNAs: Limited evidence for general uptake of dietary plant xenomiRs. *RNA Biology, 10*, 1080–1086.

Wroblewski, T., Piskurewicz, U., Tomczak, A., & Ochoa Michelmore, R. W. (2007). Silencing of the major family of NBS–LRR–encoding genes in lettuce results in the loss of multiple resistance specificities. *The Plant Journal, 51*(5), 803–818.

Wuriyanghan, H., & Falk, B. W. (2013). RNAinterference towards the potato psyllid, *Bactericera cockerelli*, is induced in plants infected with recombinant tobacco mosaic virus (TMV). *PLoS ONE, 8*, e66050.

Yin, C., & Hulbert, S. (2015). Host induced gene silencing (HIGS), a promising strategy for developing disease resistant crops. *Gene Technology, 4*, 130.

Yin, G., Sun, Z., Liu, N., Zhang, L., Song, Y., Zhu, C., & Wen, F. (2009). Production of double–stranded RNA for interference with TMV infection utilizing a bacterial prokaryotic expression system. *Applied Microbiology and Biotechnology, 84*, 323–333.

Yin, C., Jurgenson, J. E., & Hulbert, S. H. (2011). Development of a host–induced RNAi system in the wheat stripe rust fungus *Puccinia striiformis* f.sp.*tritici*. *Molecular Plant-Microbe Interactions, 24*, 554–561.

Yin, C., Downey, S. I., Klages Mundt, N. L., Ramachandran, S., & Chen, X. (2015). Identification of promising host–induced silencing targets among genes preferentially transcribed in haustoria of *Puccinia*. *BMC Genomics, 16*, 579.

Yuan, B., Latek, R., Hossbach, M., Tuschl, T., & Lewitter, F. (2004). siRNA selection server: An automated siR–NA oligonucleotide prediction server. *Nucleic Acids Research, 32*, 130–134.

Zhang, H., Guo, J., Voegele, R. T., Zhang, J., Duan, Y., Luo, H., & Kang, Z. (2012a). Functional characterization of calcineurin homologs PsCNA1/PsCNB1 in *Puccinia striiformis* f. sp. *tritici* using a host–induced RNAi system. *PLoS One, 7*, e49262.

Zhang, L., Hou, D. X., Chen, X., Li, D. H., Zhu, L. Y., & Zhang, Y. J. (2012b). Exogenous plant MIRI68a specifically targets mammalian LDLRAP1: Evidence of cross–kingdom regulation by microRNA. *Cell Research, 22*, 107–126.

Zongli, H., Urvi, P., Natsumi, M., Yuri, T., & Jose, R. B. (2015). Down regulation of *Fusarium oxysporum* endogenous genes by host–delivered RNA interference enhances disease resistance. *Frontiers in Chemistry, 3*, 1–10.

Zrachya, A., Kumar, P. P., Ramakrishan, U., Levy, Y., Loyter, A., Arazi, T., et al. (2007). Production of siRNA targettance to the virus against TYIC coat protein transcripts lead to silencing expression and resistance in virus. *Transgenic Research, 16*, 385–398.

Chapter 11
Genome Editing Technologies for Resistance Against Phytopathogens: Principles, Applications and Future Prospects

Siddra Ijaz and Imran Ul Haq

Abstract Genetic variation in crop plants is an indispensable factor for sustainable agriculture. For creating genetic variation, plant biotechnology relied on various random mutagenesis approaches such as γ-radiation, EMS generated mutagenesis etc. Now these methods are being replaced by genome editing technologies that precisely manipulate specific sequences in the genome. These technologies, based on sequence-specific nucleases (SSNs), mediated double-strand breaks in DNA at specific sites within the genome. Among them, three foundational targeted genome editing technologies are: TALENs (transcription activator-like effector nucleases), CRISPR/Cas9 (clustered regularly interspaced short palindromic repeats) and ZFNs (zinc-finger nucleases). TALENs and ZNFs are synthetic engineered bipartite enzymes, consisting of two domains, (1) DNA binding domain and (2) Fok1 domain (nuclease).CRISPR/Cas 9 is a two-component unicellular machinery, with a sgRNA first component (which targets the specific sequence in genome), and Cas 9 as a second component (an enzyme that mediates site-specific targeting of genome). The Cas9 protein is DNA nuclease whose activity is determined by target seed sequence (first 20 nucleotides) of sgRNA. Therefore, multiple sgRNAs with different target seed sequences direct Cas9 to corresponding spots. This imperative characteristic of Cas9 enables it to edit at multiple sites simultaneously and imparts potential applications in both basic and applied research. By using these approaches, disease resistance in plants is achieved by knocking out those loci contributing in disease susceptibility and negative regulator genes in genome. Thereby these techniques are becoming a new toolbox of every modern molecular biology laboratory for editing

S. Ijaz (✉)
Centre of Agricultural Biochemistry and Biotechnology (CABB),
University of Agriculture Faisalabad, Faisalabad, Pakistan

I. Ul Haq
Department of Plant Pathology, University of Agriculture Faisalabad, Faisalabad, Pakistan

© Springer Nature Switzerland AG 2020

I. Ul Haq, S. Ijaz (eds.), *Plant Disease Management Strategies for Sustainable Agriculture through Traditional and Modern Approaches*, Sustainability in Plant and Crop Protection 13, https://doi.org/10.1007/978-3-030-35955-3_11

plant genome. This chapter provides a detailed overview of genome editing approaches for developing disease resistant in plants, describing comprehensive knowledge on genome editing principles, application of these techniques, challenges and prospects.

Keywords CRISPR-Cas 9 · TALENs · ZNFs · Fok1 domain · DSBs

11.1 Introduction

In agricultural sciences, the big challenge of this century is to increase yields of crop plants with sustainability, by developing superior cultivars. The foremost objective of sustainable agricultural intensification is to surge the crop production by minimizing environmental impact. Plant pathogens are the principal factors limiting the world crop production and have become the major threat to food sustainability around the globe. Without genetic resistance exploitation, crop productions heavily relied on chemical control of phytopathogens. This control method poses disastrous impacts on environment, which has become the main concern to keep ecosystems healthy. Hence, reduction in the use of chemical control against phytopathogens by adopting secure means, which have no negative contribution to climate change globally, is the main objective of plant biologists (Tilman et al. 2002; IPCC 2007).

The rapidly evolving population of phytopathogens has brought an increase in the number of infectious diseases in plants. This scenario provokes the need of developing new strategies for disease resistance breeding. Targeting editing of plant DNA offers precise modification in the genome for modulating the expression of genetic elements involved in resistance to pathogens (Nejat et al. 2016).

Plants have diverse defense mechanisms to ward off diseases caused by phytopathogens. The response against their attacks relies on the recognition of the pathogen at the cell level, eliciting complex cellular signaling pathways (Jones and Dangl 2006; Wise et al. 2007; Andolfo and Ercolano 2015). The genome editing approaches allow the scientist to edit the genome architecture at targeted locations. Due to the highly complex architecture of the plant genome, different editing strategies have been developed. The site-directed genome editing strategies relied on the induction of double stranded breaks (DSBs) at a targeted place in the DNA, through programmable endonucleases that result in deletion and insertion at a single or multiple targeted sites, with the aid of the cellular repair mechanisms.

Among these genome editing technologies, TALENs (transcription activator-like effector nucleases), CRISPR/Cas9 (clustered regularly interspaced short palindromic repeats) and ZFNs (zinc-finger nucleases) are the foundational genome editing approaches. TALENs and ZNFs are custom-engineered proteins in which the cleavage domain of the enzyme Fok1 exonuclease is fused to a DNA binding domain

(Kim et al. 1996; Christian et al. 2010). However, CRISPR/Cas9 based on base pairing relies on RNA-guided nucleases, which impart simplicity and versatility, with high efficiency (Jinek et al. 2012). Moreover, this system is advantageous over other genome editing tools such as TALENs and ZNFs by allowing simultaneous editing at multiple sites in the genome (Cong et al. 2013). However, in comparison of zinc-finger nucleases and transcription activator-like effector nucleases, it exhibits a high rate of cleavage power (Joung and Sander 2013).

The application of genome editing technologies has been achieved in various crop plants such as *Arabidopsis thaliana*, *Nicotiana benthamiana*, *Triticum aestivum*, *Oryza sativa*, *Sorghum bicolor*, *Solanum tuberosum,* sweet orange, *Glycine max* etc. (Christian et al. 2010; Cermak et al. 2011; Li et al. 2013; Gao et al. 2015; Luo 2016; Rani et al. 2016). The emerging demand of crop yield stability urged the development of varieties with potential and ability to withstand fitness penalties induced by phytopathogens. In this perspective, genome-editing approaches are emerging molecular tools to edit key players of immunity in crop plants. In these, susceptibility associated genes (*S* genes) have been considered for modifications inducing resistance traits against pathogens.

TALENs (transcription activator-like effector nucleases), CRISPR/Cas9 (clustered regularly interspaced short palindromic repeats) are genome editing approaches that have been used successfully to edit the powdery mildew resistance-associated locus O (MLO) in *Triticum* aestivum, in order to develop powdery mildew disease resistant plant (Wang et al. 2014). The genome editing strategy has also been employed against *Xanthomonas oryzae* by editing the host factors required for the survival of the pathogen (Pyott et al. 2016).

It is well- known that the pathogens release effectors to suppress the plant immunity. Hence, to develop resistance against pathogens in plants, effector loci are the putative target candidates for genome editing tools (Gawehns et al. 2013). The effector targeted loci may be edited to deprive their molecular interaction required for pathogen activity, when establishing a disease in plants. It has been documented that single changes in polypeptide chain of effectors alter their molecular interaction, such as the EPIC1 and PmEPIC in *Phytophthora* infestans and *Phytophthora mirabilis*, respectively, as well as in their respective targets PLCP and RCR3, in tomato and potato (Dong et al. 2015; Niks et al. 2015). Despite of this, another effector targeting RIN4 has also been manipulated successfully in *Arabidopsis thaliana* through genome editing tools, resulting in resistance against *Pseudomonas syringae* and *Peronospora parasitica* (Luo et al. 2009). Another attempt in *Arabidopsis thaliana* has been made against these pathogens by modifying the negative regulators of mitogen-activated protein kinase (Petersen et al. 2000). These genome-editing technologies allow plant scientists to introduce specific mutations into target DNA sequences, reducing pleiotropic effects induced by complete deletion of a segment. These technologies also exploit gain-of-function mutations to encourage quantitative grading of resistance, as a valued method to protect crops (Mohanta et al. 2017).

11.2 Zinc Finger Nucleases (ZFNs): Principle and Applications

ZFNs are custom-made engineered proteins that mediate targeted cleavage of DNA sequences. They enable site-targeted genome editing by creating double-strand breaks (DSBs) in it. Each subunit of ZFNs has two domains, a DNA-binding and a DNA-cleaving domain. The DNA-binding presents a chain consisting of two finger modules to recognize a unique hexanucleotide sequence. The DNA-cleaving domain consists in a FokI nuclease (Carroll 2011; Carlson et al. 2012; Gupta et al. 2012). They are fused to make the zinc finger nucleases, a sort of custom-designed scissors to cleave the DNA at a specific site in the genome. A set of ZFPs consists of three Zinc finger repeats which recognize a 18 base pairs targeted DNA sequence (Durai et al. 2005). Upon the recognition of specific target sequences as DNA binding domain, the FokI endonuclease cleaves and generates double stranded breaks, in the target site (Mani et al. 2005).

The genome-editing tool, ZNFs, had been employed successfully in human, animal, zebrafish tobacco, maize and arabidopsis, since 1996. The engineering and screening are time-consuming processes, however. In addition, off targets and toxicity to host cell are other major limitations (Cathomen and Joung 2008; Hsu and Zhang 2012; Xiong et al. 2015).

11.3 Transcription Activator-Like Effector Nucleases (TALENs): Principle and Applications

Transcription activator-like effectors nucleases (TALENs) have been emerged as a plant genome editing tool that offers an alternative to Zinc finger nucleases (ZFNs). In principle, it uses double stranded breaks (DSBs) in a similar manner to ZFNs for genome editing (Joung and Sander 2013). The DSBs induces the activations of cellular DNA repair mechanisms i.e., non-homologous end joining (NHEJ) and homology-directed repair (HDR). The NHEJ is an error-prone DNA repair mechanism that develops frameshift by introducing small indel (insertion/deletion) at the point of breakage, to knock out the targeted locus. HRD is instead a template-dependent DNA repair mechanism that knocks in a DNA sequence at a targeted locus, in a genome (Lieber 2010; Chapman et al. 2012; Jankele and Svoboda 2014). TALENs also possesses FokI endonuclease domain as ZNFs (Joung and Sander 2013) that is fused with transcription activator-like effectors (TALEs). TALENs based genome editing is also a non-transgenic approach that mediates targeted mutagenesis for accurate genome editing, without the delivery of foreign genes of interest, along with selectable marker genes, as practiced in transgenic technology (Becker 2011; Xu 2012).

This technology has been employed successfully in *A. thaliana*, *Brachypodium distachyon*, *Hordeum vulgare, Oryza sativa* and *Zea mays* (Shan et al. 2013a, b;

Wendt et al. 2013; Liang et al. 2013; Char et al. 2015). In plants, TALENs-mediated genome editing has been used to dissect the effector binding elements (EBEs) in the promoter region of genes contributing to disease susceptibility to plants, for inhibiting their transcription as well as for inducing the expression of downstream disease resistance genes (R). The bacterial blight is a devastating disease caused in rice by the bacterium *Xanthomonas oryzae*, affecting rice production, worldwide. This phytopathogen activates the disease susceptibility associated gene (*S*) OsSWEET14, that encodes a SWEET sucrose transporter. The pathogen uses TALEs (transcription activator-like effectors) to take over the transcription machinery of the host plant and induces susceptibility genes, explicitly the members of the SWEET sucrose transporters family, by recognizing EBEs in the promoter regions (Zaka et al. 2018). Thus, the editing of EBE region has been achieved through TALEN, resulting in the loss of affinity of this region for binding the AvrXa7, a transcription activator-like effectors (TALEs). Because upon infection this bacterial pathogen, by using its TALEs such as, AvrXa7 targets the promoter region of *S* genes and stimulates the transcription of SWEET genes, which help the bacterial cells in promoting the disease. Hence, using TALENs approach, the promoter region of S genes has been edited that contributed resistance to this disease.

11.4 CRISPR/Cas9: Principle and Applications

Genetic variability is an important aspect in plant breeding. Several techniques are in practice for genomic manipulation and editing. Nevertheless, CRISPR-Cas9 is becoming today an emerging and promising approach. CRISPR-Cas 9 is a two-component unicellular machinery. First component is sgRNA, which targets the specific sequence in genome. The second component, Cas 9, is an enzyme, which mediates site specific targeting of genome. This enzyme knock out the sequence in genome whose match is present in sgRNA.

The CRISPR-cas9 system encompasses four major components,

- CRISPR-RNA (crRNA)
- Trans-activating CRISPR RNA (tracrRNA)
- Protospacer adjacent motif (PAM)
- Cas9 nuclease

The CRISPR system is involved in the integration of foreign DNA into the CRISPR cluster, which, upon transcription, yields crRNA. The crRNA contains PAM sequence. In custom designed CRISPR/Cas9 system the crRNA is joined with tracrRNA to make a hybrid sequence called single guided RNA (sgRNA), which stimulates Cas9 for activity. The Cas9 performs nuclease activity and induces double strand breaks in a DNA molecule that is then repaired by the cellular DNA repair systems, either the HDR or NHEJ (Sander and Joung 2014). Target specificity concerns the attribute “seed sequence” which lies 12 nucleotides upstream of the PAM motif. A perfect match between the RNA and target DNA is needed (Bortesi and

Fischer 2015). The presence of the PAM region (5'-NGG-3′ or 5'-NAG-3′) in the target DNA is pre-requisite for the cleavage activity of Cas9 (Gasiunas et al. 2012; Hsu et al. 2013).

The major criticisms faced to this approach is the off-target mutation (Hsu et al. 2013; Bortesi and Fischer 2015). However, it is very exceptional in plants, as 1.6% off-targeting was observed in rice (Xie and Yang 2013). It was initially considered that the 20 nucleotide sequence of sgRNA defines the specificity. However, later it has been found that only 8–12 nucleotides at the 3′ end (the seed sequence) is required for target regions recognition (Jinek et al. 2012; Cong et al. 2013; Jiang et al. 2013).

Simultaneous editing at different genome sites is an imperative and distinct attribute of this enzyme. So, this feature imparts advantage to this technique over other genome editing tools. By using this approach, negative regulator genes or genes with a negative impact on crop plants in terms of biotic and abiotic stresses and other agronomic traits, are knocked out from genome. Hence, pioneer researches for developing fungal disease resistance in crop plants have been accomplished and become success stories of this approach. Powdery mildew resistance in wheat, and rice blast resistance in rice, have been achieved by knocking out mildew-resistance locus O (*TaMLO*) and ERF transcription factor gene *OsERF922* respectively. Thereby this technique is becoming a new tool-box of every modern molecular biology laboratory for editing plant genome. CRISPR/Cas9 has been applied successfully for the adaptation of single or multiple targeted loci in *Nicotiana tabacum*, *Arabidopsis thaliana*, *Sorghum bicolor*, *Oryza sativa* and *Lycopersicon esculentum* (Belhaj et al. 2013; Jiang et al. 2014; Bortesi and Fischer 2015). This cutting-edge genome-editing means has conferred heritable resistance in wheat for powdery mildew disease by targeting the TaMLO-A1 locus (Jiang et al. 2013; Wang et al. 2014). In *Oryza sativa* the genetic architecture of LAZY1 and OsSPD loci was targeted using this approach, resulting in changes of the tiller angle and plant color in (Xiong et al. 2015). In rice, the ethylene pathway had been edited by down regulating the ethylene responsive factor, OsERF922, which in turn imparted resistance against *Magnaporthe oryzae* (Liu et al. 2012).

11.5 Future Prospects

Rapidly evolving plant pathogens are dreadful threats to food production and security around the globe. Current advancement made significant improvement in science and technology, through high-throughput sequencing, effector biology, complete understanding of pathogenesis as well as plant innate immunity. These ways are translated into new tools to develop sustainable disease resistance in plants. Effectoromics has unraveled effector-target genes (susceptibility and executor resistance genes) in effector-assisted breeding and has opened up new horizons to develop resistance. CRISPR/Cas9 system, TALENS and ZNFs technologies emerged as powerful breakthrough in the field of genetic engineering. These genome editing tools provide targeted modulation in the genome in a very precise and efficient manner. These approaches are a proficient and well organized way for developing crop protection strategic plans in disease resistance programs. In the end, it is noteworthy to say that this is the time to battle with phytopathogens with their own artilleries, effector targets discovery, producing novel sources of resistance in crop plants. This will reduce pesticide applications as well as cut the need for transgenic crops by applying emergent genome-editing technologies such as TALENS and CRISPR/Cas9, for targeted gene revision as novel strategies of control. This knowledge ought to be assimilated into industry to lessen production costs, sustainable food protection and augment the stability of high-yielding disease resistance in plants.

References

Andolfo, G., & Ercolano, M. R. (2015). Plant innate immunity multicomponent model. *Frontiers in Plant Science, 6*, 987.

Becker, M. (2011). Method of the year. *Nature Methods, 9*, 1.

Belhaj, K., Chaparro-Garcia, A., Kamoun, S., & Nekrasov, V. (2013). Plant genome editing made easy: Targeted mutagenesis in model and crop plants using the CRISPR/Cas system. *Plant Methods, 9*, 39.

Bortesi, L., & Fischer, R. (2015). The CRISPR/Cas9 system for plant genome editing and beyond. *Biotechnology Advances, 33*, 41–52.

Carlson, D. F., Fahrenkrug, S. C., & Hackett, P. B. (2012). Targeting DNA with fingers and TALENs. *Molecular Therapy-Nucleic Acids, 1*(1), e3.

Carroll, D. (2011). Genome engineering with zinc-finger nucleases. *Genetics, 188*, 773–782.

Cathomen, T., & Joung, J. K. (2008). Zinc-finger nucleases: The next generation emerges. *Molecular Therapy, 16*, 1200–1207.

Cermak, T., Doyle, E. L., Christian, M., Wang, L., Zhang, Y., Schmidt, C., et al. (2011). Efficient design and assembly of custom TALEN and other TAL effector-based constructs for DNA targeting. *Nucleic Acids Research, 39*, 82.

Chapman, J. R., Taylor, M. R., & Boulton, S. J. (2012). Playing the end game: DNA double-strand break repair pathway choice. *Molecular Cell, 47*, 497–510.

Char, S. N., Unger-Wallace, E., Frame, B., Briggs, S. A., Main, M., Spalding, M. H., et al. (2015). Heritable site-specific mutagenesis using TALEN s in maize. *Plant Biotechnology Journal, 13*, 1002–1010.

Christian, M., Cermak, T., Doyle, E. L., Schmidt, C., Zhang, F., Hummel, A., et al. (2010). Targeting DNA double-strand breaks with TAL effector nucleases. *Genetics, 186*(2), 757–761.

Cong, L., Ran, F. A., Cox, D., Lin, S., Barretto, R., Habib, N., et al. (2013). Multiplex genome engineering using CRISPR/Cas systems. *Science, 339*, 819–823.

Dong, X., Jiang, Z., Peng, Y. L., & Zhang, Z. (2015). Revealing shared and distinct gene network organization in Arabidopsis immune responses by integrative analysis. *Plant Physiology, 167*(3), 1186–1203.

Durai, S., Mani, M., Kandavelou, K., Wu, J., Porteus, M. H., & Chandrasegaran, S. (2005). Zinc finger nucleases: Custom-designed molecular scissors for genome engineering of plant and mammalian cells. *Nucleic Acids Research, 33*, 5978–5990.

Gao, J., Wang, G., Ma, S., Xie, X., Wu, X., Zhang, X., et al. (2015). CRISPR/Cas9-mediated targeted mutagenesis in Nicotiana tabacum. *Plant Molecular Biology, 87*(1–2), 99–110.

Gasiunas, G., Barrangou, R., Horvath, P., & Siksnys, V. (2012). Cas9–crRNA ribonucleoprotein complex mediates specific DNA cleavage for adaptive immunity in bacteria. *Proceedings of the National Academy of Sciences, 109*, E2579–E2586.

Gawehns, F., Cornelissen, B. J., & Takken, F. L. (2013). The potential of effector-target genes in breeding for plant innate immunity. *Microbial Biotechnology, 6*, 223–229.

Gupta, A., Christensen, R. G., Rayla, A. L., Lakshmanan, A., Stormo, G. D., & Wolfe, S. A. (2012). An optimized two-finger archive for ZFN-mediated gene targeting. *Nature Methods, 9*, 588.

Hsu, P. D., & Zhang, F. (2012). Dissecting neural function using targeted genome engineering technologies. *ACS Chemical Neuroscience, 3*, 603–610.

Hsu, P. D., Scott, D. A., Weinstein, J. A., Ran, F. A., Konermann, S., Agarwala, V., et al. (2013). DNA targeting specificity of RNA-guided Cas9 nucleases. *Nature Biotechnology, 31*, 827.

IPCC (Intergovernmental Panel on Climate Change). (2007). *Climate change: Impacts, adaptation and vulnerability. Contribution of working group II to the fourth assessment report of the intergovernmental panel on climate change*. Cambridge: Cambridge University Press.

Jankele, R., & Svoboda, P. (2014). TAL effectors: Tools for DNA targeting. *Briefings in Functional Genomics, 13*, 409–419.

Jiang, W., Bikard, D., Cox, D., Zhang, F., & Marraffini, L. A. (2013). RNA-guided editing of bacterial genomes using CRISPR-Cas systems. *Nature Biotechnology, 31*, 233.

Jiang, W., Zhou, H., Bi, H., Fromm, M., Yang, B., & Weeks, D. P. (2014). Demonstration of CRISPR/Cas9/sgRNA-mediated targeted gene modification in Arabidopsis, tobacco, sorghum and rice. *Nucleic Acids Research, 41*, 188.

Jinek, M., Chylinski, K., Fonfara, I., Hauer, M., Doudna, J. A., & Charpentier, E. (2012). A programmable dual-RNA–guided DNA endonuclease in adaptive bacterial immunity. *Science, 337*, 816–821.

Jones, J. D., & Dangl, J. L. (2006). The plant immune system. *Nature, 444*, 323.

Joung, J. K., & Sander, J. D. (2013). TALENs: A widely applicable technology for targeted genome editing. *Nature Reviews Molecular Cell Biology, 14*, 49.

Kim, Y. G., Cha, J., & Chandrasegaran, S. (1996). Hybrid restriction enzymes: Zinc finger fusions to Fok I cleavage domain. *Proceedings of the National Academy of Sciences, 93*, 1156–1160.

Li, J. F., Norville, J. E., Aach, J., McCormack, M., Zhang, D., Bush, J., et al. (2013). Multiplex and homologous recombination–mediated genome editing in Arabidopsis and *Nicotiana benthamiana* using guide RNA and Cas9. *Nature Biotechnology, 31*, 688.

Liang, Z., Zhang, K., Chen, K., & Gao, C. (2013). Targeted mutagenesis in *Zea mays* using TALENs and the CRISPR/Cas system. *Journal of Genetics and Genomics, 41*, 63–68.

Lieber, M. R. (2010). The mechanism of double-strand DNA break repair by the nonhomologous DNA end-joining pathway. *Annual Review of Biochemistry, 79*, 181–211.

Liu, G. H., Qu, J., Suzuki, K., Nivet, E., Li, M., Montserrat, N., et al. (2012). Progressive degeneration of human neural stem cells caused by pathogenic LRRK2. *Nature, 491*, 603.

Luo, J. (2016). CRISPR/Cas9: From genome engineering to cancer drug discovery. *Trends in Cancer, 2*, 313–324.

Luo, J., Emanuele, M. J., Li, D., Creighton, C. J., Schlabach, M. R., Westbrook, T. F., et al. (2009). A genome-wide RNAi screen identifies multiple synthetic lethal interactions with the Ras oncogene. *Cell, 137*, 835–848.

Mani, M., Kandavelou, K., Dy, F. J., Durai, S., & Chandrasegaran, S. (2005). Design, engineering, and characterization of zinc finger nucleases. *Biochemical and Biophysical Research Communications, 335*, 447–457.

Mohanta, T., Bashir, T., Hashem, A., Abd-Allah, E., & Bae, H. (2017). Genome editing tools in plants. *Genes, 8*, 399.

Nejat, N., Rookes, J., Mantri, N. L., & Cahill, D. M. (2016). Plant–pathogen interactions: Toward development of next-generation disease-resistant plants. *Critical Reviews in Biotechnology, 37*, 229–237.

Niks, R. E., Qi, X., & Marcel, T. C (2015). Quantitative resistance to biotrophic filamentous plant pathogens: Concepts, misconceptions, and mechanisms. *Annual Review of Phytopathology, 53*, 445–470.

Petersen, M., Brodersen, P., Naested, H., Andreasson, E., Lindhart, U., Johansen, B., Nielsen, H. B., Lacy, M., Austin, M. J., Parker, J. E., & Sharma, S. B. (2000). Arabidopsis MAP kinase 4 negatively regulates systemic acquired resistance. *Cell, 103*, 1111–1120.

Pyott, D. E., Sheehan, E., & Molnar, A. (2016). Engineering of CRISPR/Cas9-mediated potyvirus resistance in transgene-free Arabidopsis plants. *Molecular Plant Pathology, 17*, 1276–1288.

Rani, R., Yadav, P., Barbadikar, K. M., Baliyan, N., Malhotra, E. V., Singh, B. K., et al. (2016). CRISPR/Cas9: A promising way to exploit genetic variation in plants. *Biotechnology Letters, 38*(12), 1991–2006.

Sander, J. D., & Joung, J. K. (2014). CRISPR-Cas systems for editing, regulating and targeting genomes. *Nature Biotechnology, 32*, 347.

Shan, Q., Wang, Y., Chen, K., Liang, Z., Li, J., Zhang, Y., et al. (2013a). Rapid and efficient gene modification in rice and Brachypodium using TALENs. *Molecular Plant, 6*, 1365–1368.

Shan, Q., Wang, Y., Li, J., Zhang, Y., Chen, K., Liang, Z., et al. (2013b). Targeted genome modification of crop plants using a CRISPR-Cas system. *Nature Biotechnology, 31*, 686.

Tilman, D., Cassman, K. G., Matson, P. A., Naylor, R., & Polasky, S. (2002). Agricultural sustainability and intensive production practices. *Nature, 418*, 671.

Wang, Y., Cheng, X., Shan, Q., Zhang, Y., Liu, J., Gao, C., & Qiu, J. L. (2014). Simultaneous editing of three homoeoalleles in hexaploid bread wheat confers heritable resistance to powdery mildew. *Nature Biotechnology, 32*, 947.

Wendt, T., Holm, P. B., Starker, C. G., Christian, M., Voytas, D. F., Brinch-Pedersen, H., & Holme, I. B. (2013). TAL effector nucleases induce mutations at a pre-selected location in the genome of primary barley transformants. *Plant Molecular Biology, 83*, 279–285.

Wise, R. P., Moscou, M. J., Bogdanove, A. J., & Whitham, S. A. (2007). Transcript profiling in host–pathogen interactions. *Annual Review of Phytopathology, 45*, 329–369.

Xie, K., & Yang, Y. (2013). RNA-guided genome editing in plants using a CRISPR–Cas system. *Molecular Plant, 6*(6), 1975–1983.

Xiong, J. S., Ding, J., & Li, Y. (2015). Genome-editing technologies and their potential application in horticultural crop breeding. *Horticulture Research, 2*, 15019.

Xu, K. (2012). The next generation of biotechnology for apple improvement and beyond, the tale of TALENs and TALEs. *NY Fruit Quart, 20*, 17–20.

Zaka, A., Grande, G., Coronejo, T., Quibod, I. L., Chen, C. W., Chang, S. J., Szurek, B., Arif, M., Cruz, C. V., & Oliva, R. (2018). Natural variations in the promoter of OsSWEET13 and OsSWEET14 expand the range of resistance against *Xanthomonas oryzae* pv. *oryzae*. *PLoS One, 13*, e0203711.

Chapter 12
Plant Health Clinics (PHC) in Pakistan: Operations and Prospects

Muhammad Luqman, Saleem Ashraf, Muhammad Yaseen, and Muhammad Kaleem Sarwar

Abstract Across the world, approximate 40% crop losses pertinent to insects and pests are reported. Severity of these losses fluctuate with the varying climatic conditions. More often these losses surpass the threshold level and farmers who are already resource poor, technically unskilled and victim of impartial access to information are unsuccessful in lowering them. In Pakistan, majority of farmers are small landholders, lacking resources, less educated and vulnerable to multiple farming risks. Inadequate tendencies to adopt and implement for plant protection restrict their expectations. Poor quality chemicals further boost the damage level. Challenges associated to plant health are constant, despite of active working of extension advisory service providers. However, role of extension workers, particularly those from plant protection departments, is critical., To strengthen the extension advisory system, appropriate knowledge access to information, coordination and cooperation with other related agencies are viable options to adopt, to provide appropriate and effective plant protection services in best entrust of farmers. Among the initiatives taken by the extension service providers related to plant protection, Plant Health Clinics (PHC) is one of the most innovative steps implemented. This initiative aimed at providing farmers an opportunity to get first hand solution for their problems. Serving farmers directly is the focus of PHC. Therefore, to enhance access, PHC are organized at public places close to farms so that farmers may be able to reach easily and consult their problem with extension expert. This initiative is on-going under the umbrella of the Punjab Agriculture Department in Pakistan. There is need to unveil the answer that how establishment of PHC is feasible? Are the accountability, reporting and communication of this initiative and competency of expert understood prior implementation of PHC? In this context, this chapter aims

M. Luqman (✉) · M. Yaseen
College of Agriculture, University of Sargodha, Sargodha, Pakistan
e-mail: muhammad.luqman@uos.edu.pk

S. Ashraf
In-Service Agricultural Training Institute, Rahim Yar Khan, Punjab, Pakistan

M. K. Sarwar
Department of Plant Pathology, University of Agriculture Faisalabad, Faisalabad, Pakistan

© Springer Nature Switzerland AG 2020

I. Ul Haq, S. Ijaz (eds.), *Plant Disease Management Strategies for Sustainable Agriculture through Traditional and Modern Approaches*, Sustainability in Plant and Crop Protection 13, https://doi.org/10.1007/978-3-030-35955-3_12

at underpinning ground concerns to what extent PHC is effective and successful in attracting farmers to get their problems solved. In Pakistani context, which factors are affecting PHC performance? Furthermore, success stories of PHC in Pakistan will be compared with other countries to success guidelines. This chapter is based on a mixed method technique. One portions is based on extensive literature while success stories of PHC with primary data will be described. Through primary data collection the role of extension department in improving effectiveness of PHC will be assessed.

Keywords Plant health clinic · Rural advisory services · Rural development · Innovative agricultural extension approaches · Plant health

12.1 Introduction

12.1.1 Rural Development and Agriculture

Rural areas are mainly characterized on the basis of land utilization as a large portion of land in Pakistani rural areas is under farming. In this connection the entire development of rural areas is strongly linked to agriculture (Muzari et al. 2012). In developing countries the interaction between agriculture and rural development is very strong and clear as their national economy is mainly based on agriculture (Abdullah et al. 2005). It is reported that agricultural activities contribute significantly in developing rural areas. In Pakistan, agriculture represents the single largest sector with a major share in national Gross Domestic Product (GDP). The overall development of people residing in rural areas in Pakistan is strongly linked to the development and growth of agriculture, on sustainable basis (Nadeem and Mushtaq 2012).

12.1.2 Rural Advisory Services and Agricultural Development

Sustainable growth and development in the agricultural sector requires continuous, cost-effective and up-to-date information and transfer of knowledge towards end users (farming community). Farmers mainly need information concerning crop production and protection practices to gain maximum outputs, in the form of crop yields that ultimately improve their income level and livelihood status. Using minimum resources of manpower and materials to maximize crop yield is one of the key challenges for sustainable growth in agriculture. All growth and development of agriculture is only possible through continuous provision of advisory/extension services to farmers (Peterson et al. 2001).

The concept of "agricultural extension" is relatively old and has changed over time (Swanson 2008).It is applied to several other areas of the community. This idea was derived from the concept of "university extension". The term "extension" covers several areas of development including agriculture, education and health. The main objective of extension is to fill the gap between information source(s) and end users (Hassan et al. 2012). Several attempts were made to define the term "extension", but these definitions have been changed from time to time. It is a kind of informal education that deals with teaching and training of the community in order to uplift and improve the living standards of the community by using their available resources (Bradfield 1966).

The concepts of extension changed from informal to non-formal education during the late 1960s. Extension is a voluntary educational program for adults outside educational institutions (Agbogidi and Ofuoku 2009). It is a kind of service that provides assistance to rural community members about new trends in production, protection and marketing of agricultural products by means of an educational process. According to Christoplos (2010) it has been defined as a system that helps farmers in accessing technologies, knowledge and information, facilitate their stakeholders, interactive research, business and education to help them for their technical support and to develop their organizational as well as management skills. The term "Extension" can be defined as provision of advisory services to low income families for the utilization of local resources at their own end (Ajani and Onwubuya 2010). Such kind of services that are delivered to rural community by extension agents are termed as "rural advisory services" and their primary objective is to work for the well-being of farmers.

Finally, it is a type of voluntary education for the farming community (Kolawole and Oladele 2013). Extension services include education, protection and information dissemination on routine life issues such as environmental conservation. According to Živković et al. (2009) several extension services can be identified as optional, economical, universal and educational. In addition to educational and health-related functions of extension, one of its important functions is to work for farmers' well being especially in poverty alleviation, social inequalities and overall development of rural communities. In this regard, extension strategies, state extension departments, NGOs, private companies (extension wing), universities and research institutes through outreach programs and farmers associations or unions play a key role (Hoffmann et al. 2009). The main responsibility of agricultural extension wing is to work for the improvement of living standards of farmers. Agricultural extension service(s) have great impact on overall development of the agricultural sector. It has been proved that effective extension services are very helpful to stimulate or uplift a country's production, greatly (Akinola et al. 2011).

Agricultural extension services can work as a useful tool for the economic development of low income or developing countries whose economies are specifically based upon agriculture (Dimelu and Okoro 2011). It acts as a bridge between agriculture research and farming communities. Such services are also very helpful to transfer the scientific knowledge to farmers so that it can become part of agricultural practices, improving the overall agricultural productivity through various activities and initiatives aimed at rural development (Ifeanyi-obi et al. 2012).

Globally, one billion small land holders are being accommodated by agricultural extensionist(s), worldwide. It is also the responsibility of agricultural extensions to empower and educate rural communities for sustainable rural development (Koyenikan 2008). In Pakistan, the role of agricultural extension is to introduce new and innovative agricultural technologies or techniques to local farmers. These technologies can be adopted by local farmers and incorporated in their already existing system of crop protection and protection. Regardless of its key role in rural economy, provisions of extension services to rural communities of developing countries are very poor, regardless of their key role in rural economy (Pervaiz et al. 2013).

12.1.3 Core Functions of Rural Extension or Rural Advisory Services

The core functions of extension or advisory services of rural localities are as follows:

1. Planning
2. Contents and Methods
3. Monitoring and Evaluation
4. Problem Solving Assistance
5. Training
6. Facilitation and Animation
7. Dissemination of Information

12.2 Approaches Being Used for Extension Work and Rural Advisory Services

The type of action within an extension system is referred to as an "approach", and is the basic essence of an extension system. Different approaches are being used for effective agricultural extension work and for provision of advisory services to the rural community. These approaches are listed below:

1. General Agricultural Extension
2. Training and Visit
3. Commodity Specialized
4. Farming System Research
5. Cost Sharing
6. Participatory Extension
7. Educational Institutional
8. Project Development

12.3 Plant Clinics: An Innovative Agricultural Extension Approach

In the changing scenario regarding the role of agricultural extension services in improving livelihoods of rural people, and to meet their growing demands, there is a direct need to search for an alternative agricultural extension strategy/approach. Out of many alternative agricultural extension strategies, Plant Clinics has been designed and tested by CABI in 2001. According to an estimate, about 130 Plant Health Clinics are being established and operated in the world developing regions, including in particular Asia, Africa and Latin America. It is a type of primary health care service being provided to resource-poor farmers. It provides an opportunity for farmers to get advice and recommendations related to their crops, on the basis of field diagnosis and observations (Negussie et al. 2011). Staff members who run the Plant Health Clinics are termed as "Plant Doctors". They receive training, first related to techniques of plant disease/pest diagnosis (Boa 2009).

12.3.1 Main Features of Plant Health Clinics

Following are the main features of Plant Health Clinics:

- Operated by staff members/specialists of a local organization who is very much familiar with local agricultural problems and situations.
- Operated on the basis of problems/queries presented by farmers not staff members of research organization or Extension Field Staff (EFS).
- Location of the plant clinics should be accessible for all famers (public place). No building is required. Only furniture, shady place as well as basic equipment being used for diagnosis of plant diseases.
- Work on any problem of the crop and operates on weakly or fortnightly basis. On special occasions, i.e., agricultural exhibitions or field days, also plant clinics are operative.
- Provide relevant, regular and practical information/advisory services related to crop management.
- Record keeping are related to changing status of crop pests.
- Any farmer can approach plant health clinics to get updated information about field problems.

12.3.2 Reported Strengths

- Demand-driven and addressing farmers' priority problems.
- Accessible to farmers.

- Enhanced outreach – clinic staff solve many farmers' problems in a few hours.
- Dealing with any plant health problem.
- Update and build capacity of extension staff.
- Farmers gained confidence in extension officers.
- Enhanced interaction and collaboration among farmers and other stakeholders.
- Can be integrated into existing structure and activities – implemented by diverse local organizations.
- Help farmers gain better knowledge on pest and disease management.
- Help farmers use appropriate control options – e.g., use of the right chemicals.
- Help to obtain lists of diseases and pests.
- Better yield and quality are produced due to timely actions, hence better prices may be obtained.
- Serve as surveillance mechanism, and help to capture emerging plant pests and diseases.
- Inform research – identify priority research areas.

- Source: Negussie et al. (2011).

12.3.3 Limitations

- Resource shortage – finance, local staff and transport.
- Tendency to view the initiative as a project.
- Limited institutionalization.
- High staff turn-over among local implementing (e.g., extension) agencies.
- Workload due to other competing activities – disruption of clinics.
- Limited technical capacity among some of the local staff.
- Limited publicity for the clinics – some farmers come without samples.
- Lack of effective linkages to other services such as diagnostic labs.
- Dry season affects farmer turn out, while wet season affects mobility.
- Performance of the plant clinics at times relied on the interest and commitment of some individuals.

- Source: Negussie et al. (2011).

12.4 PHC an Innovative Step by Extension Workers to Provide First Hand Solution to Farmers

African and mostly other developing countries have major dependency on performance of agricultural sector for livelihood. Crop production, which is already challenged by various natural stressors/pests, worsens when faced with weak agricultural extension and advisory services. Farmers, especially those having small land holdings and resources, require to access latest and reliable practical extension services on a regular basis, mainly regarding market trends and upcoming management strategies for specific crops. In African countries, farmers are deprived, due to many ecological and socio-economic factors, of basic extension services. As private enterprises are more concerned about business, often in a specified crop or region, low scale farmers are not informed about unpredictable changes and their proper preventive measures, resulting in resource losses (Bentley 2009; Negussie et al. 2011). Poor advisory or extension services result in lack of information for farmer needs, that also affects the higher level of policy makers or research institutes. This situation thus seriously affects low scale farmers' recent and upcoming issues, and ultimately the rural development. As the public sector also faces serious challenges due to poor flow of information, keeping low scale farmers out of the loop, actual focus in on farmer-oriented approaches, considering extension and advisory services as major players in rural development (Boa 2007). Focus in rural development extension services is on the quality of interaction rather than just a flow of messages, to improve crop production and income of a large number of small scale holders who solely depend on their crops (Danielsen et al. 2013).

Instead of promoting a specific product or technology, plant health clinics focus on providing demand-driven solution based on field diagnosis, by utilizing available resources and information on a regular basis. Demand in PHC is driven by farmer queries rather than researchers, and are preferably run by local workers, expert in diagnosis and management of plant health, and familiar with the local conditions. PHC have to adjust to social values and institutional structures already working in specific area, and get adapted to local conditions (Boa 2009; Danielsen et al. 2013). The PHC personnel mostly operate in open public places once a week, where farmers can bring samples for diagnosis of their crops or management queries regarding any crop. The PHC is equipped with all basis devices required for diagnosis of field problems. If these are difficult to diagnose then samples may be sent to a nearby laboratory, and the farmer is guided to keep in mind affordability and availability of management options (Danielsen and Fernández 2009). The PHC focus is on latest trends in disease and insect pest management, incorporating traditionally successful and scientific methods for any specific problem, with minimum use of synthetic chemicals, through an approach known as Integrated Disease Management (IDM) (Danielsen et al. 2013). All necessary data of client and diagnosis including appropriate solutions, farmer priority and changing trends of pests and diseases, are registered, to be used for research purposes. PHS is the best platform for collaboration among different institutes and networks, with various expertises to achieve a betterment of low scale farmers conditions. As clinics provide benefits yielding solutions to farmers' problems, next step is analyzing and utilizing the information collected for collective actions. These are timely informing about possible epidemics in the next season, formulating management strategies and activating regulatory agencies before time (Negussie et al. 2013).

Due to the diversity of biotic and abiotic factors responsible for many diseases with confusing symptoms, choosing the best management strategy for a particular issue produces constant challenges for extension workers and farmers. in the form i.e. of a potential loss of crop production and quality. PHC is a practical way to improve management strategies for farmers through extension workers powered by plant health experts. In the USA, plant health clinics have joint ventures in 42 states by linking agricultural experts to university scientists for better extension and research responsibilities. India has 3000 PHC which are commercially funded, mostly by input sales or other services, blending research and extension, although the farmers access is less on the national level. (Boa et al. 2016; NPDN). Institutional PHC have their own diagnostic laboratories, whereas extension plant health clinics provide direct services to farmers, mostly at their door step or workplace.

PHC is a demand-driven service provided during daily routine extension and advisory services. They are most effective when a complete package if offered with expertise and knowledge, regarding local conditions and applicable solutions (Danielsen and Matsiko 2016).

Limitations faced by the plant health management system were overcome by CABI, through a plantwise program in which plant health clinics were established across Africa, Asia and some parts of America, during 2003 (Romney et al. 2013). PHC is an innovative idea based on a community service with focus on small land holders. It is available to everyone, being often established on a place easily accessible to farmers such as markets or farmer cooperatives. (Boa 2009).

12.5 Accessibility and Consultancy to Farmer Problems

In Uganda, PHC established in 2005 and 2006 have improved the extension services, increased outreach of farmers and improved disease assessment in local districts. Success of this approach provoked the government to include a PHC program in its 5-year development and investment plan. PHC was properly activated by CABI, in 2010 and 2011, under plantwise program by government and NGOs (Danielsen et al. 2012). PHCs improve efficiency of agri-extension departments through innovative methods of helping farmers to deal with insect pests, diseases and other on farm problems, effectively. Plant health clinics in Teso were operating on the same principle and intentions as were in Bolivia (Bentley et al. 2009, 2011) and Bangladesh (Harun-Ar-Rashid et al. 2010). Vision of plant clinics was to cover all the aspects of crop production by proposing local solutions to them. Major issues related to farmers regarding their crop is plant health, so diagnosis of diseases and their proper management strategies became the focus of this PHC system. The management procedure starts with educating farmers on the importance of proper germplasm and application of inputs necessary for plant health and growth, which will improve plant resistance and mitigate losses. Ultimately, improved yield will increase farmer's profitability and quality of produce. If producers are bringing samples of viral disease for which there is no treatment, the solution relies on management strategies. In 2011, by utilizing the experience of Global Plant Clinic (GPC) and activities in Uganda, Plantwise program was used to scale up plant clinics. For effective and long term success of this program proper monitoring, evaluation, assessing impact, improvement with time and according to local conditions, are necessary (Brubaker et al. 2013). Participation in sessions of plant clinics based on agri-extension and research have positive effects on rural development by focusing farmers and improving production, knowledge and adoption to latest trends and disease management strategies (Birkhaeuser et al. 1991). Farmers attending agri- extension and research related projects such as PHC show reasonably good rates of return, varying according to design and purpose of the project (Alston et al. 2000). Results of Alston et al. (2000), further expanded by Anderson (2008), highlighted various measurement issues regarding extension work evaluation. These Authors concluded that there are mixed results in term of return rates. The same conclusions — lack of strong impact evidence in extension projects — were attained by Davis (2008).

In Nepal PHC was established in 2008, for an early detection and warning system of plant diseases, discovery of new diseases/pests and to support quarantine program. PHC in its latest form provides best surveillance, make extension workers technically stronger, improving and providing services to low scale farmers (Adhikari et al. 2013). PHC gained tremendous popularity in Nepal as an excellent extension services provider. In 2013 the Nepal Government signed an agreement with CABI for structured implementation of the system which was considered a success. Today 40 PHC are working in various districts of the country (CABI 2015). The number of sessions, queries and farmers increased in 2014 as compared to the previous year. PHC is fulfilling its core value of increasing access, but still it has not a wide spread across the country.

Important points to increase the access and interest of farmers in PHC are easily accessible location, advertisements for attraction, quality of the service, expert plant doctors and proper coordination with different farmer stakeholders. For a specific solution to the problems of a better location, PHCs must be built in areas where they are previously absent, in order to improve livelihood of small land holding farmers (Adhikari et al. 2018).

PHC system in Rawanda was a success and was appreciated by the Agricultural board and Government, due to its major role in technically strengthening farmers, disease vigilance and reduction in crop losses. This was proved by a survey in 2015 in which more than 90% of PHC users were satisfied with this system and referred it as an effective service provider (Nsabimana et al. 2015). A comparative study of attendance within and among countries in PHC sessions showed great variability, when data of Rawanda, Ghana, Malawi and Zambia were compared. This differences in attendance may be due to many reasons. The case study of Uganda shows that clinics which are operated by collaboration of NGOs and local government are successful as they have ability to withstand unconducive policies in the institute's environment, as compared to plant clinics which are under local Government control (Danielsen and Matsiko 2016). Significant variation in performance of plant clinics was observed despite the fact that staff, fund and logistics conditions were similar across the districts. All PHCs were facing problem of funds and shortage of staff. However, PHC with higher return rates and attendance studied proved that commitment, publicity, motivation, accountability of staff, proactive communication among PHC staff and farmers, including local leaders, were key factors. Factors such as crop grown and purpose, distance to PHC, problem being faced, farmer's social and economic status may also influence the effectiveness of the clinics.

In conclusion, PHC significantly improve farmers' ability to tackle the major insect pest and disease problems in their fields, when attended regularly, as proved by several studies and testimonies (Majuga et al. 2018).

References

Abdullah, A., Gillani, W. A., Naveed, S., Amanullah, K., & Kashif, H. (2005). *Computerized farm guide: Using ICT for better dissemination of agriculture extension information*. Islamabad: Center for Agro-Informatics Research, National University of Computer and Emerging Sciences.

Adhikari, R. K., Regmi, P. P., Boa, E., Dhoj, G. Y., & Thapa, R. B. (2013). Innovation in plant health extension services: The case of plant clinics in Nepal. *Economia Agro-alimentare, 15*, 235–245.

Adhikari, D., Sharma, D. R., Pandit, V., Schaffner, U., Jenner, W., & Dougoud, J. (2018). Coverage and access of plant clinic in Nepal. *Journal of Agriculture and Environment, 18*, 51. https://doi.org/10.3126/aej.v18i0.19889.

Agbogidi, O. M., & Ofuoku, A. U. (2009). Forestry extension: Implications for forest protection. *International Journal of Biodiversity and Conservation, 1*, 98–104.

Ajani, E. N., & Onwubuya, E. A. (2010). Farmer field school (FFS) and junior farmer field and life school (JFFLS) as challenges to agricultural extension development and practice in Nigeria. *Journal of Agricultural Extension, 14*, 53–61.

Akinola, M. O., Issa, F. O., & Sanni, S. A. (2011). Agricultural extension Systems in West Africa: Adoptable strategies for Nigeria's agricultural extension reform agenda. *Journal of Agricultural Extension, 15*, 1–7.

Alston, J. M., Chan-Kang, C., Marra, M. C., Pardey, P. G., & Wyatt, T. J. (2000). *A meta-analysis of rates of return to agricultural R&D: Ex pede Herculem?* (Vol. 113). Washington: IFPRI.

Anderson, J. R. (2008). *Agricultural advisory services*. Washington, DC: World Bank. 44 pp.

Bentley, J. W. (2009). Impact of IPM extension for smallholder farmers in the tropics. In R. Peshin & A. K. Dhawan (Eds.), *Integrated pest management: Dissemination and impact* (pp. 333–346). Dordrecht: Springer.

Bentley, J. W., Boa, E., Danielsen, S., Franco, P., Antezana, O., Villarroel, B. (2009). Plant health clinics in Bolivia. 2000—2009: Operations and preliminary results. *Food Security, 1*, 371–386.

Bentley, J., Boa, E., Almendras, F., Franco, P., Antezana, O., Díaz, O. (2011). How farmers benefit from plant clinics: An impact study in Bolivia. *International Journal of Agricultural Sustainability, 9*, 393–408.

Birkhaeuser, D., Evenson, R. E., & Feder, G. (1991). The economic impact of agricultural extension: A review. *Economic Development and Cultural Change, 39*, 607–650.

Boa, E. (2007). Plant healthcare for poor farmers: An introduction to the work of the global plant clinic. *APSnet*, October 2007.

Boa, E. (2009). How the global plant clinic began. *Outlooks on Pest Management, 20*, 112–116.

Boa, E., Franco, J., Chaudhury, M., Simbalaya, P., & Van Der Linde, E. (2016). Plant health clinics. Note 23. In *GFRAS good practice notes for extension and advisory services*. Lausanne: GFRAS.

Bradfield, D. J. (1966). *Guide to extension training* (Vol. 11). Rome: FAO.

Brubaker, J., Danielsen, S., Olupot, M., Romney, D., & Ochatum, N. (2013). Impact evaluation of plant clinics: Teso, Uganda. *CABI working paper 6*, 92 pp.

CABI. (2015). *Plantwise annual report 2015*. https://www.plantwise.org/impact/annual-reports/.

Christoplos, I. (2010). *Mobilizing the potential of rural and agricultural extension*. Rome: Neuchatel Group.

Danielsen, S., & Fernández, M. (2009). Nicaragua-public plant health services for all. *Results and lessons learned 2005–2007*, 84 pp.

Danielsen, S., & Matsiko, F. B. (2016). Using a plant health system framework to assess plant clinic performance in Uganda. *Food Security, 8*, 345–359.

Danielsen, S., Matsiko, F., Mutebi, E., & Karubanga, G. (2012). *Second generation plant health clinics in Uganda: Measuring clinic performance from a plant health system perspective 2010–2011*.

Danielsen, S., Centeno, J., López, J., Lezama, L., Varela, G., Castillo, P. (2013). Innovations in plant health services in Nicaragua: From grassroots experiment to a systems approach. *Journal of International Development, 25*, 968–986.

Davis, K. (2008). Extension in sub-Saharan Africa: Overview and assessment of past and current models and future prospects. *Journal of International Agricultural and Extension Education, 15*, 15–28.

Dimelu, M. U., & Okoro, C. P. (2011). Prospects of Farmer Field School Extension Approach in Enugu State. *Journal of Agricultural Extension, 15*, 79–89.

Harun-Ar-Rashid, M., Bentley, J., & Boa, E. (2010). *Benefits of plant health clinics to farmers; bigger harvests in Bangladesh*.

Hassan, M. Z. Y., Butt, T. M., Khalid, M. F., Mahmud, K., Mahr-un-Nisa, A. R. (2012). Impact of poultry extension services for the rural women. *African Journal of Agricultural Research, 7*, 1893–1900.

Hoffmann, V., Gerster-Bentaya, M., Christinck, A., & Lemma, M. (Eds.). (2009). *Handbook: Rural extension volume 1. Basic issues and concepts*. Weikersheim: Margraf Publishers GmbH.

Ifeanyi-obi, C. C., Etuk, U. R., & Jike-wai, O. (2012). Climate Change, Effects and Adaptation Strategies; Implication for Agricultural Extension System in Nigeria. *Greener Journal of Agricultural Science, 2*(2), 53–60.

Kolawole, A. E., & Oladele, O. I. (2013). Non-agricultural information services provided by extension agents in Oyo state, Nigeria. *Life Science Journal, 10*, 628–632.

Koyenikan, M. J. (2008). Issues for Agricultural Extension Policy in Nigeria. *Journal of Agricultural Extension, 12*, 52–62.

Majuga, J. C. N., Uzayisenga, B., Kalisa, J. P., Almekinders, C., & Danielsen, S. (2018). "Here we give advice for free": The functioning of plant clinics in Rwanda. *Development in Practice, 28*, 858–871.

Muzari, W., Gatsi, W., & Muvhunzi, S. (2012). The impacts of technology adoption on smallholder agricultural productivity in Sub-Saharan Africa: A review. *Journal of Sustainable Development, 5*, 69–70.

Nadeem, N., & Mushtaq, K. (2012). Role of agricultural research and extension in enhancing agricultural productivity in Punjab, Pakistan. *Pakistan Journal of Life and Social Sciences, 10*, 67–73.

Negussie, E., Karanja, P., Romney, D., Muriithi, C., Kamau, R., & Boa, E. (2011). Role of plant health clinics in enhancing adaptive capacity to climate-induced plant health problems: Experiences from Kenya. In *Africa adapt climate change symposium*, 9–11.

Negussie, E., Karanja, P., Day, R., Romney, D., Reeder, R., Boa, E. (2013). Role of plant health clinics in meeting the needs of small-scale farmers for advisory services: Experiences from Eastern Africa. In *Proceedings, international conference on innovations in extension and advisory services, 15–18 Nov. 2011* (pp. 1–9). Nairobi: The Technical Centre for Agricultural and Rural Cooperation (CTA).

Nsabimana, J. D., Uzayisenga, B., & Kalisa, J. P. (2015). *Learning from plant health clinics in Rwanda. Plantwise-CABI*. Kigali: Rwanda Agriculture Board.

Pervaiz, U., Khan, F., Jan, D., & Zell-e-Huma Zafarullah, M. (2013). An analysis of sugarcane production with reference to extension services in union council Malakandher – Peshawar. *Serhad Journal of Agriculture, 29*(1), 145–150.

Peterson, W., Galleno, V., Epomas, T., Wuyts-Fivawo, A., & Wilks, M. (2001). Methods for planning effective linkage. In *International service for national agricultural research. Briefing paper 45*, www.laolink.org.

Romney, D., Day, R., Faheem, M., Finegold, C., LaMontagne-Godwin, J., & Negussie, E. (2013). Plantwise: Putting innovation systems principles into practice. *Agriculture for Development, 18*, 27–31.

Swanson, B. E. (2008). *Global review of good agricultural extension and advisory services practices*. Rome: FAO.

Živković, D., Jelić, S, & Rajić, Z. (2009). The role of knowledge, innovation and human capital in multifunctional agriculture and territorial rural development. In *113th EAAE Seminar, Dec. 9–11*. Belgrade, Republic of Serbia.

Chapter 13
Precision Agriculture Technologies for Management of Plant Diseases

Siva K. Balasundram, Kamlesh Golhani, Redmond R. Shamshiri, and Ganesan Vadamalai

Abstract Plant diseases contribute 10–16% losses in global harvests each year, costing an estimated US$ 220 billion. Abundant use of chemicals such as bactericides, fungicides, and nematicides to control plant diseases are causing adverse effects to many agroecosystems. Precision plant protection offers a non-destructive means of managing plant diseases based on the concept of spatio-temporal variability. Global Navigation Satellite System (GNSS) and Geographic Information System (GIS) allow for assessment of field heterogeneity due to disease problems and can enable site-specific intervention. Similarly, hyperspectral remote sensing is a cutting-edge spectral approach for plant diseases detection. The main aim of precision plant protection is to significantly reduce the injudicious use of chemical inputs and hence the adverse impact of chemicals to the environment. This chapter provides some insights into the deployment of site- and time-specific approaches to manage plant disease problems in a balanced and optimized manner.

Keywords Precision agriculture · Global positioning systems · Geographic information systems · Remote sensing · Spectroradiometer

S. K. Balasundram (✉) · K. Golhani · R. R. Shamshiri
Department of Agriculture Technology, Universiti Putra Malaysia (UPM), Serdang, Selangor, Malaysia

G. Vadamalai
Department of Plant Protection, Universiti Putra Malaysia (UPM), Serdang, Selangor, Malaysia

© Springer Nature Switzerland AG 2020

I. Ul Haq, S. Ijaz (eds.), *Plant Disease Management Strategies for Sustainable Agriculture through Traditional and Modern Approaches*, Sustainability in Plant and Crop Protection 13, https://doi.org/10.1007/978-3-030-35955-3_13

13.1 Introduction

13.1.1 *Plant Disease Management*

Crop stress represents a major challenge in agricultural production. Stress sources are classified into two categories: biotic and abiotic. Biotic stresses are caused by diseases (Carter and Knapp 2001), insect pests (Riedell et al. 2000) and weeds (Pinter et al. 2003), while abiotic stresses are caused by physical factors such as water deficit (Steele et al. 2008), salinity overload (Pinter et al. 2003), and nutrient deficiency (Blackmer et al. 1995). Particularly, plant diseases result in changes in crop physiology (transpiration, photosynthesis), morphology (tissue shape or color) and crop density (West et al. 2010). Changes in host plants are also caused by hypersensitive reactions (Chaerle 2004) and cell wall degradation (Blackburn 2007).

At present, plant diseases represent a major threat to the global economy. Severe economic losses have been incurred in the agriculture industry due to diseases. Therefore, effective disease monitoring and early detection system should be facilitated to reduce their incidence and spread (Martinelli et al. 2014). It has been reported that many developed countries have established disease surveillance system. In developing countries, however, where dense populations reside and extensive agricultural operations take place, disease surveillance systems are lacking (Lemon et al. 2007). Precision agriculture could be a very useful and effective approach to enable disease surveillance at the field scale.

13.1.2 *Precision Agriculture vis-à-vis Plant Protection*

Precision agriculture comprises management strategies that use information technology to process high resolution spatial and temporal data related to crop production. From tillage to harvesting, precision agriculture provides different packages of operations to reduce inputs, increase profits, and protect the environment. Recently, under the umbrella of precision agriculture, a new branch popularly known as "precision plant protection" has emerged. Precision plant protection means taking the right action in the right place at the right time, to protect plants from biotic and abiotic stresses. Precision plant protection comprises disease management strategies based on precision agriculture tools. Remote sensing, Geographic Information System (GIS) and Global Navigation Satellite System (GNSS), Unmanned Aerial Vehicle (UAV) and machine learning techniques such as Artificial Neural Network (ANN), with Partial Least Square (PLS) are the major drivers being applied.

Oerke et al. (2010) showed that the strategies for precision plant protection should be planned on the basis of information obtained from the previous crops. Only then, a proper precision plant protection program can be implemented. Over the last decade, most studies focused on the application of remote sensing for early disease detection (Lee et al. 2010). Some examples include diseases caused by fungi

(West et al. 2010), viruses (Grisham et al. 2010; Krezhova et al. 2015) and viroids (Beltrán-Peña et al. 2014; Golhani et al. 2017a, b, 2019a; Selvaraja et al. 2013). Rumpf et al. (2010) highlighted the use of different data mining techniques with hyperspectral data for plant disease detection.

13.2 Orange Spotting Disease in Malaysia

Orange Spotting (OS) is an emerging disease of oil palm (*Elaeis guineensis* Jacq.; Arecaceae) in Malaysia. Coconut cadang-cadang viroid (CCCVd) is the causal agent of OS disease. CCCVd is one of the known species of viroids, which thus far have only been found in plants. They are single-stranded, low molecular weight, circular RNAs between 246 and 401 nucleotides that lack a protective protein coat (Diener 1999). Recent reports indicate that OS disease can result in an epidemic that could bring significant economic losses to oil palm production in Malaysia. In 2006 for the first time, three CCCVd variants (OP_{297}, OP_{293}, and OP_{270}) were reported in an asymptomatic oil palm in Malaysia. This was the first incident outside the Philippines in a species other than coconut palm (Vadamalai et al. 2006). Then, Wu et al. (2013) reported the first incident of a variant of CCCVd (OP_{246}) with clear orange color spots. In a recent investigation, an oil palm variant (OP_{293}) showed low accumulation of viroid load with no symptoms, 1 year after inoculation (Thanarajoo 2014).

Selvaraja et al. (2013) reported OS disease has similar foliar symptoms to that of potassium deficiency in oil palm. Symptomatic separability between OS disease and potassium deficiency is also very difficult to achieve via visual assessment. It proves that symptom expression is not a necessary outcome of CCCVd infection. It is challenging to scout for healthy palms from the diseased palms due to the lack of visible symptoms. It is believed that the use of hyperspectral sensor (spectroradiometer) can serve as a useful tool for preliminary screening of CCCVd infected seedlings, at the nursery stage. Therefore, it is important to identify OS disease at an early stage, most preferably at the leaf scale (nursery stage). The current approach of remote sensing can serve as a useful tool for preliminary screening of CCCVd infected seedlings.

Real-time detection of OS disease has become possible now using a spectroradiometer. Existing molecular marker techniques typically take a longer time (3–6 months), from sampling to laboratory analysis, for detection of CCCVd infection. Instead, precision plant protection may provide rapid and non-invasive detection of OS disease. Recently, Golhani et al. (Golhani et al. 2017a, b; 2019a, b) used an Analytical Spectral Device (ASD) spectroradiometer for non-invasive detection of OS disease in oil palm. In this research, oil palm seedlings were inoculated with a CCCVd oil palm variant (OP_{246}). The research was designed to observe the spectral changes between CCCVd-inoculated and healthy oil palm seedlings followed by the development of spectral signatures, selection of red edge wavebands, selection of red edge indices and development of an Orange Spotting Disease Index (OSDI) using red edge parameters.

13.3 Objectives

The main objective of this chapter is to discuss major drivers of precision agriculture in the context of precision plant protection. This chapter describes hyperspectral remote sensing in general and Visible/Near-infrared (VNIR) spectroscopy and Spectral Disease Index (SDI) in particular. The advantages of UAV, GIS and GNSS are also discussed. Few machine learning techniques are also reviewed. In addition, a successful case study on OS disease detection comprising the use of remote sensing and machine learning techniques is reported.

13.4 Major Drivers in Precision Plant Protection

13.4.1 Remote Sensing

In order to efficiently apply remote sensing in precision plant protection, it is very important to understand the fundamental interaction of radiant energy with the earth surface (Huete 1989). The radiant energy (electromagnetic radiation) propagates through the atmosphere to the earth surface in the form of electromagnetic waves. These waves are well distributed across the electromagnetic spectrum comprising several spectral regions, viz. ultraviolet, blue, green, red, red edge and Near-infrared (NIR) (Fig. 13.1). The electromagnetic radiation interacts with the atmosphere in different ways via absorption, transmission, diffusion, scattering, and reflection. In this process, approximately 40% of the solar flux is received by the earth surface (Lacis and Hansen 1974).

The spectral composition of solar flux interacting with the earth surface provides information about the physical properties of soil, water, and vegetation. In vegetation (plant leaf and canopy), the reflectance from radiation results in diffuse and specular characteristics. Spectral diffusion takes place due to multiple scattering,

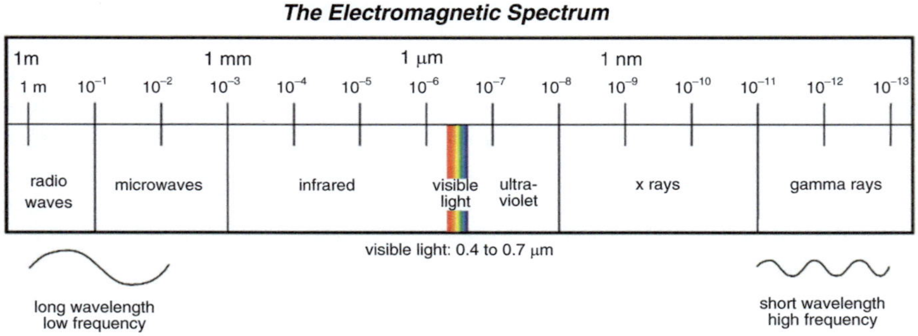

Fig. 13.1 The electromagnetic spectrum showing different segments of the spectrum comprising γ rays, X rays, ultraviolet, visible light, infrared, microwaves, and radio waves

depending on the different physical and structural design of the leaves. The topography of the cuticular wax and hair at the leaf surface affects specular characteristics of leaf reflectance. In remote sensing, spectral signatures are developed from leaf reflectance, which is often found to be sensitive to these changes. Light is scattered in all directions when interacting with unhealthy plant tissue, while the light is scattered in a diffused manner when interacting with healthy tissues, next to small symptomatic tissue present in the epidermis layer. Thomas et al. (2017) has recently studied plant-pathogen interaction using hyperspectral imaging reflectance and transmission measurements.

The measurement of reflectance has a significant role in detecting crop diseases as well as in quantifying the complex spatio-temporal dynamics of plant-pathogen interactions. Mahlein (2010) investigated foliar sugar beet diseases (Cercospora leaf spot, sugar beet rust, and powdery mildew) at the canopy and leaf scales using spectral signatures derived from hyperspectral sensors. The term "hyperspectral" refers to the use of hundred over contiguous narrow spectral bands. Hyperspectral sensors can be either active or passive. The sensor system equipped with its own source of radiation is called an active sensor, while sensors that depend on solar radiation are called passive sensors. Thus, the basic mechanism of remote sensing completely depends on the type of sensors (active or passive) being employed (Schellber et al. 2008).

13.4.2 *Hyperspectral Remote Sensing*

Hyperspectral remote sensing is also known as reflectance spectroscopy. Hyperspectral wavebands measure the reflectance from the leaf surface. On the basis of the percentage of reflectance of wavebands, different types of stress such as diseases, nutrient deficiency, and water scarcity can be differentiated at their corresponding spectral regions. Clevers et al. (2004) reported that leaf pigments dominate in the visible region (400–700 nm), while cell structure and leaf water contents dominate in the NIR (700–1000 nm) and Short-wave infrared (SWIR, 1000–2500 nm) regions, respectively. Changes in reflectance characteristics have been observed due to alterations in plant biochemistry and cellular composition of leaves. Unhealthy vegetation (senescent and stressed) has more reflection in the red region and lower reflectance in the NIR region (Li et al. 2005). In a landmark study, Knipling (1970) stated that low reflectance in the NIR region has been often associated with an advanced stage of disease attack, where the breakdown of leaf cells has taken place.

A practical hyperspectral sensor is the hand-held variant (called a spectroradiometer), which is available within a spectral range of 400–1100 nm (Visible/NIR or VNIR) and 400–2500 nm (Visible-NIR-SWIR). Slonecker (2011) recommends a multiscale spectroradiometer for laboratory-to-field scale experiments. Hand-held spectroradiometers can be used repeatedly in order to understand changes in spectral reflectance of plants. They are also called non-invasive crop sensors and classified into imaging and non-imaging sensors, depending on their detection specifications. These sensors are primarily used for real-time stress detection.

According to Kuska and Mahlein (2018), hyperspectral sensors are potentially powerful tools in protecting crops against diseases. Hyperspectral sensors can facilitate a proximate and objective detection. Thomas et al. (2018) mentioned that hyperspectral sensors can play an important role for measuring pathogen-induced changes. They showed that hyperspectral sensors are practically diagnostic as well as valuable for disease investigation at different scales, from the tissue to the canopy level. Several recent studies have included early disease detection using hyperspectral imaging sensors (Mahlein 2016), disease forecasting from meteorological parameters (Grinn-Gofroń et al. 2019), disease warning (Gillespie and Sentelhas 2008), and estimation of disease stages using the Spectral Disease Index (SDI) (Ashourloo et al. 2016).

13.4.3 Visible/Near-Infrared (VNIR) Spectroscopy

The VNIR region is being widely explored in the field of precision plant protection. As an effort to increase the effectiveness of the VNIR region for studying plant diseases, different wavebands within this range have been examined. Ayala-Silva and Beyl (2005) described an unambiguous relationship between VNIR region and green parts of the plants. The maximum absorption was found in the blue (400–500 nm) and red (600–700 nm) regions via chlorophyll content, while maximum reflection was found in NIR region due to scattering in leaf mesophyll. VNIR spectra typically characterize the baseband frequency range of organic compounds, therefore they give potentially pertinent and inherent information about an object (Zhao 2012). Several recent studies (Ashourloo et al. 2016; Mahlein 2016; Rumpf et al. 2009) have explained the importance of the VNIR range in discriminating diseased plants.

Also, there is evidence of qualitative and quantitative changes of chlorophyll content in the course of plant growth, coupled with biotic and abiotic stresses. Merzlyak et al. (2003) developed algorithms for pigment analysis within the VNIR region. They showed that two spectral wavebands (550 nm and 700 nm) were more sensitive to chlorophyll content. According to Chappelle et al. (1992), chlorophyll absorption is reduced when plant growth is under stress due to the effect of biostressors. Carter and Miller (1994) demonstrated the importance of spectral wavebands and spectral ratios within 690–700 nm, which can provide early detection of stress-induced chlorosis.

13.4.4 Spectral Disease Index (SDI)

An SDI is developed from a combination of disease sensitive hyperspectral wavebands represented as a waveband ratio (Golhani et al. 2018). The SDI has the potential to discriminate and differentiate between diseased and healthy plants. Development of SDI comprises different methods of algebraic or mathematical analyses such as normalizing, differentiation, summation, and linear combinations. The preferred methods for development of SDI are continuum-removal and band ratioing.

Mahlei et al. (2013) described how each disease affects the leaf reflectance spectrum in a specific way. Therefore, an SDI should be developed based on the progression of disease symptoms and their spectral characteristics. Ashourloo et al. (2014) developed two SDIs on the basis of disease progression for detection of leaf rust in wheat. Most of the SDIs have been developed with a general interest to diagnose a plant disease at an early stage. As a matter of fact, different SDIs represent specificity, sensitivity and severity of the vegetation at different stages of infection. In general, a combination of red and NIR wavebands are used for the development of the vegetation index.

Clevers et al. (2004) and Broge and Leblanc (2001) reported that the common vegetation indices developed from combinations of red and NIR wavebands have also been useful for plant stress detection, such as Normalized Difference Vegetation Index (NDVI) (Rouse et al. 1974). For a very long time, NDVI has served as a plant stress indicator. However, recent studies conceptually differ from this notion. Most vegetation indices and spectral ratios provide information about a specific phenomenon only, such as crop vitality and greenness, which are not widely tested for disease diagnosis. Jimenez and Landgrebe (1999) reported that the selection of optimal wavebands for crop stress detection is very important. For development of SDI, wavebands must be free from redundant information without losing the ability for discrimination and class separability.

13.4.5 Unmanned Aerial Vehicle (UAV)

Over the years, the idea of developing vegetation indices has been popularized with the use of different airborne and satellite imageries. Processing airborne or satellite imageries is time consuming and very costly. Using an advanced hyperspectral camera onboard an UAV can be comparably more useful at the field scale. UAV allows for lower flight altitude and light-weight platform for hyperspectral camera. Recently, Behmann et al. (2018) described the disadvantage of not providing high-quality correction signals.

13.4.6 Geographic Information System (GIS) and Global Navigation Satellite System (GNSS)

For precision plant protection, GIS can play an important role in managing plant diseases in the field. A GIS usually needs GNSS coordinates, soil, crop, weather data and satellite imageries to serve as a decision support system. These data inputs are processed using several data analysis tools. Currently, GIS has emerged as a valuable tool to achieve the goal of precision plant protection. For plant disease management, GIS can be employed from field scale up to country scale. Remotely sensed vegetation indices with GIS can result in good outcomes for on-site plant disease assessment. GIS can provide an environment for utilizing the indices for studying crop health. For example, the application of NDVI data with phenological characteristics of plants can assess the suitability of remote sensing for estimating biotic and abiotic stress. This approach has become popular for analysis of phenological phenomena to facilitate disease detection, monitoring and diagnosis (Bolton and Friedl 2013; Chen et al. 2012; Granados-Ramírez et al. 2004). Damm et al. (2015) reported that NDVI along with photochemical reflectance index and sun-induced chlorophyll fluorescence can measure plant functional properties and detect infected vegetation canopies.

13.5 Machine Learning Techniques

13.5.1 Artificial Neural Network (ANN)

In recent years, there has been an increasing interest in applying ANN techniques for plant disease management. ANN techniques are capable of processing information similar to the way neurons process information in a human brain (Wasukar 2014). ANN requires a series of mathematical expressions, commonly referred to as algorithms. ANN is applicable for non-parametric regression, non-linear function, clustering, and data classification. Typically, ANN first analyzes the sample data and then makes a prediction from them (Paydipati 2004). Recently, Golhani et al. (2018) summarized relevant details of ANN mechanism, types, models, and classifiers in the context of plant disease detection using hyperspectral data. Basically, hyperspectral data provide near-continuous narrow-bands, which are helpful in developing spectral signatures and SDIs. The use of hyperspectral data has established ANN as an essential tool, particularly for large volume data processing. ANN has a powerful discriminating capability for plant disease classification, as it combines the best trainer sets for accurate classification (Golhani et al. 2018). For example, Zhu et al. (2016) processed the hyperspectral image for presymptomatic detection and classification of tobacco mosaic virus in tobacco leaves using back propagation neural network and PLS (see below).

13.5.2 Partial Least Square (PLS)

PLS is an extension of econometric path modeling which was developed during the late 1970s (Wold 1975). PLS was developed for solving chemometric problems, specifically for analyzing multivariate chemometric data (Martens 2001). Sundberg (1999) highlighted that PLS is appropriate for explaining a dependent factor where independent variables are also defined. Major advantages of PLS are the ability to reduce matrix dimensions, ability to find the number of relevant components, and ability to identify latent structure models in the data matrix (Helland 2001; Lingjaerde and Christophersen 2000). Therefore, PLS is known as a technique for analyzing spectroscopic data (Balabin and Smirnov 2011). PLS regression helps in obtaining stress sensitive wavebands from the VNIR spectrum. Van Maanen and Xu (2003) pointed out that an accurate regression model may reveal distinguished spectral patterns either before or after disease infection. Principally, PLS regression establishes the relationship between the independent variables (spectra) and the dependent variables (attribute information) (Indahl 2014). This method is superficially similar to principal component analysis where principal components are extracted from independent variables, and a regression model is established to predict the attribute information of unknown samples. PLS is often referred to as the analysis of multi co-linearity spectral data comprising a high degree of co-linearity among neighboring wavebands (Jones et al. 2010).

13.5.3 Cluster Analysis

Cluster analysis allows for grouping within spectral samples, also called dendrogram. A dendrogram is represented as a tree of spectral data which does not only identify similar groups of variables but successfully merges them (Ahmed et al. 2010; Iounousse et al. 2015). Krafft et al. (2009) explained that reflectance spectra are just like fingerprints which have different types of pattern. Cluster analysis works like a key to classify the pattern of fingerprints. In cluster analysis, the difference between subsets of clusters is minimized whereas the difference between groups of clusters is maximized. Lee et al. (2005) investigated a range of applications of cluster analysis for studying hardness and proximate constitutes of maize kernel. They organized a total of 248 maize samples into 7 and 10 subgroups by cluster analysis. The groups resulting from cluster analysis had unique physical and chemical properties showing the different levels of hardness measurement. In a recent study, Golhani et al. (2017a) demonstrated the use of a spectroradiometer for reflectance measurement and cluster analysis to construct dendrograms of measured data. Their study was focused on a real-time screening of CCCVd-inoculated seedlings at the leaf scale.

13.6 Case Study of OS Detection

13.6.1 Experiment Details

13.6.1.1 Experiment Setup for Years 2015 and 2017

A research group comprising experts of precision agriculture, viroids and ANN has conducted a study to screen CCCVd-inoculated oil palm seedlings in a glass house. A highly infective CCCVd variant (OP_{246}) was used to inoculate three-months-old oil palm seedlings under a glasshouse facility in the Universiti Putra Malaysia (UPM), at Serdang, Selangor, Malaysia. Fifteen inoculated and ten healthy oil palm seedlings were evaluated throughout a 4-month experiment. The study was designed for two experimental years, 2015 and 2017. Reflectance data collected in the year of 2015 were used for calibration while data collected in 2017 were used for validation.

13.6.1.2 Reflectance Measurement at the Leaf Scale

The reflectance of inoculated and healthy seedlings were collected fortnightly from 15 through 120 days after inoculation (dai) using a spectroradiometer with hyperspectral capacity. A VNIR range spectroradiometer (325–1075 nm), ASD FieldSpec-2, was used in 2015. While a full range spectroradiometer (350 nm −2500 nm), FieldSpec®4, was used in 2017. Spectroradiometers were employed at leaf scale using an ASD plant-probe containing a 100 W halogen reflectorized lamp. A 10 mm diameter portion of an oil palm leaf was clipped using the leaf-clip holder during spectral measurement. A total of twenty spectral readings were collected and averaged from each oil palm seedling.

13.6.1.3 SPAD Measurement

The chlorophyll content is a key indicator to assess the stress caused by OS disease in oil palm seedlings. Chlorophyll content was measured using a Minolta SPAD-502 (Konica Minolta, Inc. Japan), popularly known as SPAD meter, which measures chlorophyll content non-destructively at leaf scale within the range 0–100. The SPAD meter was used to measure chlorophyll content from the oil palm seedlings throughout the experiment. It measures a chlorophyll absorbance based on absorbance data collected at 650 nm and 940 nm (Castro and Sanchez-Azofeifa 2008). The chlorophyll content is measured in just a fraction of a second when a leaf is clamped between two Light Emitting Diodes (LEDs) positioned at the tip of the SPAD meter (Benetoli da Silva et al. 2012). The readings of chlorophyll content were taken between 10:00 am and noon, recording an average of five SPAD readings from the midrib of the third leaf of each seedling. SPAD readings were measured at an interval of 15 days through 120 dai.

13.7 Orange Spotting (OS) Detection

This case study was aimed at investigating reflectance data of Coconut cadang-cadang viroid (CCCVd)-inoculated seedlings at the VNIR region of the spectrum, especially at the red edge region (680–780 nm), located between the far red and the Near-infrared (NIR) wavelengths. The red edge region is able to extract precise and detailed information on crop stress. In the first step of this work, sensitive and insensitive wavebands were identified within the red edge region using cluster analysis. In the second step, the VNIR region was investigated using PLS for selecting the efficient wavebands, while four red edge indices were also evaluated using ANN. In the third and final step, the work generated a spectral index specifically for OS disease, i.e., OSDI by focusing on the red edge and twenty noble red edge parameters (Li et al. 2016). Details are given below.

13.7.1 Selection of Spectral Signature Using Cluster Analysis

Multivariate statistical techniques are widely applied to analyze hyperspectral remote sensing data. The application of hyperspectral data has been rapidly increasing thus far with the help of multivariate statistical technique such as cluster analysis. We used cluster analysis to extract the relevant spectral signature from reflectance of healthy and inoculated seedlings. This process typically involves identification of sensitive wavebands within the reflectance spectra, followed by determination of reflectance sensitivity.

Technically, cluster analysis produces groups of similar spectral reflectance. Similar reflectance spectra are closer to each other than dissimilar spectra. Joining these groups or clusters progressively results in a tree-like structure known as dendrogram. The scale at the top of the dendrogram is the normalized Euclidean distance among observations or clusters (Köksal 2011). By this process, reflectance spectra were archived from each interval of inoculation (i.e., days after inoculation, dai) of inoculated oil palm seedlings corresponding to 15, 30, 45 and 60 dai. At the same intervals, the reflectance of control seedlings was also collected. The dendrograms obtained from spectral readings of inoculated and healthy seedlings were used to compute the minimum Euclidean distance measured within each interval of spectral measurement. For example, Fig. 13.2a,b shows the dendrograms obtained from spectral readings of inoculated and healthy seedlings, measured at 30 dai, in which the nearest clusters based on minimum Euclidean distance were selected. Cluster A was found to be the nearest in both dendrograms. Its member spectra were averaged. As such, dendrograms of different inoculation intervals (15, 30, 45, 60 dai) were identified and their mean spectra were averaged to get the representative spectral signatures of inoculated and control oil palm seedlings (Fig. 13.3).

The representative spectra of inoculated and control seedlings were plotted against the VNIR region, specifically to understand changes in the red edge region.

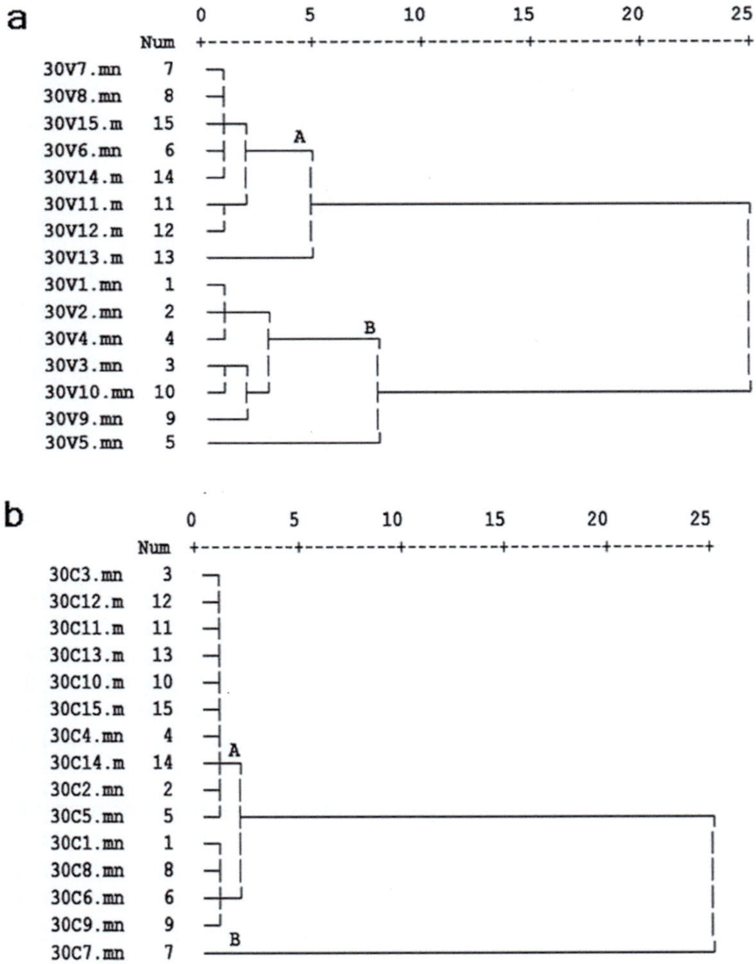

Fig. 13.2 Dendrogram structure of spectral reflectance obtained from fifteen inoculated (30 V1. mn – 30 V15.mn, **a**) and fifteen control seedlings (30C1.mn – 30C15.mn, **b**) at 30 days after inoculation

It was observed that the beginning point of red edge (680 nm), which is a relative chlorophyll absorption maxima, and a second point (754 nm), which is the first steep slope, were located in the red edge region (680–780 nm). This region has less reflectance due to chlorophyll absorption, while the NIR region is typically characterized by a high percentage of reflectance, due to the scattering of light in the intercellular mesophyll volume of leaves. Finally two spectral bands, 680 nm and 754 nm, were identified. A sharp change was also observed between 680 nm and 754 nm, which characterized a transition from chlorophyll absorption to leaf scattering in the red edge region.

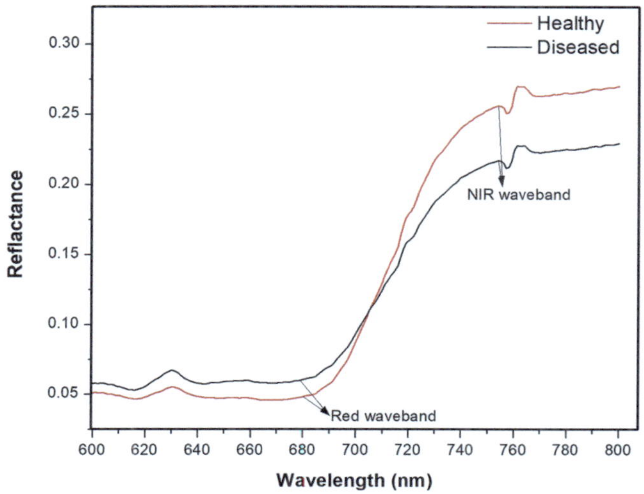

Fig. 13.3 Representative spectra of control and inoculated seedlings

Reflectance Sensitivity (RS), as proposed by Riedell et al. (2000), was applied to derive stress-sensitive and insensitive wavebands from these corresponding representative spectra (Fig. 13.3). As a result, an appreciable increase in RS (20%) was observed at 680 nm, while RS decreased up to 18% at 754 nm. To the authors' knowledge, these two red edge wavebands have not been previously studied for OS disease diagnosis in oil palm. These selected wavebands will be useful to screen infected seedlings prior to destructive sampling for biomolecular investigation. It is recommended that these wavebands be employed for evaluation of existing red edge indices and/or development of an SDI. This study provides a useful lead for canopy level diagnosis of OS disease in mature oil palm stands.

13.7.2 Estimating Chlorophyll Content Using PLS

The VNIR region (400–1050 nm) appears as most sensitive to chlorophyll stress. In this study, PLS regression was used to estimate chlorophyll content specifically from VNIR reflectance of CCCVd-inoculated and healthy oil palm seedlings. Information about chlorophyll stress could aid in diagnostics and decision-making for OS detection. Non-destructive estimation of chlorophyll content can be helpful in precision plant protection. In the PLS regression, independent and dependent variables need to be pre-defined before executing the program. For this, VNIR spectra were selected as independent variables and SPAD meter readings were selected as dependent variables. A Matlab freeware tool, Interval Partial Least-Squares Regression (iPLS), was used for estimating the chlorophyll content. Raw spectral data were pre-processed using first-order derivative, Savitzky-Golay (SG) smoothing, Multiplicative Scatter Correction (MSC) and Standard Normal Transformation (SNV) methods.

Five datasets were prepared including the raw dataset. For each dataset, 80 samples were selected, in which 60 calibration samples were obtained from CCCVd-inoculated seedlings and 20 prediction samples were obtained from healthy oil palm seedlings. MSC pre-processed spectra gave outstanding performance with a root mean square error of prediction of 3.70% and a correlation coefficient for prediction of 0.72. Thirty sensitive wavebands (601–630 nm) were identified from VNIR reflectance using MSC pre-processed spectra. See technical and experimental details in Golhani et al. (2019b).

13.7.3 Selection of Red Edge Wavebands Using Artificial Neural Network (ANN)

A Multilayer Perceptron Neural Network (MLPNN) model was used to establish a relationship between red edge bands and spectral indices. Two spectral bands (700 nm and 768 nm) were identified from reflectance spectra of CCCVd-inoculated and healthy seedlings. The bands were used for evaluation of spectral indices, namely: simple ratio, red edge normalized difference vegetation index, two-band enhanced vegetation index 2 (EVI 2). In MLPNN model, identified spectral bands were used as input, and values of spectral indices were used as target. The EVI 2 resulted as best spectral index which resulted in zero errors at the training, testing, and validation datasets. In this work, the highest coefficient of correlation ($r = 1$) was recorded by EVI 2. Golhani et al. (2019a) mentioned that identified spectral bands and spectral index could be evaluated, using airborne or space-borne hyperspectral sensor platforms, for detection of OS disease in mature oil palm stands.

13.7.4 Development of Orange Spotting Disease Index (OSDI)

The main purpose of this biennial experiment was to develop the OSDI, which could specifically be used for early detection of OS disease at the leaf scale. The OSDI was developed from reflectance spectra obtained from inoculated and healthy oil palm seedlings. It is believed that the OSDI values will give a reliable indication of OS disease, prior to confirmation by biomolecular marker techniques.

During the first experimental year (2015), twenty-four red edge parameters which were developed from First Derivative Reflectance (FDR) of the electromagnetic spectrum were used to develop the OSDI. Then, the OSDI values were verified with a repeated experiment in 2017. In Fig. 13.4, mean reflectance (30–120 dai) of diseased and healthy FDR was plotted in the red edge region. Four red edge parameters, viz. Red Edge Position (REP), mid-point (P), Right-side peak area (RSDR),

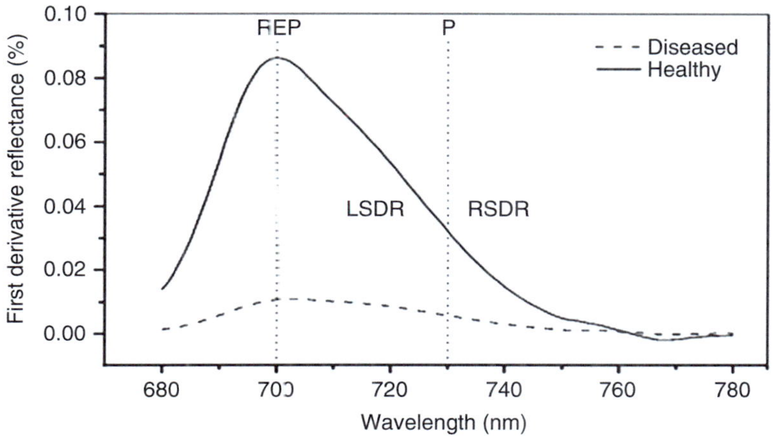

Fig. 13.4 A spectral plot of average spectra of healthy and diseased oil palm seedlings within the red edge region (680–780 nm) showing red edge parameters, viz. REP, LSDR, mid-point (P), and RSDR

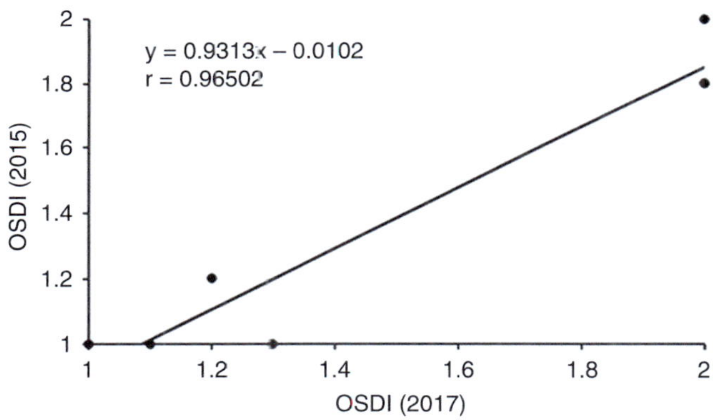

Fig. 13.5 Correlation between OSDI values computed in the years of 2015 and 2017

and Left-side peak area (LSDR), were studied to observe basic changes between diseased and healthy spectral signatures. In both spectra, reflectance increased in LSDR region between 680 nm and mid-point P nm, and decreased in RSDR region between P nm and 780 nm. REP was found at 700 nm, which could not be shifted towards shorter or longer wavelengths. This figure, while preliminary, suggested the need for exploration on other red edge parameters in order to find the most tangible criterion for comparison. Yang et al. (2010) mentioned that reflectance around the REP has been found to be most sensitive to plant stress.

Finally, a simple ratio representing the sum of the FDR of the right side – Red Edge Point (REP) to the sum of the FDR of the left side – REP of the red edge region, was identified as OSDI. The validation results showed a strong correlation (r = 0.96) between OSDI values from experimental years of 2015 and 2017 (Fig. 13.5). In the future, OSDI values will be analyzed using ANN. OSDI is the first spectral index developed for detection of both symptomatic and asymptomatic OS-infected oil palms, at the leaf scale.

13.8 Conclusion

In this chapter, important drivers in precision agriculture have been discussed. A case study on OS detection was described. A precision approach to plant protection will expedite disease control and save financial resources and valuable time. In the shown case study, an attempt of quantifying reflectance data was made to augment effectiveness of OS phytopathometry appraisal in oil palm. Basically, CCCVd damages oil palm seedlings by crippling the chlorophyll apparatus. This case study showed that selected red edge wavebands, red edge indices, with the newly developed OSDI, are good predictors of chlorophyll stress caused by viroid attack. Development of the OSDI is the best outcome from this attempt, although to date, the newly developed OSDI has only been tested under a glasshouse environment. The verification of OSDI values under a wide range of growing conditions is recommended. Future work should be directed at investigating the efficacy of OSDI for diagnosis of OS disease at the canopy scale.

References

Ahmed, S. S. S. J., Santosh, W., Kumar, S., & Thanka Christlet, T. H. (2010). Neural network algorithm for the early detection of Parkinson's disease from blood plasma by FTIR micro-spectroscopy. *Vibrational Spectroscopy, 53*, 181–188. https://doi.org/10.1016/j.vibspec.2010.01.019.

Ashourloo, D., Mobasheri, M., & Huete, A. (2014). Developing two spectral disease indices for detection of wheat leaf rust (*Puccinia triticina*). *Remote Sensing, 6*, 4723–4740. https://doi.org/10.3390/rs6064723.

Ashourloo, D., Matkan, A. A., Huete, A., Aghighi, H., & Mobasheri, M. R. (2016). Developing an index for detection and identification of disease stages. *IEEE Geosc Remote Sens Lett, 13*, 851–855. https://doi.org/10.1109/LGRS.2016.2550529.

Ayala-Silva, T., & Beyl, C. A. (2005). Changes in spectral reflectance of wheat leaves in response to specific macronutrient deficiency. *Advances in Space Research, 35*, 305–317.

Balabin, R. M., & Smirnov, S. V. (2011). Variable selection in near-infrared spectroscopy: Benchmarking of feature selection methods on biodiesel data. *Analytica Chimica Acta, 692*, 63–72. https://doi.org/10.1016/j.aca.2011.03.006.

Behmann, J., Acebron, K., Emin, D., Bennertz, S., Matsubara, S., Thomas, S., et al. (2018). Specim IQ: Evaluation of a new, miniaturized handheld hyperspectral camera and its application for plant phenotyping and disease detection. *Sensors, 18*, 441. https://doi.org/10.3390/s18020441.

Beltrán-Peña, H., Soria-Ruiz, J., Téliz-Ortiz, D., Ochoa-Martínez, D. L., Nava-Díaz, C., & Ochoa-Ascencio, S. (2014). Molecular and satellite spectral imaging detection of Avocado Sunblotch Viroid (ASBVd). *Revista Fitotecnia Mexicana, 37*, 21–29.

Benetoli da Silva, T. R., Reis de Sousa, A. C., & de Goés Maciel, C. D. (2012). Relationship between chlorophyll meter readings and total N in crambe leaves as affected by nitrogen topdressing. *Industrial Crops and Products, 39*, 135–138. https://doi.org/10.1016/j.indcrop.2012.02.008.

Blackburn, G. A. (2007). Hyperspectral remote sensing of plant pigments. *Journal of Experimental Botany, 58*, 855–867. https://doi.org/10.1093/jxb/erl123.

Blackmer, T. M., Schepers, J. S., & Meyer, G. E. (1995). Remote sensing to detect nitrogen deficiency in corn. In: *Site-specific management for agricultural systems* (pp. 505–512). https://doi.org/10.2134/1995.site-specificmanagement.c35.

Bolton, D. K., & Friedl, M. A. (2013). Forecasting crop yield using remotely sensed vegetation indices and crop phenology metrics. *Agricultural and Forest Meteorology, 173*, 74–84. https://doi.org/10.1016/j.agrformet.2013.01.007.

Broge, N. H., & Leblanc, E. (2001). Comparing prediction power and stability of broadband and hyperspectral vegetation indices for estimation of green leaf area index and canopy chlorophyll density. *Remote Sensing of Environment, 76*, 156–172. https://doi.org/10.1016/S0034-4257(00)00197-8.

Carter, G. A., & Knapp, A. K. (2001). Leaf optical properties in higher plants: Linking spectral characteristics to stress and chlorophyll concentration. *American Journal of Botany, 88*, 677. https://doi.org/10.2307/2657068.

Carter, G. A., & Miller, R. L. (1994). Early detection of plant stress by digital imaging within narrow stress-sensitive wavebands. *Remote Sensing of Environment, 50*, 295–302. https://doi.org/10.1016/0034-4257(94)90079-5.

Castro, K. L., & Sanchez-Azofeifa, G. A. (2008). Changes in spectral properties, chlorophyll content and internal mesophyll structure of senescing *Populus balsamifera* and *Populus tremuloides* leaves. *Sensors, 8*, 51–69. https://doi.org/10.3390/s8010051.

Chaerle, L. (2004). Thermal and chlorophyll-fluorescence imaging distinguish plant-pathogen interactions at an early stage. *Plant & Cell Physiology, 45*, 887–896. https://doi.org/10.1093/pcp/pch097.

Chappelle, E. W., Kim, M. S., & McMurtrey, J. E. (1992). Ratio analysis of reflectance spectra (RARS): An algorithm for the remote estimation of the concentrations of chlorophyll A, chlorophyll B, and carotenoids in soybean leaves. *Remote Sensing of Environment, 39*, 239–247. https://doi.org/10.1016/0034-4257(92)90089-3.

Chen, X., Han, W., & Li, M. (2012). Spectroscopic determination of leaf water content using linear regression and an artificial neural network. *African Journal of Biotechnology, 11*, 2518–2527. https://doi.org/10.5897/AJB11.2733.

Clevers, J. G. P. W., Kooistra, L., & Salas, E. A. L. (2004). Study of heavy metal contamination in river floodplains using the red-edge position in spectroscopic data. *International Journal of Remote Sensing, 25*, 3883–3895. https://doi.org/10.1080/01431160310001654473.

Damm, A., Guanter, L., Verhoef, W., Schläpfer, D., Garbari, S., & Schaepman, M. E. (2015). Impact of varying irradiance on vegetation indices and chlorophyll fluorescence derived from spectroscopy data. *Remote Sensing of Environment, 156*, 202–215. https://doi.org/10.1016/j.rse.2014.09.031.

Diener, T. O. (1999). Viroids and the nature of viroid diseases. In *100 Years of Virology* (pp. 203–220). Vienna: Springer. https://doi.org/10.1007/978-3-7091-6425-9_15.

Gillespie, T. J., & Sentelhas, P. C. (2008). Agrometeorology and plant disease management: a happy marriage. *Science in Agriculture*. https://doi.org/10.1590/S0103-90162008000700012.

Golhani, K., Balasundram, S. K., Vadamalai, G., & Pradhan, B. (2017a). Use of reflectance spectroscopy as a tool for screening viroid-inoculated oil palm seedlings. *Open Access Journal of Agricultural Research, 2*, 1–5. https://doi.org/10.23880/OAJAR-16000145.

Golhani, K., Balasundram, S.K., Vadamalai, G., & Pradhan, B. (2017b). *Red-edge indices to diagnose orange spotting disease of oil palm in Malaysia*. In: 38th Asian Conference on Remote Sensing - Space Applications: Touching Human Lives, ACRS 2017.

Golhani, K., Balasundram, S. K., Vadamalai, G., & Pradhan, B. (2018). A review of neural networks in plant disease detection using hyperspectral data. *Information Processing in Agriculture, 5*, 354–371. https://doi.org/10.1016/j.inpa.2018.05.002.

Golhani, K., Balasundram, S. K., Vadamalai, G., & Pradhan, B. (2019a). Selection of a spectral index for detection of orange spotting disease in oil palm (*Elaeis guineensis* Jacq.) using red edge and neural network techniques. *Journal of the Indian Society of Remote Sensing, 47*, 639–646. https://doi.org/10.1007/s12524-018-0926-4.

Golhani, K., Balasundram, S. K., Vadamalai, G., & Pradhan, B. (2019b). Estimating chlorophyll content at leaf scale in viroid- inoculated oil palm seedlings (*Elaeis guineensis* Jacq.) using reflectance spectra (400 nm–1050 nm). *International Journal of Remote Sensing, 40*, 647–7662. https://doi.org/10.1080/01431161.2019.1584930.

Granados-Ramírez, R., Reyna-Trujillo, T., Gómez-Rodríguez, G., & Soria-Ruiz, J. (2004). Analysis of NOAA-AVHRR-NDVI images for crops monitoring. *International Journal of Remote Sensing, 25*, 1615–1627. https://doi.org/10.1080/0143116031000156855.

Grinn-Gofroń, A., Nowosad, J., Bosiacka, B., Camacho, I., Pashley, C., Belmonte, J., et al. (2019). Airborne *Alternaria* and *Cladosporium* fungal spores in Europe: forecasting possibilities and relationships with meteorological parameters. *Science of the Total Environment, 653*, 938–946.

Grisham, M. P., Johnson, R. M., & Zimba, P. V. (2010). Detecting sugarcane yellow leaf virus infection in asymptomatic leaves with hyperspectral remote sensing and associated leaf pigment changes. *Journal of Virological Methods, 167*, 140–145. https://doi.org/10.1016/j.jviromet.2010.03.024.

Helland, I. S. (2001). Some theoretical aspects of partial least squares regression. *Chemometrics and Intelligent Laboratory Systems, 58*, 97–107. https://doi.org/10.1016/S0169-7439(01)00154-X.

Huete, A. R. (1989). Soil influences in remotely sensed vegetation-canopy spectra. In G. Asrar (Ed.), *Theory and applications of optical remote sensing* (pp. 107–141). New York: Wiley-Interscience.

Indahl, U. G. (2014). The geometry of PLS1 explained properly: 10 key notes on mathematical properties of and some alternative algorithmic approaches to PLS1 modelling. *Journal of Chemometrics, 28*, 168–180. https://doi.org/10.1002/cem.2589.

Iounousse, J., Er-Raki, S., El Motassadeq, A., & Chehouani, H. (2015). Using an unsupervised approach of Probabilistic Neural Network (PNN) for land use classification from multitemporal satellite images. *Applied Soft Computing, 30*, 1–13. https://doi.org/10.1016/j.asoc.2015.01.037.

Jimenez, L. O., & Landgrebe, D. A. (1999). Hyperspectral data analysis and supervised feature reduction via projection pursuit. *IEEE Transactions on Geoscience and Remote Sensing, 37*, 2653–2667.

Jones, C. D., Jones, J. B., & Lee, W. S. (2010). Diagnosis of bacterial spot of tomato using spectral signatures. *Computers and Electronics in Agriculture, 74*, 329–335. https://doi.org/10.1016/j.compag.2010.09.008.

Knipling, E. B. (1970). Physical and physiological basis for the reflectance of visible and near-infrared radiation from vegetation. *Remote Sensing of Environment, 1*, 155–159. https://doi.org/10.1016/S0034-4257(70)80021-9.

Köksal, E. S. (2011). Hyperspectral reflectance data processing through cluster and PCA for estimating irrigation and yield related indicators. *Agricultural Water Management, 98*, 1317–1328. https://doi.org/10.1016/j.agwat.2011.03.014.

Krafft, C., Steiner, G., Beleites, C., & Salzer, R. (2009). Disease recognition by infrared and Raman spectroscopy. *Journal of Biophotonics, 2*, 13–28. https://doi.org/10.1002/jbio.200810024.

Krezhova, D., Stoev, A., & Maneva, S. (2015). Detection of biotic stress caused by apple stem grooving virus in apple trees using hyperspectral reflectance analysis. *Comptes Rendus de l'Académie Bulgare des Sciences, 68*, 175–182.

Kuska, M. T., & Mahlein, A. K. (2018). Aiming at decision making in plant disease protection and phenotyping by the use of optical sensors. *The European Journal of Plant Pathology*, 1–6.

Lacis, A. A., & Hansen, J. (1974). A parameterization for the absorption of solar radiation in the earth's atmosphere. *Journal of the Atmospheric Sciences, 31*, 118–133.
Lee, K. M., Herrman, T. J., Lingenfelser, J., & Jackson, D. S. (2005). Classification and prediction of maize hardness-associated properties using multivariate statistical analyses. *Journal of Cereal Science, 41*, 85–93. https://doi.org/10.1016/j.jcs.2004.09.006.
Lee, W. S., Alchanatis, V., Yang, C., Hirafuji, M., Moshou, D., & Li, C. (2010). Sensing technologies for precision specialty crop production. *Computers and Electronics in Agriculture, 74*, 2–33. https://doi.org/10.1016/j.compag.2010.08.005.
Lemon, S. M., Hamburg, M. A., Sparling, P. F., Choffnes, E. R., & Mack, A. (2007). *Global infectious disease surveillance and detection : assessing the challenges — finding solutions, Workshops summary* (p. 284). Washington, DC: National Academies Press. doi:978-0-309-11114-0.
Li, L., Ustin, S. L., & Lay, M. (2005). Application of AVIRIS data in detection of oil-induced vegetation stress and cover change at Jornada, New Mexico. *Remote Sensing of Environment, 94*, 1–16. https://doi.org/10.1016/j.rse.2004.08.010.
Li, L., Ren, T., Ma, Y., Wei, Q., Wang, S., Li, X., et al. (2016). Evaluating chlorophyll density in winter oilseed rape (*Brassica napus* L.) using canopy hyperspectral red-edge parameters. *Computers and Electronics in Agriculture, 126*, 21–31. https://doi.org/10.1016/j.compag.2016.05.008.
Lingjaerde, O. C., & Christophersen, N. (2000). Shrinkage structure of partial least squares. *Scandinavian Journal of Statistics, 27*, 459–473. https://doi.org/10.1111/1467-9469.00201
Mahlei, A.-K., Rumpf, T., Welke, P., Dehne, H.-W., Plümer, L., Steiner, U., & Oerke, E.-C. (2013). Development of spectral indices for detecting and identifying plant diseases. *Remote Sensing of Environment, 128*, 21–30. https://doi.org/10.1016/j.rse.2012.09.019.
Mahlein, A. (2010). *Detection, identification and quantification of fungal diseases of sugar beet leaves using imaging and non-imaging hyperspectral techniques*. Bonn: University of Bonn.
Mahlein, A. K. (2016). Plant disease detection by imaging sensors – parallels and specific demands for precision agriculture and plant phenotyping. *Plant Disease, 100*, 241–251. https://doi.org/10.1094/PDIS-03-15-0340-FE.
Martens, H. (2001). Reliable and relevant modelling of real world data: a personal account of the development of PLS regression. *Chemometrics and Intelligent Laboratory Systems, 58*, 85–95.
Martinelli, F., Scalenghe, R., Davino, S., Panno, S., Scuderi, G., Ruisi, P., et al. (2014). Advanced methods of plant disease detection. A review. *Agronomy for Sustainable Development, 35*, 1–25. https://doi.org/10.1007/s13593-014-0246-1.
Merzlyak, M. N., Gitelson, A. A., Chivkunova, O. B., Solovchenko, A. E., & Pogosyan, S. I (2003). Application of reflectance spectroscopy for analysis of higher plant pigments. *Russian Journal of Plant Physiology, 50*, 704–710.
Oerke, E. C., Gerhards, R., Menz, G., & Sikora, R. A. (2010). *Precision crop protection-the challenge and use of heterogeneity* (p. 441). Dordrecht: Springer.
Paydipati, R. (2004). Evaluation of classifiers for automatic disease detection in citrus leaves machine vision. Master Thesis, Univ. FL. In *USA*.
Pinter, P. J., Hatfield, J. L., Schepers, J. S., Barnes, E. M., Moran, M. S., Daughtry, C. S. T., & Upchurch, D. R. (2003). Remote sensing for crop management. *Photogrammetric Engineering and Remote Sensing, 69*, 647–664. https://doi.org/10.14358/PERS.69.6.647.
Riedell, W.E., Osborne, S.L., Hesler L.S., & Blackmer, T.M. (2000). Remote sensing of insect damage in wheat. In: *Proceedings of the 5th International Conference on Precision Agriculture, Bloomington, Minnesota, USA* (16–19 July, 2000), 1–11.
Rouse, J. W., Haas, R. H., Deering, D. W., & Harlan, J. C. (1974). *Monitoring the vernal advancement and retrogradation (green wave effect) of natural vegetation*. Greenbelt: NASA/GSFC type III final report.
Rumpf, T., Mahlein, A., Dörschlag, D., & Plümer, L. (2009). Identification of combined vegetation indices for the early detection of plant diseases. In: C.M.U. Neale, and A. Maltese (Eds.). *SPIE Europe Remote Sensing*, 747217. doi:https://doi.org/10.1117/12.830525.

Rumpf, T., Mahlein, A.-K., Steiner, U., Oerke, E.-C., Dehne, H.-W., & Plümer, L. (2010). Early detection and classification of plant diseases with support vector machines based on hyperspectral reflectance. *Computers and Electronics in Agriculture, 74*, 91–99. https://doi.org/10.1016/j.compag.2010.06.009.

Schellber, J., Hill, M. J., Gerhards, R., Rothmund, M., & Braun, M. (2008). Precision agriculture on grassland: applications, perspectives and constraints. *European Journal of Agronomy, 29*, 59–71.

Selvaraja, S., Balasundram, S. K., Vadamalai, G., & Husni, M. H. A. (2013). Site-specific disease management: a preliminary case with orange spotting in oil palm. In J. V. Stafford (Ed.), *Precision Agriculture 13* (pp. 577–584). Dordrecht: Precision Agriculture '13. Wageningen Academic Publishers. https://doi.org/10.3920/978-90-8686-778-3_71.

Slonecker, E. (2011). Analysis of the effects of heavy metals on vegetation hyperspectral reflectance properties. In *Hyperspectral remote sensing of vegetation* (pp. 561–578). Boca Raton: CRC Press. https://doi.org/10.1201/b11222-33.

Steele, M., Gitelson, A. A., & Rundquist, D. (2008). Nondestructive estimation of leaf chlorophyll content in grapes. *American Journal of Enology and Viticulture, 2*, 299–305. https://doi.org/10.2307/2445170.

Sundberg, R. (1999). Multivariate calibration — direct and indirect regression methodology. *Scandinavian Journal of Statistics, 26*, 161–207.

Thanarajoo, S. S. (2014). *Rapid detection, accumulation and translocation of Coconut cadang-cadang viroid variants in oil palm*. Serdang: Universiti Putra Malaysia.

Thomas, S., Wahabzada, M., Kuska, M. T., Rascher, U., & Mahlein, A. K. (2017). Observation of plant--pathogen interaction by simultaneous hyperspectral imaging reflection and transmission measurements. *Functional Plant Biology, 44*, 23–34.

Thomas, S., Kuska, M. T., Bohnenkamp, D., Brugger, A., Alisaac, E., Wahabzada, M., et al. (2018). Benefits of hyperspectral imaging for plant disease detection and plant protection: A technical perspective. *The Journal of Plant Diseases and Protection, 125*, 5–20.

Vadamalai, G., Hanold, D., Rezaian, M. A., & Randles, J. W. (2006). Variants of Coconut cadang-cadang viroid isolated from an African oil palm (*Elaeis guineensis* Jacq.) in Malaysia. *Archives of Virology, 151*, 1447–1456. https://doi.org/10.1007/s00705-005-0710-y.

Van Maanen, A., & Xu, X. M. (2003). Modelling plant disease epidemics. In *Epidemiology of Mycotoxin Producing Fungi* (pp. 669–682). Dordrecht: Springer. https://doi.org/10.1007/978-94-017-1452-5_2.

Wasukar, A. R. (2014). Artificial neural network – an important asset for future computing. *International Journal of Emerging Trends in Science, 1*, 28–34.

West, J. S., Bravo, C., Oberti, R., Moshou, D., Ramon, H., & McCartney, H. A. (2010). Detection of fungal diseases optically and pathogen inoculum by air sampling. In *Precision crop protection - the challenge and use of heterogeneity* (pp. 135–149). Dordrecht: Springer. https://doi.org/10.1007/978-90-481-9277-9_9.

Wold, H. (1975). Soft modeling by latent variables: the nonlinear iterative partial least squares (NIPALS) approach. *Journal of Applied Probability, 12*(S1), 117–142.

Wu, Y. H., Cheong, L. C., Meon, S., Lau, W. H., Kong, L. L., Joseph, H., & Vadamalai, G. (2013). Characterization of Coconut cadang-cadang viroid variants from oil palm affected by orange spotting disease in Malaysia. *Archives of Virology, 158*, 1407–1410. https://doi.org/10.1007/s00705-013-1624-8.

Yang, F., Li, J., Gan, X., Qian, Y., Wu, X., & Yang, Q. (2010). Assessing nutritional status of *Festuca arundinacea* by monitoring photosynthetic pigments from hyperspectral data. *Computers and Electronics in Agriculture, 70*, 52–59. https://doi.org/10.1016/j.compag.2009.08.010.

Zhao, X. (2012). Advances and technology in infrared spectroscopy. *Journal of Anqing Teachers College, 18*, 94–97.

Zhu, H., Cen, H., Zhang, C., & He, Y. (2016). Early detection and classification of tobacco leaves inoculated with tobacco mosaic virus based on hyperspectral imaging technique. *ASABE Annual International Meeting, 1*. https://doi.org/10.13031/aim.20162460422.

Chapter 14
Quarantine and Regulations

Yasir Iftikhar and Ashara Sajid

Abstract Quarantine is one of the approaches adopted for the management of plant diseases. It restricts the movement of pathogens between geographical areas. Different plant pathogens including bacteria, fungi, virus and nematodes can be ruled out from different economically important crops, through quarantine regulations. Actually several countries are implementing the quarantine rules and regulations at air and sea ports. Different international agencies related to plant health published lists of quarantine pests to exclude. Plant protection and health are the major aims of quarantine and help in formulating and implementing of quarantine procedures and regulations.

Keywords Quarantine Regulation · Phytosanitation · Plant Protection · Plant biosecurity · World Trade Organization

14.1 Introduction

Different approaches have been adopted to manage plant pathogens. Integrated plant disease management (IPDM) strategies are under practice, worldwide. Some of the plant diseases caused by fungi and bacteria are of great importance, due to their devastating nature and effects. Quarantine is a basic and important component of IPDM. We can attempt to prevent our crops from plant pathogenic contaminations by applying quarantine rules and regulations. Some devastating fungal infections such as smuts, rusts and bunts of wheat can be controlled up to a great extent by adopting quarantine measures. Similarly, some bacterial diseases also of great importance may be prevented such as citrus canker, which caused considerable losses to the citrus industry in the USA. Plant viruses being mesobiotic and com-

Y. Iftikhar (✉) · A. Sajid
Department of Plant Pathology, College of Agriculture, University of Sargodha, Sargodha, Pakistan
e-mail: yasir.iftikhar@uos.edu.pk

© Springer Nature Switzerland AG 2020
I. Ul Haq, S. Ijaz (eds.), *Plant Disease Management Strategies for Sustainable Agriculture through Traditional and Modern Approaches*, Sustainability in Plant and Crop Protection 13, https://doi.org/10.1007/978-3-030-35955-3_14

prised of nucleic acid and proteins in combination, have a latent periods of different lengths depending on the host and environment. Like other pathogens, plant viruses are also seed borne and very difficult to diagnose. Due to globalization international trade, the chances of invasion of alien plant viruses have been increased. Therefore, there is a direct need to implement effective quarantine rules and regulations, established at every point of entry in the country.

14.1.1 Quarantine

The word Quarantine originated from an Italian word "*Quarantina*", which means "about 40". The existence of this term originated around the arrival of Black death in Europe in the XIV century. The incubation period of disease was about 40 days, therefore this duration was implemented on the ships carrying suspected material (MacKenzie 2001). Precautionary measures were taken when suspected material was found in the ship or port, by refusing the entry or access. This practice helped in preventing the entry and spread of unknown diseases in the receiving countries.

Plant quarantine addresses most of the subjects covered by plant health, including eradication, prevention, certification and registration processes. There were certain legislative measures taken to implement the national quarantine regulations. Application, registration and regulation of pesticides also come under the term of plant health. Plant protection and quarantine are hence inter-related subjects.

14.2 Plant Health and Phytosanitation

Plant health plays an important role in yield and productivity of crops, and ultimately in any country's economy. The discipline of plant health pertains to the applied biology, being closely associated with plant pathological and entomological disciplines. Nematological and biological studies of other invertebrates and pests are fall within this discipline. Plant health is also affected by the other biological factors, such as those investigated in pesticide science and chemistry.

There are many problems to face while maintaining plant health. Controlling different plant pathogens, insects and other invertebrates attacking plants is one of the practices to improve plant health. There are several comprehensive glossaries related to plant health and phytosanitary aspects, published previously (Hopper 1995; Anon 1996, 2002).

Plant health has received a remarkable attention due to its importance, and now several biological and chemical journals deal with this topic, i.e. the *EPPO Bulletin*, published by the European and Mediterranean Plant Protection Organization (EPPO) both in French and English or *Crop protection* and *Plant protection*, addressing plant health. Moreover, journals related to plant pathology and entomology also give space to articles related to plant health. The *FAO Plant Protection*

Bulletin was formerly a leading journal in this field, especially in covering legal and administrative matters, but ceased publication after 1994. However, within the United Nations Food and Agriculture Organization (FAO), the Secretariat to the International Plant Protection Convention (IPPC) publishes *International Standards for Phytosanitary Measures* and many other plant health documents, available on the informative website www.ippc.int/.

Some plant species may cause damage to other plants and are not controlled by normal methods applied for disease management. Plants which are in competition with other plants by different means, including nutrition, light and space, are indicated as "plant pests". They not only affect the nutritional requirements of plants but also cause injuries leading to huge losses. Commonly, phytosanitary authorities are responsible for the control of such plants that may act as parasitic species or weeds. They are called invasive plants or serious weeds. A common example of such invasive plants is dwarf mistletoes (e.g. *Arceuthobium*). Different countries have a range of legislative control measures, adopted against invasive plants, however the phytosanitary measures do not represent the permanent concern for control. In some countries the invasive plants are not found or rare, but still they have their own legislative control measures to avoid damages. The administration of these measures is always under the responsibility of the phytosanitary authorities. Alternative methods such as the concern of the government dealing with the environment, agriculture and some other extensions agencies may also apply (Ebbels 2003).

14.3 Menace of Foreign Pests

As human are moving from one area to other areas, also the forage plants, food and livestock move along with the human population (Diamond 1998). This practice has been facilitated by the conquests and empires involved in establishment of populations. The cultivated plants were historically introduced within the ancient empires of Greeks, Middle East, Romans and Egyptians. When plants migrated from one place to the other their associated pests were kept along with them. Some host species, such as apple and orange, have been travelling from east to west whereas others such as barley and wheat have been transported from west to east, along the pathway known as "silk route", established between the Asia and Europe. China, South-East Asia, the Middle East and East Africa have been involved in the transport of crops and their associated pests, including sugarcane, rice, banana and soybean, by a trade route system.

In the early XIX century seeds became a trade commodity, an innovation as before the exchange included the plant products, while the seeds were occasionally traded. As time passed, seeds of other crops as well as ornamental plants were also collected by the specialists and transported to Europe and North America. Long distance transport of growing plants became more common during the middle of XIX century and in turn increased the risk of pests transportation through their specific hosts, from one area to the other.

The invention of the Wardian case by Nathaniel Bagshaw Ward in 1830 greatly facilitated the long-distance transport of living plants (Hobhouse 1992). In this way the long distance transport of different commodities increased day by day after the introduction of the steam ships. However, the transportation of growing plants increased the risk of spreading pests from one country to the other within short periods of time. Containers developed to maintain the product quality as well as of the use of refrigeration helped growing plants to be maintained and kept fresh, favoring as well the pests associated with them. Trading is easy in countries where commercial partnerships have been established from many decades, as partners facilitate the exchange of commodities. Climatic conditions also favor the trading system, if they are similar in the origin and destination countries. In these conditions the trading product and their associated pests can easily survive in the destination country. This mutual matching between partner countries during trading presents great plant health risks (McRae and Wilson 2002). There are severe pests that are yet present in a country but did not spread. It is hence necessary to avoid the spread of such invasive pests during the trading of living plants or products. Main objective at this point is to avoid further spreading of pathogens by eradication procedures, adopting preventive measures as well. To maintain plant health along with preventive measures it is necessary to enhance the awareness of farmers and growers, in order to use disease-free planting material so that healthy plants are obtained.

14.4 Legislation and Phytosanitary Measures

In order to minimize the health risk of plants it is necessary to establish an appropriate phytosanitary legislation. First phytosanitary legislation was established and implemented in 1660 by France against common barberry (*Berberis vulgaris*) bushes, due to their likely association with blast (later called black stem rust) of wheat and other cereals. These legislative measures were effective and implemented by New England and German authorities in mid XVIII and early XIX centuries, respectively, to control the wide spreading of the cereal blast causal agent, especially in barberry vicinity. Association of barberry with the wheat blast pathogen (*Puccinia graminis*) was confirmed and explained, two centuries after the first observations, by the mycologist Antone de Bary who reported the occurrence of the alternative host and thus solved the mystery. As the scientific basis of the association was provided, barberry eradication campaigns started in many European countries, not forced but encouraged by the respective agricultural authorities. Phytosanitary legislation in USA started in 1912 with the Plant Quarantine Act presented by the Congress, later endorsed in 1918 by other 13 states which passed barberry eradication laws in wheat growing areas. In response to this action, the shipment of 35 species of barberry and *Mahonia* altogether was prohibited (Large 1940; Fulling 1942, 1943; Ebbels 2003).

14.5 Developing Stories on Phytosanitary Measures

Phytopathogens cause huge losses in particular to farmer with limited resources, as well as threatening food security, disturbing national economies and, in severe cases, cause starvation or even death for million people. These problems can be minimized by proper legislation and agreements among trading countries. When the wine industry was threatened in France by plant pests during 1865–1875, the first international quarantine and phytosanitary legislations were implemented against different pests. During this period the French wine industry was threatened by the American vine louse (*Viteus vitifolii*) in all vineyards across the country. This American pest, formerly known as *Phylloxera vastatrix*, rapidly spread to France and all European wine production regions in close countries, causing in 1875 losses of about 50 million £ to France only (Large 1940). Vine industry is the backbone of many European growers' economy, and protecting this commercially important industry was a first priority. This kind of problem must be determined and dealt with by finding the sources of infection and checking the import material.

Humans also represent a potential source of pathogens spread within and across countries. Berne and Switzerland authorities acted together, by inviting interested countries to discuss the role of import/export of planting material in pathogens' spread, across countries including the role of human travels. The purpose of the meeting held on 17 September 1878 was to avoid the spread of pathogens hosted by Switzerland through the "International Convention on Measures to be taken against *Phylloxera vastatrix*". In the presence of Austria, France, Germany, Italy, Portugal, Spain and Switzerland representatives, this first international agreement was passed to check the spread of phytopathogens across countries (Anon 1914). The principles set in this convention are still internationally recognized today for protection of plant health. Terms decided in the convention were followed in the next 3 years. During this period it was concluded that some principles and terms had some deficiencies and were not clearly defined. By keeping in mind these deficiencies, and aiming at their improvement, a second International meeting was conducted in 1881, also signed in Berne. This convention worked for eight years successfully. Again, further improvements were identified in certain aspects and in 1889 a third Convention was held in Berne, attended by different countries. At the start of the XX century, with much progress achieved in plant protection with the discovery of new chemicals to manage diseases and of improved techniques for diagnosis, the needs for legislations and phytosanitary measures increased. National and international trade is still increasing today the risk of pathogen spread across and within country, so national legislations have to be followed strictly.

14.6 Quarantine: Food Security and Food Safety

For biosecurity measures, plant quarantine regulations have been designed to prevent the introduction and spread of economically important pests to previously unaffected areas, or where the pests or disease causal agents are present, but at a limited scale or under official control (FAO 2004). The laws concerning plant quarantine regulations can be implemented at the national, international or even regional levels. All countries have right to implement plant quarantine laws to regulate the movement of material to protect crops, ecosystem and environment. These laws are also helpful to restrict pests occurrence within limited areas. Plant quarantine regulations also can prevent trade of a specific plant material and products and help to reduce the risk of introduction of pests, therefore limiting the movement of vegetative parts and material. However, movement of germplasm is very important in the economic development through the export, trade, and capacity building for industries and markets at global, national and regional levels, in order to ensure food security (Beale et al. 2008; Stack 2006).

For the World Trade Organization (WTO) member countries, the WTO Agreement on the application of Sanitary and Phytosanitary (SPS) measures emphasizes that the quarantine regulations and measures should be appropriate and follow the standards developed by the International Plant Protection Convention (IPPC). Moreover, plant quarantine should not be used as a technical barrier to stop the trade, without any reason (FAO 2004; World Trade Organization 1994). It is also important to note that no country can operate or violate quarantine regulations, otherwise it will have to face sanctions that will be imposed upon its trade (Beale et al. 2008).

It is important to recognize a policy of risk management based on scientific knowledge, and that all quarantine measures have to be tailored in correspondence with the risk related to a specific pest (Naim et al. 1996; Ochoa-Corona 2011). Countries can analyze pest threats by performing pest risk analyses (PRA) to assess the potential for a pest to entry, establish and spread, as well as to assess potential biological and economic consequences on such entry [see International Standards for Phytosanitary Measures (ISPM) No. 2 (FAO 2007) and ISPM No. 11 (FAO 2004)].

PRA is used to determine and prepare the biosecurity measures required to minimize all risks of introduction and to mitigate the consequences of introduction, when it happens. The plant quarantine import conditions depend on the genus and species of plant and the type of imported material. In undertaking a PRA, the first step is to determine whether a pest is present in a region and, if so, whether it has a restricted distribution and is under official control. Surveillance should be performed in a region to determine the presence or absence of a pest, before implementing plant quarantine regulations. The potential for establishment and spreads and its economic consequences should also undertaken to determining the quarantine status of a disease or pest.

When plants or plant products reach the border, they may be kept in quarantine for an interval of time dependent on the product and its risk of introducing a pest into that area or region. Some plant propagation material and seed can be inspected, treated, and made pest-free and/or pathogen-tested upon arrival, which can provide a rapid and safe release. However, many plant propagation materials require much time for growth. Disease screening and/or pathogen testing may be applied in a post-entry quarantine (PEQ) facility to ensure pest-free material and certify the absence/presence of pathogens and diseases. This process should be adopted because some pests, especially viruses and vascular-limited bacteria, exist in very low concentrations and can require a long time to reach detectable levels, or symptoms may appear lately as well (Constable et al. 2013).

In some cases, PEQ has been imposed for several years to minimize the risk of undetectable pathogens, which can represent a significant delay in germplasm to become available in new markets, thus reducing its competitiveness. It has been reported many times that several export countries guarantee the absence of pests within the regions from which the material originates. Importing material from approved suppliers who have processes in place to minimize the incidence of pests and diseases in propagation material can reduce PEQ time in some countries (Constable et al. 2016). For example, for strawberry (*Fragaria annassa*) Australia have two approved sources for provision of meristem-cultured plantlets, produced from mother plants that have been maintained in an insect-proof environment and tested using acceptable detection procedures.

Candidatus Liberibacter is one of the most serious threats to potato producing countries and worldwide market access, and is a high priority quarantine pathogen for Australia (Plant Health Australia 2013). Studies revealed that it may be seed-borne in carrots (Bertolini et al. 2015), and infected seeds are a potential pathway for the bacterium introduction into Australia. Therefore, in 2015, Australia introduced phytosanitary measures against the import of carrot seeds that included heat treatment prior to entry onshore or offshore, in order to minimize the risk of the presence of pathogen in the seed, or onshore or offshore testing, to provide clear evidence for absence of the bacterium in the seed. After using offshore treatment or testing for *C. iberibacter*, carrot seed arriving at the Australian border may be allowed immediately and released as long as they comply with all measures, meaning a faster access of growers to seeds and meeting planting deadlines.

14.7 Modern Trends in Certification

Phytosanitary certification protocols are mostly targeted toward specific agents, especially those which represent significant threats to production and ecosystems. They are mostly targeted towards insect pests, viruses, viroids, vascular bacteria and endophytic fungi, that are able to spread in propagation material and whose eradication is much difficult through chemical or physical control treatments, including chemical dips, sprays, fumigation, or heat treatments. For several crops, testing is required for multiple pests and pathogens. Detection methods in quarantine regulation and certification programs include visual inspection and observation, isolations onto growth media, bioassays in which a sensitive indicator is inoculated and inspected for specific symptoms, microscopy, serology. Molecular methods such as polymerase chain reaction (PCR), reverse transcription-PCR (RT-PCR), quantitative-PCR (q-PCR), or nucleic acid blot assays are more sensitive and often applied.

In every phytosanitary certification protocol the main issue is its diagnostic reliability: protocols should be characterized by high levels of repeatability, reproducibility and diagnostic sensitivity in order to minimize false readings and results (EPPO 2014). However, the detection methods used in certification greatly vary in their processes and analytical sensitivity, and there is not any single method that is completely reliable for detection. All methods are affected by the biology of the pest/pathogen and its interaction with the host and the environment, which can affect symptoms' expression and pathogen titer. Many of the methods above mentioned were designed for the specific detection of a target pest or disease (Constable et al. 2013). Most of them were developed using the best available knowledge of the pest/pathogen biology and genetics, a knowledge was limited at the time of the test development, therefore bearing the risk of being limited in its detection capability.

Another limitation is that the methods were developed using plant materials grown under greenhouse conditions in which the pathogens' titres are often higher than those encountered in plants under field conditions. Additionally, virus inocula can be greatly affected in plants having mixed infections, which is the normal occurrence in perennial crops. All the detection methods are sensitive to the pest/pathogen concentration, their genetic variability, and the similarities between the target and other organisms. Moreover, they may not be helpful for the discovery of unknown noxious organisms (Bebber 2015).

Visual inspection of plant material and selection of healthy germplasm (not showing obvious symptoms of any specific disease) is one of the simplest methods used for detection, being the basis for early certification programs (Frost et al. 2013). However, visual inspection on germplasm or in a bioassay is influenced by the pathogen concentration in the host and by environmental and biological factors that have their effect on the expression of symptoms (Constable et al. 2013, 2016; Martin et al. 2013; Schaad et al. 2003). Some pests that are less virulent, asymptomatic, or symptomless in some plant varieties, clones, or cultivars may get missed in visual detection. For example, many host plants which are infected by the vascular bacterium *Xylella fastidiosa* remain symptomless (Hopkins and Purcell 2002; Purcell and Saunders 1999). Visual inspection of germplasm and bioassays may not be able to differentiate among species and strains of pathogens that cause similar symptoms, e.g. all *Grapevine leaf roll virus* species cause nearly identical leafroll disease symptoms, irrespective of the cultivars and sensitive indicator plants such as Cabernet franc and Pinot noir. Detection either by visual inspection of germplasm or bioassays also require the skill in identifying symptoms and differentiating them from those caused by other factors (Riley et al. 2002). Bioassays, specially indexing for viruses on woody indicators for different crops hosts such as grapevine, stone or pome fruits, are dependent on successful inoculation (Constable et al. 2013). Bioassays can take time, ranging from several weeks to many years to complete (Rowhani et al. 2005). Therefore, more active diagnostic and detection methods may improve the chance of detection and can reduce the time taken for diagnosis procedures.

Microscopic methods may be important steps for diagnosis, based on morphology, such as fungi and insects, and provide a useful support, programmed to prove the presence of some viruses. However, they require a high level of skill and knowledge of use. For some pests, such as viruses from same family, genus, and species with identical particle morphology, microscopy can be non-specific. It can also lack sensitivity for pathogens detected in plants that occur in low amounts, such as some viruses and phytoplasmas.

Serological methods such as enzyme-linked immunosorbent assays (ELISA) rely on antibodies to detect the presence of an antigen, such as a viral coat protein, present in the sample (Clark 1981, Clark et al. 2012). ELISA is a simple and easy technique and is useful for high-throughput testing. The reliability of the polyclonal antiserum may be, however, compromised by the poor quality of purification, due to contaminating proteins of plant origin or proceeding from co-infecting organisms, leading to false and non-specific detection, and false positive results (Hsu et al. 1988).

Poly- and monoclonal antisera may be affected by the degree of conservation of a protein sequence, and may possibly cross-react with other species or pest that produce protein-like antigens (Gugerli 2009). Occasionally, significant variations in the structure of the protein between closely related species of pest and targeted organisms may result in non-recognition, which may lead to a false negative result (Bertazzon and Angelini 2004; Fajardo et al. 2007). ELISA may also lack the sensitivity of other methods, such as PCR and some bioassays, that multiply the target nucleic acid up to a level at which they are easily detectable (Chevalier et al. 1995; Huttinga 1996; Sefc et al. 2000).

Molecular methods, such as PCR, have become the best choice for plant pest/pathogens detection programmes. PCR molecular tests are highly specific and can target individual strains or species, or may be generic, with multiple targets (Christensen et al. 2004; Bertolini et al. 2015; Maliogka et al. 2007; Zheng et al. 2010). They can be used in combination with sequencing methods to confirm their target identity (Martin et al. 2000; McCartney et al. 2003; Weisberg et al. 1991).

Genetic variability of targets may affect the test reliability, reducing the binding capacity of the primers even in conserved regions of the target genome or inducing mispriming, because of genetic similarity in genomes to other organisms at the primer binding sites (Powney et al. 2011; Vincelli and Tisserat 2008). Reverse transcription and polymerase enzymes are sensitive to compounds that may be mixed with extracted nucleic acids and may result in false negative consequences (López et al. 2009; Wilson 1997). PCR methods have a sensitivity greater than that of other detection methods. However, there is still a problem of lower detection limits of, and its reliability may depend upon a range of appropriate sampling techniques.

14.8 Quarantine Detection

The biosecurity of many crops can be compromised by the natural dispersal of pests and pathogens, which can prevent quarantine measures to accomplish their goals. The dispersal of airborne pests and diseases is impacted by multiple factors, the weather patterns being the most significant and yet uncontrollable ones. Example of such cases are the emergence of the soybean rust in North America and the spread of the wheat stem rust strain Ug99 around the globe (Isard et al. 2005; Krupa et al. 2006; Singh et al. 2011).

Human factors affect the emergence and recurrence of pests and diseases by the induced climate changes, limiting the production potential of crops in several areas across the globe (Singh et al. 2011; Vermeulen et al. 2012). Still, there are many other examples of disease agents and/or vector dispersal, correlated with human activities at the individual level (Baker et al. 1993; Gergerich et al. 2015); Knobler et al. 2006). Grafting onto susceptible indicator plants is the gold standard for virus detection in quarantine facilities. A mix-up in the choice of a plant cultivar used as virus indicator, leading to false negative results, could lead to the spread of important pathogens across the globe.

Advanced technology allowed for major expansion in the list of pathogens infecting crops, providing another challenge to maintain quarantine regulation and certification lists and detection protocols. For example, in the 2014 International Committee on Virus Taxonomy list there were more than 1200 plant viruses formally recognized, compared with 980 in 2005 and 380 in 1991, with most viruses associated with a disease (http://ictvonline.org/). Many recent discoveries in virology have occurred with the help of new technologies, to identify viruses-associated diseases (Adams et al. 2015; Barba et al. 2014; MacDiarmid et al. 2013). Ecological prospecting projects revealed that the number of plant viruses is likely far greater than known today, with many being cryptic or even possibly granting a kind of benefit to their hosts (Roossinck 2005; Roossinck et al. 2010; Wren et al. 2006).

The number of previously recognized plant-pathogenic bacteria is also increasing, especially because of the introduction of new unculturable species such as *Ca.* Liberibacter and *Ca.* Phytoplasma spp., with further pathovars and strains (Bull et al. 2012, 2014). Along with viruses, there are many unique uncultured bacteria observed in different samples, especially soils, based on genome fragments such as the 16S rRNA gene, whose biology is unknown.

Many plant-inhabiting pathogens, such as viruses, viroids, and phytoplasmas, can be unevenly distributed and in low quantity, making their detection difficult (Constable et al. 2003; Constable et al. 2016; López et al. 2009). It is a prerequisite, for any technology or method used, to identify best sampling strategies to improve the chance of detection and decrease the risk of false results. Regardless of how sensitive a test is, the sampling strategy imposes an impact on the assay reliability.

As technology evolves, new detection methods develop. In general, they are becoming faster and more sensitive, allowing us to monitor and detect targets in real time (Boonham et al. 2014; Gundersen and Lee 1996; Martin et al. 2000; Seyrig et al. 2015). If we look at the most common methods used for detection by quarantine facilities, it becomes evident that the newer methods are more specific than those previously used. For example, PCR is more specific than ELISA, which is more specific than grafting or mechanical inoculation onto indicators plants (Martin et al. 2000).

It is a kind of natural law: the pathogen evolves faster than the host. Viruses are an excellent example, as they evolve more often to overcome host resistance, improve vector transmissibility, or increase their host range using both micro- and macroevolution. They rely on genetic drift via the quasispecies distribution of mutations, to recombine and reassort in multipartite viruses (Hull 2013). It has hence become evident that the definition of a virus isolates, or species, is a fluid concept. The population structure and diversity of most pathogens are grossly understudied. As few isolates are characterized at the molecular and biological level, they are used for the development of laboratory detection tests. Still, this is only a small fraction of a species population, where the studied isolates may be outliers, rather than the most representative of their species. In such a case, the newly established tests may provide us less reliable results because they are too much specific and lack identification of many isolates. There are several examples showing this effect. Monoclonal antibodies recognize a single epitope, which may result in a test with a minimal background, as they are screened to react to the pathogen and not towards the host antigen (Chamberlain et al. 2003; Pleško et al. 2009).

A very useful genus-specific monoclonal antibody has been developed towards potyviruses for a first screening, when examining a unknown target, although it must be cautioned that its coverage is partial Jordan and Hammond (1991). Current diagnostic methods provide opportunities for disease diagnosis and pathogen detection, but a further challenge for plant quarantine is the ability to determine the cause of a disease or symptoms, especially the unfamiliar cases or those for which a pathogen is not known. If a pest cannot be detected by any of the current methods, should that diseased material be released? A plant protection organization is often conservative in this situation and stops the release of germplasm exhibiting unusual symptoms, for which a cause cannot be unequivocally identified. This is where new unbiased detection technologies could play an important role, to identify all known and, most importantly, unknown agents, in a timely fashion. Quarantine plays an important role in biosecurity and prevents invasions by foreign pathogens. Therefore, regulations should be implemented and coupled with new and updated detection technologies, also suitable for unknown pathogens and pests.

References

Adams, M. J., Hendrickson, R. C., Dempsey, D. M., & Lefkowitz, E. J. (2015). Tracking the changes in virus taxonomy. *Archives of Virology, 160*, 1375–1383.

Anon. (1914). *Actes de la Conférence Internationale de Phytopathologie* (pp. 237–241). Rome: Institut International d'Agriculture.

Anon. (1996). *Glossary of Phytosanitary terms*. Paris: European and Mediterranean Plant Protection Organisation.
Anon. (2002). Glossary of Phytosanitary Terms 2002. In *International standards for Phytosanitary measures no 5*. Rome: Food and Agriculture Organization of the United Nations.
Baker, C. R. B., Barker, I., Bartlett, P. W., & Wright, D. M. (1993). Western flower thrips, its introduction and spread in Europe and role as a vector of tomato spotted wilt virus. In D. L. Ebbels (Ed.), *Plant Health and the European Single Market* (Monograph No. 54) (pp. 355–360). Farnham Royal: British Crop Protection Council.
Barba, M., Czosnek, H., & Hadidi, A. (2014). Historical perspective, development and applications of next generation sequencing in plant virology. *Viruses, 6*, 106–136.
Beale, R., Fairbrother, J., Inglis, A., & Trebeck, D. (2008). *One biosecurity: a working partnership. Independent review of Australia's quarantine and biosecurity arrangements.* Rep. Aust. Gov., Dep. Agric. Fish. For., Sydney.
Bebber, D. P. (2015). Range-expanding pests and pathogens in a warming world. *Annual Review of Phytopathology, 53*, 335–356.
Bertazzon, N., & Angelini, E. (2004). Advances in the detection of grapevine leafroll associated virus 2 variants. *Journal of Plant Pathology, 86*, 283–290.
Bertolini, E., Teresani, G. R., Loiseau, M., Tanaka, F. A. O., Barbé, S., et al. (2015). Transmission of "*Candidatus* Liberibacter solanacearum" in carrot seeds. *Plant Pathology, 64*, 276–285.
Boonham, N., Kreuze, J., Winter, S., van der Vlugt, R., Bergervoet, J., et al. (2014). Methods in virus diagnostics: from ELISA to next generation sequencing. *Virus Research, 186*, 20–31.
Bull, C. T., De Boer, S. H., Denny, T. P., Firrao, G., Fischer-Le Saux, M., et al. (2012). List of new names of plant pathogenic bacteria (2008–2010). *Journal of Plant Pathology, 94*, 21–27.
Bull, C. T., Coutinho, T. A., Denny, T. P., Firrao, G., Fischer-Le Saux, M., et al. (2014). List of new names of plant pathogenic bacteria (2011–2012). *Journal of Plant Pathology, 96*, 223–226.
Chamberlain, C. J., Kraus, J., Kohnen, P. D., Finn, C. E., & Martin, R. R. (2003). First report of Raspberry bushy dwarf virus in *Rubus multibracteatus* from China. *Plant Disease, 87*, 603.
Chevalier, S., Greiff, C., Clauzel, J. M., Walter, B., & Fritsch, C. (1995). Use of immunocapture-polymerase chain reaction procedure for the detection of grapevine virus A in Kober stem grooving-infected grapevines. *Journal of Phytopathology, 143*, 369–373.
Christensen, N. M., Nicolaisen, M., Hansen, M., & Schultz, A. (2004). Distribution of phytoplasmas in infected plants as revealed by real time PCR and bioimaging. *Molecular Plant-Microbe Interactions, 17*, 1175–1184.
Clark, M. F. (1981). Immunosorbent assays in plant pathology. *Annual Review of Phytopathology, 19*, 83–106.
Clark, C. A., Davis, J. A., Abad, J. A., Cuellar, W. J., Fuentes, S., et al. (2012). Sweetpotato viruses: 15 years of progress on understanding and managing complex diseases. *Plant Disease, 96*, 168–185.
Constable, F. E., Gibb, K. S., & Symons, R. H. (2003). The seasonal distribution of phytoplasmas in Australian grapevines. *Plant Pathology, 52*, 267–276.
Constable, F. E., Connellan, J., Nicholas, P., & Rodoni, B. C. (2013). The reliability of woody indexing for detection of grapevine virus-associated diseases in three different climatic conditions in Australia. *Australian Journal of Grape and Wine Research, 19*, 74–80.
Constable, F. E., Nancarrow, N., Kelly, G., Milinkovic, M., Ko, L., et al. (2016). Improved diagnostic testing and on farm biosecurity plan to support Australian strawberry certification programs. *Acta Horticulturae*, (1117), 171–176.
Diamond, J. (1998). *Guns, Germs and Steel. A Short History of Everybody for the Last 13,000 Years* (pp. 104–113). London: Vintage Random House.
Ebbels, D. L. (2003). *Principals of plant health and quarantine*. Wallingford: CAB International.
EPPO. (2014). PM 7/76 (3) Use of EPPO diagnostic protocols. *EPPO Bulletin, 44*, 335–337.
Fajardo, T. V. M., Barros, D. R., Nickel, O., Kuhn, G. B., & Zerbini, F. M. (2007). Variability of the coat protein gene of *Grapevine leafroll-associated virus 3* in Brazil. *Fitopatol Brasil, 32*, 335–340.

FAO. (2004). *Pest risk analysis for quarantine pests, including analysis or environmental risks and living modified organisms*. Int. Stand. Phytosanit. Meas. Number 11, Food Agric. Organ., Rome.

FAO. (2007). *Framework for pest risk analysis*. Int. Stand. Phytosanit. Meas. Number 2, Food Agric. Organ., Rome.

Fulling, E. H. (1942). Plant life and the law of man. *III. Barberry eradication. Journal of the New York Botanical Garden, 43*, 152–157.

Fulling, E. H. (1943). Plant life and the law of man. IV. Barberry, currant and gooseberry, and cedar control. *The Botanical Review, 9*, 483–592.

Frost, K. E., Groves, R. L., & Charkowski, A. O. (2013). Integrated control of potato pathogens through seed potato certification and provision of clean seed potatoes. *Plant Disease, 97*, 1268–1280.

Gergerich, R. C., Welliver, R. A., Gettys, S., Osterbauer, N. K., Kamenidou, S., et al. (2015). Safeguarding fruit crops in the age of agricultural globalization. *Plant Disease, 99*, 176–187.

Gugerli, P. (2009). 25 years of serological identification of grapevine leafroll-associated viruses: antiserum and monoclonal antibodies to GLRaV-1 to GLRaV-9. *ICVG, 16*, 24–28.

Gundersen, D. E., & Lee, I.-M. (1995). Ultrasensitive detection of phytoplasmas by nested-PCR assays using two universal primer pairs. *Phytopathologia Mediterranea, 35*, 144–151.

Hobhouse, H. (1992). *Seeds of Change. Five Plants that Transformed Mankind*. London: PaperMac.

Hopkins, D., & Purcell, A. (2002). Xylella fastidiosa: Cause of Pierce's disease of grapevine and other emergent diseases. *Plant Disease, 86*, 1056–1066.

Hsu H.T., Franssen J.M., van der Hulst C.T.C., Derks A.F.L.M., Lawson R.H. 1988. Factors affecting selection of epitope specificity of monoclonal antibodies to tulip breaking potyvirus. Phytopathology 78:1337–1340.

Hopper, B. E. (1995). *NAPPO compendium of Phytosanitary terms. north American plant protection organisation*. Ontario: Nepean.

Hull, R. (2013). *Plant virology* (5th ed). New York: Academic.

Isard, S. A., Gage, S. H., Comtois, P., & Russo, J. M. (2005). DATE principles of the atmospheric pathway for invasive species applied to soybean rust. *Bioscience, 55*, 851–861.

Huttinga, H. (1996). Sensitivity of indexing procedures for viruses and viroids. *Advances in Botanical Research, 23*, 59–71.

Jordan, R., & Hammond, J. (1991). Comparison and differentiation of potyvirus isolates and identification of strain-, virus-, subgroup-specific and potyvirus group-common epitopes using monoclonal antibodies. *The Journal of General Virology, 72*, 25–36.

Knobler, S., Mahmoud, A., Lemon, S., & Pray, L. (2006). *The Impact of Globalization on Infectious Disease Emergence and Control*. Washington, DC: The National Academies Press.

Krupa, S., Bowersox, V., Claybrooke, R., Barnes, C. W., Szabo, L., et al. (2006). Introduction of Asian soybean rust urediniospores into the midwestern United States: a case study. *Plant Disease, 90*, 1254–1259.

Large, E. C. (1940). *The Advance of the Fungi*. London: Jonathan Cape.

Lopez, M. M., Llop, P., Olmos, A., Marco-Noales, E., Cambra, M., & Bertolini, E. (2009). Are molecular tools solving the challenges posed by detection of plant pathogenic bacteria and viruses? *Current Issues in Molecular Biology, 11*, 13–46.

MacDiarmid, R., Rodoni, B., Melcher, U., Ochoa-Corona, F., & Roossinck, M. (2013). Biosecurity implications of new technology and discovery in plant virus research. *PLoS Pathogen, 9*, e1003337.

MacKenzie, D. (2001). All fall down. *New Scientist, 172*, 34–37.

Maliogka, V. I., Dovas, C. I., & Katis, N. I. (2007). Demarcation of ilarviruses based on the phylogeny of RNA2- encoded RdRp and a generic ramped annealing RT-PCR. *Archives of Virology, 152*, 1687–1698.

Martin, R. R., MacFarlane, S., Sabanadzovic, S., Quito, D., Poudel, B., & Tzanetakis, I. E. (2013). Viruses and virus diseases of Rubus. *Plant Disease, 97*, 168–182.

Martin, R. R., James, D., & Levesque, C. A. (2000). Impacts of molecular diagnostics on plant disease management. *Annual Review of Phytopathology, 38*, 207–239.
McCartney, H. A., Foster, S. J., Fraaije, B. A., & Ward, E. (2003). Molecular diagnostics for fungal plant pathogens. *Pest Management Science, 59*, 129–142.
McRae, C. F., & Wilson, D. (2002). Plant health as a trade policy issue. *Australasian Plant Pathology, 31*, 103–105.
Nairn, M. E., Allen, P. G., Inglis, A. R., & Tanner, C. (1996). *Australian Quarantine: A Shared Responsibility*. Canberra: Dep. Prim. Ind. Energy.
Ochoa-Corona, F. M. (2011). Biosecurity, microbial forensics and plant pathology: Education challenges, overlapping disciplines and research needs. *Australasian Plant Pathology, 40*, 335–338.
Plant Health Australia. (2013). *Industry biosecurity plan for the potato industry. Version 2* (p. 0). Plant Health Aust: Canberra, Aust. http://www.planthealthaustralia.com.au.
Ples˘ko, I. M., Marn, M. V., S˘ irca, S., & Urek, G. (2009). Biological, serological and molecular characterisation of *Raspberry bushy dwarf virus* from grapevine and its detection in the nematode *Longidorus juvenilis*. *European Journal of Plant Pathology, 123*, 261–268.
Powney, R., Beer, S., Plummer, K., Luck, J., & Rodoni, B. (2011). The specificity of PCR-based protocols for detection of *Erwinia amylovora*. *Australasian Journal of Plant Pathology, 40*, 87–97.
Purcell, A. H., & Saunders, S. R. (1999). Fate of Pierce's disease strains of Xylella fastidiosa in common riparian plants in California. *Plant Disease, 83*, 825–830.
Riley M. B., Williamson M. R., & Maloy O. 2002. Plant disease diagnosis. *Plant Health Instructor*. doi:https://doi.org/10.1094/PHII.
Roossinck, M. J. (2005). Symbiosis versus competition in plant virus evolution. *Nat. Rev. Microbiol., 3*, 917–924.
Roossinck M, J., Saha, P., Wiley, G. B., Quan, J., White, J. D., et al. (2010). Ecogenomics: Using massively parallel pyrosequencing to understand virus ecology. *Molecular Ecology, 19*, 81–88.
Rowhani, A., Uyemoto, J. K., Golino, D., & Martelli, G. P. (2005). Pathogen testing and certification of *Vitis* and *Prunus* species. *Annual Review of Phytopathology, 43*, 261–278.
Schaad, N. W., Frederick, R. D., Shaw, J., Schneider, W. L., Hickson, R., & Petrillo, M. D. (2003). Advances in molecular-based diagnostics in meeting crop biosecurity and phytosanitary issues. *Annual Review of Phytopathology, 41*, 305–324.
Sefc, K. M., Leonhardt, W., & Steinkellner, H. (2000). Partial sequence identification of *Grapevine-leafroll associated virus-1* and development of a highly sensitive IC-RT-PCR detection method. *Journal of Virological Methods, 86*, 101–106.
Seyrig, G., Stedtfeld, R. D., Tourlousse, D. M., Ahmad, F., Towery, K., et al. (2015). Selection of fluorescent DNA dyes for real-time LAMP with portable and simple optics. *Journal of Microbiological Methods, 119*, 223–227.
Singh, R. P., Hodson, D. P., Huerta-Espino, J., Jin, Y., Bhavani, S., et al. (2011). The emergence of Ug99 races of the stem rust fungus is a threat to world wheat production. *Annual Review of Phytopathology, 49*, 465–481.
Stack, J. P. (2006). Challenges to crop biosecurity. In J. ML Guillino, A. G. Fletcher, & J. P. Stack (Eds.), *Crop biosecurity: Assuring Our Global food supply (NATO science for peace and security series-C: Environmental security)* (pp. 15–23). Dordrecht: Springer.
Vermeulen, S. J., Campbell, B. M., & Ingram, J. S. (2012). Climate change and food systems. *Annual Review of Environment and Resources, 37*, 195–222.
Vincelli, P., & Tisserat, N. (2008). Nucleic acid–based pathogen detection in applied plant pathology. *Plant Disease, 92*, 660–669.
Weisberg, W., Barns, S., Pelletier, D., & Lane, D. (1991). 16S ribosomal DNA amplification for phylogenetic study. *Journal of Bacteriology, 173*, 697–703.
Wilson, I. G. (1997). Inhibition and facilitation of nucleic acid amplification. *Applied and Environmental Microbiology, 63*, 3741–3751.

World Trade Organization. (1994). Agreement on the application of sanitary and phytosanitary measures. In *Results of the Uruguay Round of Multilateral Trade Negotiations: The Legal Texts* (pp. 69–84). Geneva, Switz: WTO.

Wren, J. D., Roossinck, M. J., Nelson, R. S., Scheets, K., Palmer, M. W., & Melcher, U. (2006). Plant virus biodiversity and ecology. *PLoS Biology, 4*, e80.

Zheng, L., Rodoni, B. C., Gibbs, M. J., & Gibbs, A. J. (2010). A novel pair of universal primers for the detection of potyviruses. *Plant Pathology, 59*, 211–220.

Chapter 15
Development and Implementation of IDM Program for Annual and Perennial Crops

Qaiser Shakeel, Guoqing Li, Yang Long, and Hafiz Abdul Samad Tahir

Abstract In developing countries, more than a quarter of potential food and fibre crop yield is routinely lost to pests, weeds and diseases. Minimizing the damages caused by these organisms in a sustainable manner cannot be achieved without considering the whole farming system in which the crops are produced. These challenges require the development of integrated crop management systems for pests, diseases and weeds, which are environmentally sustainable and socio-economically appropriate. Participatory approaches are needed for the development and extension of integrated crop management technologies and strategies. Hence, there is need to develop/utilize crop management strategies that cause very little environmental pollution. Management of crops in an integrated manner could be quite useful in this direction. This will include studies ranging from land preparation through seeding, propagation and management, up to harvesting.

Keywords Integrated disease management · Nutrition · Pest · Weed · Crop certification

Q. Shakeel (✉)
Discipline of Plant Pathology, University College of Agriculture and Environmental Sciences, The Islamia University of Bahawalpur, Bahawalpur, Pakistan
e-mail: qaiser.shakeel@iub.edu.pk

G. Li · Y. Long
State Key Laboratory of Agricultural Microbiology and Hubei Key Laboratory of Plant Pathology, Huazhong Agricultural University, Wuhan, China

H. A. S. Tahir
Pakistan Tobacco Board, Tobacco Research Station, Okara, Ministry of Commerce, Government of Pakistan, Islamabad, Pakistan

© Springer Nature Switzerland AG 2020

I. Ul Haq, S. Ijaz (eds.), *Plant Disease Management Strategies for Sustainable Agriculture through Traditional and Modern Approaches*, Sustainability in Plant and Crop Protection 13, https://doi.org/10.1007/978-3-030-35955-3_15

15.1 Introduction

The systems of cropping was developed since the ancient times in the India subcontinent, matching climate, crops and seasons. The knowledge on the plants growing period led to the methodical development of cropping systems. The biological activity was maintained by the incorporation of crop residues in soil, ultimately adding much required organic matter. With the increase in human population, agriculture was intensified. New, high yielding varieties and species moved the agriculture away from a methodical system. In recent times, however, it is considered that farmers also need to recall the knowledge gained during the development of ancient methods. Emphasis must be placed upon the value of increasing biodiversity in the agricultural system. In this regard, practices sustaining soil fertility should be integrated by aboveground management approaches based on either a single crop, or considering crops integrated with trees, in order to enhance natural microflora and add green manure, including even animals in local biodiversity.

To sustain the economic growth, increasing the agricultural productivity is prerequisite for development. In developing countries, especially in Pakistan, agricultural productivity is a dominant part in the economy. It plays a significant role in elevation from poverty. It provides livelihood to hundred million poorest people in Pakistan. With increasing environmental concerns in Pakistan, environment management should also be achieved through agriculture. Well-managed agriculture may help in fact in poverty elevation, providing livelihood, conserving soil and water resources, increasing preservation of trees and biodiversity. Contaminated and polluted environments, depletion of natural resources, food scarcity and ultimately nutrition and health problems, are all the results of a not methodically managed agriculture (Kumar and Shivay 2008).

In developing countries like Pakistan almost half of the produce is lost due to the impact of diseases and insect pests. To minimize the damages caused by these organisms, agricultural systems must be managed in a sustainable manner. One of the most appropriate ways to manage disease problems is by the integration of all factors active in crop management. The challenges that need to be addressed are as follows:

- Developing Integrated disease management (IDM) systems, which must be environment friendly and economically appropriate.
- Minimizing the use/misuse/overuse of synthetic chemicals by developing disease management systems based on innovative and safer chemicals.
- Developing alternative disease management technologies, i.e. biological control, phytobiochemicals.
- Participation of different agencies for the development of technologies to properly implement suitable IDM approaches (Kumar and Shivay 2008).

15.2 Concepts and Definitions

IDM is a practice using all possible ways to manage a disease, with synthetic chemicals as last option. IDM balances traditional and modern technologies with a focus on economics and environmental impact. In other terms, we can say that IDM is a whole-farm approach. The main components of IDM are: crop, nutrition, and insect-pest managements, without ignoring the financial status of the farming community. The farm management and IDM relationship is indeed very dynamic (Mishra et al. 2016). For example, soil sickness is a renowned phenomenon in cultivation of crops. Soil-borne pathogens usually persist if we follow a monoculture, year after year. This will ultimately lead towards a population increase of soil pathogens, and making that field unfit for the cultivation of a given crop.

Another approach is the cultivation of resistant crops for 4–5 years, as it is expected that the pathogen population will decline. This may be due to some biochemical substances released by a particular crop, or by the pathogen starvation. Mechanisms behind this approach also include an increase in the population of antagonistic microorganism, killing the pathogen.

Soil invaders can be obligate pathogens or saprotrophs. The latter can survive in the soil for at-least 5–6 years. In this regard, different disease management strategies can be implemented. Among them crop rotation is an effective method especially against soil pathogens (Agrios 2005). If we can rotate the crop for more than 3–4 years with a non-host variety or cultivar, the pathogen population will decline gradually. Most of the pathogens will infect only one or more particular species of plant or sometime a few within their families. The weakness of this approach is that the pathogen produces hard resting propagules or structures such as durable spores, in this case the crop rotation will not achieve a long-term control of the pathogen (Kumar and Shivay 2008).

Another approach, known as soil solarization, is feasible in areas with a hot summer. In such areas, the field is ploughed fallowed by a fallowing for 1 to 1.5 years. The increase in soil temperature will have a negative impact on the pathogen population. This system is helpful for a number of obligate parasites, especially nematodes (Agrios 2005).

Keeping in view these conditions, the definition of an appropriate IDM requires the definition of some ideas and goals which need to be implemented by the farmers. IDM considers the management of different crops, planted in different fields, in order to get benefit from their interactions. The benefits may include management of diseases and/or insect-pests, maintaining soil fertility etc. However, inclusion of modern technologies are far more important to maximize the IDM benefits (Kumar and Shivay 2008).

15.3 Underlying Principles of IDM

Basic objectives to be achieved through IDM are:

1. Improving crop production for either quantity and quality.
2. Reducing the introduction of disease into crops.
3. Avoiding conditions suitable for a disease establishment and spread.
4. Sustainability with regard to biological-environmental and financial aspects.
5. Balancing local and external inputs with more emphasis on encouraging the use of local resources.
6. Focusing mainly on environmental and human health protection.
7. Farmers information should be combined with science-based knowledge to manage the agro-ecological zones with respect to human and economic aspects.
8. IDM practices should be developed by keeping in mind the type and stages of crops development.

Keeping in mind the major agro-ecological zones of Pakistan, one single ideal IDM system will never work efficiently, As the conditions of climate, soil, pathogens' prevalence are different. The aim is to devise some basic guideline, principles and know-how that can be communicated to farmers, showing them how to optimize their own IDM using the provided information. One of the most important aspect of IDM is its detailed application and testing. This system should start with practices supporting the exploitation of friendly microorganism, resistance, with the precise application of biological control agents, disease forecasting model, reliable diagnostic techniques, considering chemical application as the last possible alternative (Khoury and Makkouk 2010; Kumar and Shivay 2008).

15.4 Plant Disease Management Strategies

Whetzel (1929) was the first to define some terms for managing diseases, i.e. exclusion, eradication, protection and immunization, with the addition of two new approaches namely avoidance and therapy. Definitions follow.

15.4.1 Avoidance

Creating an unfavourable environment by changing the site and season of the crop. This will change the environment and also reduce the effective inoculum in a particular area.

15.4.2 Exclusion

Restriction in inoculum entry.

15.4.3 Eradication

Eradicate, extinguish, or deactivate the inoculum.

15.4.4 Protection

Direct application of chemicals to prevent infection.

15.4.5 Resistance

Exploring the genes or group of genes that will counteract or halt infection, or act as a therapy rehabilitating the infected parts.

15.4.6 Antagonism

Suppressing the activity/population density of one or more pathogens by enhancing the activity of its/their natural regulating organisms. Antagonists are important for the development of biological control.

15.5 Integrated Disease Management

Cost effective and environment-friendly methods will ideally keep the plant pathogen population below any economic threshold or injury level, with the benefit of increase productivity and economic growth (Nutter 2007; Zadoks 1985).

It may also be defined as the use of multiple manoeuvres in such a successive way that managing the pathogen results naturally and thriftily (Mills and Nutter Jr. 1991; Nutter and Guan 2001). IDM main traits are:

1. Effective in managing a vast majority of pathogens.
2. Use of up-to-date information about the current status of the pathogen(s) and antagonist(s) development.

3. Monitoring the chemicals application with respect to environment and economy.
4. Integrated use of numerous, effective tactics.

15.6 Guidelines of IDM

The guidelines for IDM can be divided into (1) a general IDM strategy and (2) a specific strategy. Both must be followed by all farmers. The general IDM strategy is more concerned to the practices that are being followed at the farm level. It is applicable to all types of disease and pathogens. The specific strategy level usually deals with epidemic diseases (Fry 1982; Agrios 2005), that require automated IDM practices for individual cases.

First actions to begin with IDM are:

1. Knowing the current scenario of the disease at farm level.
2. Being familiar with the basic concepts of plant pathology and IDM
3. Starting with level 1, i.e. apply basic IDM strategies.
4. With the passage of time, monitoring the disease development at the field and farm levels.
5. If a specific problem persists, apply specific management strategies for that particular disease.
6. For proper diagnosis and further guideline on disease management consult with expert and eventually act properly.

15.7 Factors Affecting Occurrences

The plant disease causal factors are divided into two categories, i.e. biotic and abiotic. Biotic factors include microorganism such as some fungi, bacteria, nematodes and higher parasitic plants. Abiotic factors include high and low temperatures, soil conditions, soil pH, aeration, deficiency and excess of micro- and macro-nutrients, soil moisture etc. Both factors continuously affect plants tissues and cells (Agrios 2005), with a harmful impact on plant physiology, also eliciting a physiological response towards harmful processes. The physiological response that arouse as a result of disruption in plant physiology originates particular appearance with characteristic symptoms. Based on the causal factors, plant diseases are infectious (biotic) and non-infectious (abiotic) (Nutter and Mills 1990; Zadoks 1985). Management of non-infectious diseases are somewhat easier, while for infectious diseases there is the need to develop a proper strategy for management.

15.7.1 *Biotic Factors*

Micro-organism can by saproptrophs or parasites. Saprotrophs obtain their nutrients by decomposing dead plants or parts. Parasites feed on living plants or parts and may or may not cause a disease. All plant pathogens are parasites. The mechanism by which these microorganism obtain their nutrients is exocytic (they produce enzymes to digest their food outside their body and then absorb the resulting digested nutrients) or phagotrophic (member of oomycetes have a phagotrophic mood of nutrition, engulfing cells or their parts). In either way of nutrition a substantial disruption in the plant metabolic activity takes place (Agrios 2005; Fry 1982).

15.7.2 *Parasites*

There are two types of parasites, i.e. facultative and obligate. Facultative parasites are those that can become parasites if a living host is available. Obligate parasites are those that can only survive on a living host. Parasites can spread from one plant to another, being "infectious" a term used to denote a diseases caused by biotic factors (Fry 1982).

15.7.3 *Inoculum for Propagation*

Parasites have different ways of reproduction. Pathogens or parts of a pathogen (i.e. spores) used for propagation are generally referred to "*inoculum*". The inoculum can be provided by any part of the pathogen devoted to the function of disease transmission to a new host. For example in case of fungi, the spores are often an inoculum. Single cells for bacteria, virions, prions or viroids in case of viruses, second stage juveniles in case nematodes, all are *inocula* (Latin, plural) of the concerned pathogens (Fry 1982). Although plants can directly be infected by an inoculum, the development of a pathogenic process is often needed, before infection can proceed.

15.7.4 *Inoculation*

The process of passive transportation (or the active movement) of the inoculum from its source to a given, susceptible host plant or part of it, is known as inoculation. The inoculum must be transported to the susceptible host to start an infectious disease (Fry 1982). Different vectors may carry the pathogens, i.e. insects, wind, and part of infected parts, animals and/or water splashes. Insects can transmit several pathogens such viruses, mycoplasma-like organisms, fungi and bacteria. The same applies to wind and water splashes (Zadoks and Schein 1979). Active inocula include parasitic invertebrates, such as insects and nematodes, which reach plant tissues by active movements.

15.7.5 Pathogens' Entry

Once the pathogen reaches its host, it must enter the plant to cause a disease. There are different ways of entry, including natural openings, wounds as entry points or direct penetration through the host surface. Natural openings include stomatas, hydathodes and lenticels. Bacterial pathogens usually gain entry through natural openings and wounds. Fungi have structures that can penetrate directly through the protective covering of the plant. They may also soften the tissues by enzymatic reactions. For viruses, the vectors are important, as they usually inject them into the host, during their feeding activity (Strange 2003; Chaube and Singh 2004).

15.7.6 Establishment of Pathogens

The establishment and colonization of pathogens in the host complete the infection. Once the pathogens got entry, they induce changes that make the nutrients available for the pathogens. Most parasites attack the host cell by damaging its cytoplasmic membrane, to make the membrane "spongy", so that the nutrients are readily available (Agrios 2005; Strange 2003; Chaube and Singh 2004). This process is followed by the attack on carbohydrates, proteins and lipids by the parasite secreted enzymes. As result of all this process, complex compounds are converted into simpler ones, i.e. sugars and amino acids, that diffuse through the cytoplasmic membrane to be assimilated by the parasite (Pathak et al. 2006).

15.8 Plant Disease Concept

The disease development results by the interaction of three factors. These are: host, pathogen and environment. A susceptible host, a virulent pathogen and a conducive environment must meet in these states to cause a disease. The host that can counteract a pathogen's effect is regarded as "resistant". If it can withstand the pathogen attack then it is considered as immune. Pathogens hence need to develop a mechanism by which they can breach through resistance, which is specific to each host. Once the pathogen is able to bypass the barriers deployed as resistance, then the host is termed "susceptible" and the pathogen is regarded as "virulent" to that particular host. Once a virulent pathogen spreads to the susceptible host, the environment must be suitable for the infection to be successful (Pathak et al. 2006; Agrios 2005; Strange 2003; Chaube and Singh 2004).

To develop a plant disease management strategy, one must keep in mind the triangle formed by the host, the pathogen and the environment. Success of IDM lies in the selection of a range of strategies that increase productivity and reduce losses due to biotic and abiotic factors (Pathak et al. 2006; Agrios 2005; Strange 2003; Chaube and Singh 2004).

The IDM strategies that place emphasis on host are: (1) resistance (2) activation of host defence mechanism either by plant defence activator or by biological agents and (3) maintenance of vigorous plant growth through application of proper and balanced nutrition.

Considering the pathogen, the control strategies that must be followed are: (1) clearing of a particular area from the pathogen (exclusion), (2) destruction of alternative hosts, (3) cultivation of non-host crops (crop rotation), (4) elimination, eradication and management of crop residues that may provide opportunity for the pathogen to survive and pass the unfavourable seasons, (5) seed treatment, (6) application of chemicals and (7) enhancing the pathogens' natural enemies (biological control).

IDM strategies for environment management includes: (1) proper seed bed preparation, (2) irrigation at proper time and rate, (3) management of row to row and plant to plant distances, (4) proper decisions about early or late sowing to avoid early infection and (5) harvest at proper time and maturity (Pathak et al. 2006; Agrios 2005; Strange 2003; Chaube and Singh 2004).

15.9 Integrated Disease Management (IDM)

The goal of an integrated disease management (IDM) is to keep the disease under its economic threshold level, by using a wide range of management practices that also are environment friendly. In other words, crop management practice should be so integrated that the profit is maximize and the loss is minimum. In the end, in any biological system the population of any organism must be in a balance with its resources. If somehow there is any disturbance in balance, it should be reinstated by integration of different practices in such a way to achieve an acceptable level of damaging factor(s).

IDM practices are general or specific. General practices focus on all the diseases that can infect a specific crop, while specific management has a specific target, to manage a single infectious disease in a crop (Overton 1996).

15.10 Principles of IDM

15.10.1 Exclusion

The concept of exclusion means keeping the pathogen away from plant. This practice is usually followed in such area where the pathogen is absent. Crops growing in that particular area are free of diseases. They are not genetically explored to identify a gene or a set of genes effective against the pathogen. If the latter gains entry in that particular area it can cause an epidemic. National and state level laws are implemented to restrict the entry and spread of particular plant pathogens in a pathogen-free area. Different strategies including exclusion are: quarantine, inspection and crop certification (Pathak et al. 2006; Agrios 2005; Strange 2003; Chaube and Singh 2004).

15.10.1.1 Quarantine and Inspection

If a new plant pathogen got entry in area where the plant has been grown in the absence of that particular pathogen (Hall 1995; Dhaliwal et al. 2007). The introduction of a pathogen in an cultivated area in which it was previously absent can lead to huge losses as compared to those induced by existing pathogen. This happen because:

1. the crop has never been tested against that specific pathogen, so no activation of resistance factor against the pathogen occurs;
2. there is no competition towards the pathogen nor antagonistic activity by any microorganism;
3. an unrestrained development and reproduction of the pathogen may occur, due to huge availability of susceptible hosts.

Historically, there are examples of severe plant disease epidemics due to the entry of unknown pathogens, such as Late blight of potato, bacterial blight of rice, downy mildew of grapes, citrus cankers and so on. A number of diseases and pathogens even changed the course of human history (see Chap. 14, this Volume) (Razdan and Sabitha 2009).

In 1912, the United States Congress passed the 'Plant Quarantine Act' to prevent the entry of foreign pathogen into new localities. The act was then superseded by 'Plant Protection Act' 2000, which regulates the import and export of plant or plant parts in and out of the USA. Similar quarantine laws also exists nowadays in other parts of the world. Quarantine application monitors the materials that may carry plant pathogens that do not exist in a particular area (Khetarpal and Gupta 2005). For quarantine, experienced inspectors are positioned at each and every entry point of the country. The responsibilities of quarantine department include (Mathys 1975):

1. observation of growing plant for a specific period of time for detection of any pathogenic problem, before being released to the importer;

2. inspection of the material at farmers' fields level whether there is any chance of introduction of any pathogen. In case of seed, using replicated molecular and serological analyses for pathogens detection. If after all necessary actions and inspection the material is found free of diseases or pathogens, a certificate is issue, thereby allowing the entry of that material into the country;
3. the laws apply to inter- or intrastate movements of material, as authorities regulate the movement and sale of materials among states;
4. for spread of plant pathogens through natural means, the plant quarantine regulations are not applicable.

15.10.1.2 Crop Certification

A further form of quarantine or exclusion is crop certification, which is sometimes compulsory and, in several cases, voluntary. For the import or export of seed, seed certification is necessary. If the seed is not properly inspected and even do not have inspection certificate, the lot may be rejected or not allowed to enter the country (Fry 1982). The farmers who are interested in import/export of plants, may voluntary submit their product for inspection and certification, in field or storage conditions. After submission the plants are inspected through recommended procedures. Once the plant or planting material is found free from all kind of pathogens and diseases, the inspection authorities will issue a certificate for the concerned pathogens or diseases. After issuance of the certificate, the farmers can advertise or sale their product to the market. It is worth mentioning here that the certificate will refer only to the specific problem or disease for which the plants were tested (Akoroda 2010; Khoury and Makkouk 2010).

15.10.2 Avoidance

After exclusion, next option is to avoid the disease. Measures such as the shifting of the sowing time, selection of field, alternation in cultivation areas, aim at prohibiting any contact between susceptible hosts and pathogen. Changing the environmental conditions may be helpful to avoid disease development. The main procedures are:

15.10.2.1 Choice of Geographical Area

Management of crop diseases can also be achieved by focusing on crop environmental factors such as humidity, dry and wet conditions (Palti 1981; Thurston 1992). The location in which a crop is cultivated may represent an issue, and changing the geographical area can solve it. In a given location, the pathogen may adopt to local conditions and even become virulent. Wet and dry conditions can affect the development of fungal and bacterial diseases. If the development of a disease is severe in a wet area, we can cultivate that particular crop in an irrigated, dry area to reduce or even avoid the disease (Palti 1981; Thurston 1990, 1992). Simply changing the cultivation of bajra from wet area to dry area, smut and ergot may be avoided and profit can be increase. Similarly, *Colletotrichum lindemuthianum* (bean anthracnose), *Xanthomonas phaseoli* (bacterial blight) and *Pseudomonas phaseolicola* prefer wet conditions, and the bean cultivation in dry areas can remarkably reduce the incidence of these pathogens (Palti 1981; Thurston 1992).

15.10.2.2 Selection of Field

Selection of a disease-free land is key for a successful crop. Many soil-borne diseases can be avoided by proper site selection. If the causal organism of a soil-borne disease is present in a field it is always advisable to avoid the cropping for some years. Red rot fungus (*Colletotrichum falcatum*) can persist in soil for few months. Therefore, if sugarcane is planted in the same field immediately after harvest of the preceding (diseased) crop, many chances of disease exacerbation exist, and the crop may fail (Palti 1981; Thurston 1992). Similarly, bacterial wilt of potato, smut of bajra, ergot of bajra, ear cockle of wheat, root knot or other nematodes, etc. are diseases whose causal organisms persist in soil for varying periods. Such fields should not be selected for susceptible crops. The management of drainage plays a major role in field selection. Many diseases, such as red rot of sugar cane and downy mildew of bajra, occur more severely in fields where water logging is common. In orchards, the site selection is of special importance, as fruit trees may remain on the same land for 40–50 year. If proper land selection is not made at the time of planting, the trees might show signs of abnormalities and diseases (i.e. "greening", fungal diseases or bacterial cankers) will spread, after a few years.

15.10.2.3 Choice of Sowing Time

If the location of a crop is not in question, to avoid a disease next comes the timing of cultivation. Early or late sowing will be the option. The objective behind this choice is to avoid the matching of a disease favourable environmental condition with the susceptible stage of a crop. The sowing time of each crop should be adjusted according to prevailing environmental conditions, so that little or no active inoculum is present in the field. The factors that need to be considered for sowing are the time of maximum activity of the pathogen, the suitable epidemiological conditions for the pathogen, soil conditions, and last but not least, the crop susceptibility. Any or all of these conditions should not coincide. Early or late sowing will enable vigorous crop growth by avoiding most pathogens' favourable conditions. For example, cultivation of gram (*Cicer arietinum*) and pea (*Pisum sativum*) in soil with high temperature and moisture due to rain is not feasible, because these conditions favour the development of blight and root rot diseases in both crops. Alternatively, cultivation of gram or pea in November or December is more suitable, as late sowing will help in reducing blight or root rot diseases, due to low moisture and temperature (Palti 1981; Thurston 1992).

15.10.2.4 Disease Escaping Varieties

Varieties with growth characters that help to avoid or escape diseases should be considered to develop disease management strategies. The disease escaping capability in this case is not due to the plants genetic resistance, but to their characteristics of growth and time of maturity. For example, powdery mildew and rust of pea may be escaped by early maturing varieties. The diseases normally become serious in January or later. If pods have developed before serious disease incidence, the losses are considerably reduced (Ciancio and Mukerji 2007).

15.10.2.5 Seed and Planting Stock

Many diseases are introduced in the field only through seed or other propagating material (cuttings, grafts, bulbs, tubers, etc.). Most vascular pathogens (protozoa, phytoplasma, viroids and viruses) including fungi and bacteria are transmitted through infected propagating material (rootstocks, rhizomes, buds, grafting, bulbs, etc.). Pathogens i.e. nematodes, spores of fungi, bacterial oozes may also be present on the surface of these propagating material. If plants are propagated through belowground parts, i.e. root cuttings, tubers etc., these may carry an inoculum of soil borne pathogens, especially nematodes, internally or externally. Viruses can be also carried through infected seeds. Diseases such as anthracnose, wilt, smuts, and spots, either bacterial or fungal, can be avoided by proper selection of propagating material. Planting of such disease-free seeds in a disease-free field are often the most effective method of managing epidemics of certain diseases such as smuts, red rot of sugarcane, black scurf of potato, etc. Various activities must be included in IDM strategies to evade pathogens, such as use of disease-free seed, early or late sowing in disease-free area, plant-plant and row-row distance maintenance, cultivation of trap crops, soil drainage, maintaining plant protection and wind breaks. The results of these activities will allow avoiding pathogens, vigorous growth keeping the damage below economic threshold levels (Palti 1981; Thurston 1992).

15.10.2.6 Epidermal Coating to Avoid Pathogens

Avoiding the contact of plant and pathogens is fundamental to control a disease. There are many compounds that can be used to spray plants that will form a membrane on their surface, originating a barrier avoiding the host and pathogens connection. The pathogen penetration, and ultimately the establishment of infection, are thus avoided. One of such compound is dodecyl alcohol, forming lipid membranes of high quality, that can last for 15 days. This membrane blocks water diffusion, but allows oxygen and carbon dioxide diffusion. This compounds helps in water conservation contributing in yield increase. Similarly, effectiveness of Kaolin-based films has also been reported. These films are quite helpful in protecting i.e. apple, tomato, wheat, cucumber, rice etc. against many diseases. Still the commercial application and success of these compounds is still in question (Agrios 2005).

15.10.3 Eradication

Pathogens can still enter crops or fields bypassing the mentioned precautions, as it is possible that an inoculum is already present in the field neighbourhood. Under these situations eradication of the pathogen from the field or crop is necessary. This practice aims at: decreasing, deactivating, eliminating or destroying the inoculum source at the plant, field or whole geographic areas levels. Eradication can be through biological methods, crop rotation, elimination of diseased plants, physical or chemical treatments, etc. (Pathak et al. 2006; Agrios 2005).

15.10.3.1 Cultural Methods that Eradicate or Reduce the Inoculum

15.10.3.1.1 Host Eradication

Even after strict quarantine measures, if a pathogen gain entry into an area, a disease epidemic may occur, with an inoculum that is continuously being generated through infected plants (Agrios 2005). The presence of diseased plants in the field or orchard is a source of continuous release of inoculum. Therefore, to reduce disease spreading as far as practicable, such plants or their affected organs should be removed and destroyed. On the same basis, eradication of alternate and collateral hosts is also recommended. The burning or removal of infected plant or parts becomes inevitable. This practice will not only reduces the disease epidemic but also eliminates the pathogens, in such a way that further spreadings are evaded (Palti 1981; Thurston 1992).

15.10.3.1.2 Rouging

The procedure known as rouging involves field sanitation by eliminating diseased plants, to halt disease spreading to healthy plants and helping in production of disease-free seed. In orchards, where removal of the entire tree is not feasible unless it is very badly damaged, the affected organs can be cut (tree-surgery) and burnt. Rouging is successful against smuts of wheat, barley, maize sorghum etc. The procedure is practical only when the size of the plots is small and the number of diseased plants is not very high (Palti 1981; Thurston 1992). Best example of rouging in the human history is the attempt to eliminate Citrus canker from Florida, made during the XIX century at four different occasions i.e. 1915, 1984, 1992 and 1995. The eradication costed million citrus trees, removed even from homeyards and gardens. But still the disease persists in the State. Similar attempts were made to eliminate pathogens of coffee rust, pear and apple fire blight, witchweed and plum pox virus of stone fruits. Unfortunately these pathogens still persists and spread (Palti 1981; Thurston 1992; Ciancio and Mukerji 2007).

15.10.3.1.3 Crop Rotation

Soil infection is a prominent problem in farming. Growing of the same crop in the same field successively, year after year, facilitates the increase of soil borne pathogens in soil, where they multiply. Heavy pathogens densities can make soil unsuitable for sowing of a certain crop. If resistant or non-host crops are cultivated for a short period after a susceptible one, the pathogen will not survive because of the unavailability of required host and nutrients. Moreover, various metabolites released from the roots excretion can affect the pathogen or stimulate the establishment of antagonistic microorganisms. This phenomenon makes crop rotation a successful approach for control of root diseases. Crop rotation is unsuitable or less efficient for soil borne pathogens that can persist as a saprotrophs in soil for about 5–6 years. In some cropping patterns, after tillage, the field fallowing for about a year results in the reduction of pathogens. Population of nematodes and other pathogens can be reduced considerably by burning or removing the root residues or heating and drying the fallowed field soil, during summer. Although some diseases can be reduced effectively, others can be however enhanced such as stalk rot of rain sorghum and corn, barley scab and septoria leaf blotch of wheat (Palti 1981; Thurston 1992; Ciancio and Mukerji 2007).

15.10.3.1.4 Sanitation

Sanitation includes all activities aiming at elimination, eradication or reduction in pathogenic populations in a field, plant and storage houses and at prevention of pathogen dispersal to healthy fields and/or plants. After harvesting, ploughing of field to cover remaining infected plant parts such as stems, leaves, infected fruits etc. with soil simultaneously encourages the breakdown or damage of pathogens that are present on these infected parts. In houses or gardens, the inoculum can be reduced by removing or pruning the infected plant parts i.e., leaves, stems, branches and fruits. Such activities result in the reduction of a pathogen population that will progress later. Sanitation also includes burning of leftover infected plant parts, washing or disinfecting the hands before handling some particular crops, disinfestation of equipment used for pruning or cutting of propagative material, washing off the soil from farming equipment, washing the containers and warehouses walls with chlorinated water. Various diseases can be inhibited through sanitation e.g., red rot of sugarcane, downy mildew of maize and peas, powdery mildew of barley, peas, and wheat (Salamanca 2015).

15.10.3.1.5 Polyethylene Traps and Mulches

Airborne aphid vectors disseminate several plant viruses, i.e. cucumber mosaic virus, to healthy crops like peppers. A large population of aphids can be distracted by yellow polyethylene sheets that are sticky in nature, standing vertically across the boundaries of susceptible crops. The virus inoculum to healthy crops can be reduced by some degrees through this method, by tricking the incoming insects that play an important role in virus dissemination. Field mulching between rows and plants with whitish-grey, black, or coloured polyethylene sheets or reflectant aluminium, repels incoming insect vectors such as thrips, aphids etc., keeping them away. The outcome of mulching and use of polyethylene traps is that a few insect vectors carrying virus can reach plants and infect them. However, under the crop canopy, the reflecting mulches stop performing this function (Kasirajan and Ngouajio 2012).

15.10.3.1.6 Suppressive Soil

Biological control is an effective and eco-friendly method for various pathogens, based on antagonistic organisms that can damage the pathogen, completely or partly. Antagonistic microorganisms are present in some soil known as suppressive soil. Higher plants, known as trap plants, restrict pathogens inocula by tricking the pathogens or by secreting detrimental metabolites in soil. The suppression mechanism over various pathogens is not completely understood. However, abiotic or biotic factors are involved whose effect can vary with the pathogens' density. It seems that they mostly function through the action of various antagonistic species in the soils. Addition of suppressive soil to a conducive one may result in the reduction of a pathogen population due to the introduction of antagonistic microorganisms. This effect was shown for e.g. potato scab disease, effectively reduced by addition of suppressive soil with *Streptomyces* spp. to conducive soil containing *Streptomyces scabies*, the causal agent of potato scab. Similarly, cultivation of papaya seedlings with addition of suppressive soil in the holes of garden soil containing *Phytophtora* sp. reduced the root rot of papaya. Furthermore, in some cases, disease incidence can also be reduced due to repeated cultivation of the same crop in conducive soil, year after year. After severe infections in first years the disease may be reduced ultimately because of the production of a large number of antagonistic take all species. Damping-off of cucumber and wheat caused by *Rhizoctonia* was reduced because of the successive plantation of cucumber and wheat in the same field. Likewise, successive cultivation of 'Crimson Sweet' a variety of watermelon results in reduction of Fusarium wilt due to an increase of other *Fusarium* spp. antagonistic to Fusarium wilt. Such type of soils appear suppressive to development of upcoming diseases. Pasteurization of suppressive soil for 30 minutes at 60 °C totally removed the suppressiveness due to the presence of antagonistic microorganisms (Chandrashekara et al. 2012).

15.10.3.1.7 Control Through Trap Plants

Plantation of tall plants such as corn or rye in a few rows around peppers, beans and squash reduced the pathogen infection as the incoming insect virus vectors, i.e. aphids, first nourish themselves on taller plants. As the viruses carried by the aphids are non-persistent, during feeding on taller plants the vectors usually transmit the virus. Hence, pathogen populations, or their viral loads, can be reduced before reaching the particular crop to protect. In some cases, a different method is used to control nematodes through trap plants, e.g. some plants that are resistant against sedentary, endoparasitic nematodes produce metabolites that encourage hatching of eggs. After hatching, the juveniles die ultimately as they can enter the plants but are not able to develop till adults. These type of plants are also referred to as "trap crops" (Palti 1981; Thurston 1992). Nematode populations can be remarkably reduced by planting trap crops in a rotation. For instance, juveniles of the root-knot nematode *Meloidogyne* spp. are trapped by crotalaria plants. Population of *Globodera rostochiensis* (potato cyst golden nematode) can be reduced by *Solanum nigrum* (black nightshade). Nematodes can also be eliminated through the plantation of extremely susceptible plants that are removed or burned after root attack and penetration, in order to restrict the nematodes from reaching their maturity stage, so that they cannot reproduce (Palti 1981; Thurston 1992).

15.10.3.1.8 Control through Antagonistic Plants

Some plants, i.e. marigolds and asparagus, show antagonistic effects on nematodes that they affect by excreting toxic exudates in soil. Nematodes populations can be reduced in susceptible roots as well as in soil through inter-plantation of these antagonistic plants with crops that are susceptible. Like trap plants, nematodes control through antagonistic plants is not suitable on a large scale (Wood and Tveit 1955). Moreover, some root knot nematode species are able to multiply on asparagus (Murga Gutierrez et al. 2012).

15.10.3.2 Physical Methods That Eradicate or Reduce the Inoculum

Air, temperature, light and different radiation types are the physical factors that are generally used to control the plant diseases.

15.10.3.2.1 Soil Sterilization by Heat

Steam in pressure is used to sterilize soil. In glasshouse benches, steam is provided through channels that diffuse it through soil.

Nematodes, some water moulds and few oomycetes are killed at 50 °C, while 60–72 °C is the effective temperature to kill virulent bacteria and fungi, together with some slugs, centipedes and worms (Palti 1981; Thurston 1992). Plant viruses that persist in crop litter, the remaining bacteria and insects are killed at approximately 82 °C. Boiling temperature (95–100 °C) is used to kill some plant viruses such as tobacco mosaic virus (TMV) and some weeds that can tolerate heat. Mostly, soil sterilization is performed maintaining the temperature at 82 °C for 30 min, as this temperature can kill nearly all the soil borne pathogens. In alternative to hot water or steam, electric current is also used to produce heat for soil sterilization. An important point to keep in mind is that extremely high temperatures or extensive sterilization time periods should not be provided to sterilize soil (Palti 1981; Thurston 1992).

15.10.3.2.2 Mulching

Soil borne plant pathogens can also be decreased through mulching. During strong sunny summer days, a slightly wet soil covered with polyethylene sheet may enhance its temperature in upper 5 cm at about 52 °C. This temperature is much higher than that of soil without mulching, in which its maximum may reach up to 37 °C. The temperature rises in mulching by taking heat from sun, so its application during sunny weather for some days or weeks appears suitable to kill soil borne pathogens (Shtienberg et al. 2010).

15.10.3.2.3 Hot-Water Treatment of Propagative Organs

Hot-water treatment is used to kill any pathogen inside or on the external surface of nursery seedlings, bulbs, seeds etc. For some seed borne diseases such as loose smut of cereals, hot water seed treatment was the only method available to control the disease. It was used for many years, as the disease could not be controlled through chemicals because of their inability to reach inside the seedcoat. Likewise, the method applies also to some nematodes that persist inside bulbs or nursery seedlings, such as *Ditylenchus dipsaci* and *Radolpholus similis* occurring in different ornamental plants and citrus rootstocks, respectively (Agrios 2005).

To maintain the fruit quality for long time, the treatment of melon fruit by rinsing or brushing for just 15 sec. With hot water at about 58–60 °C shows significant result, as it washes the soil or dust particles and the spores of the fungi from the external surface. This also makes the openings in the epidermis partly or completely impenetrable (Nelson 2004).

15.10.3.2.4 Disease Control by Refrigeration

Post-harvest diseases of fleshy fruits of plants can be controlled effectively by using a refrigeration method, which is the most useful. Refrigeration does not kill any pathogen as it maintains fruit at a freezing temperature, or to some extent, at more than a freezing point (Beattie et al. 1989). Refrigeration halts the pathogens' growth and all its activities, resulting in the reduction of a prevailing disease spread. After harvesting, the perishable fruit and vegetables are transported in refrigerated vehicles or preserved under refrigeration as early as possible, until consumption (Champ et al. 1984).

15.10.3.3 Chemical Methods That Eradicate or Reduce the Inoculum

Pesticides are used to protect the external surface of plants from pest attacks or reduce prevailing infections. Some chemicals are considered most effective in decreasing the inoculum before it starts its interaction with the plant. Sanitation of storage houses, disinfection of handling equipment, fumigation, soil treatment and restriction of insect vectors can be achieved through chemical treatments (Thomas and Waage 1996).

15.10.3.3.1 Soil Treatment with Chemicals

Chemical treatment of soil may be applied to overcome certain diseases of ornamentals, trees, strawberries, high-value crops, or vegetables. Seedling blights, damping-off, crown and root rots, and other diseases can be controlled through soil treatment based on various fungicides. Their application may be done through different methods including liquid drenches, dusts or granules, with irrigation water, i.e. in sprinkler irrigation. Captan, metalaxyl, diazoben, and penchloronitrobenzene (PCNB) are used for soil treatment. However, the former two are mainly used as seed treatments (Hewitt 1998).

15.10.3.3.2 Fumigation

Fumigation is the most potent method to overcome various soil borne diseases by using fumigants. Some of them, including metam sodium, dazomet, methyl bromide and chloropicrin are pre-plant fumigants that diffuse in soil and/or convert into gases. These fumigants are used for a general-purpose, as they are effective in controlling a broad range of microorganisms, i.e., many fungi, nematodes, various bacteria, insects, and weeds. They are applied to soil by broadcasting in the field or applying just in the rows in which the crop is sown (row treatment). After pre-plant soil fumigation, the crop or seedlings must be planted in the treated field after numerous days or weeks, in order to evade plant damage due to phytotoxic residues.

The potential of fumigants in fields where control of nematodes or other pathogens is aimed, depends on their dispersion in a gaseous form through the soil pores. Size and continuity of pores, soil moisture content (80% of field capacity is the best), soil type (organic matter rich soils require more amounts of fumigants), soil temperature (10–20 °C is best range) and characteristics of fumigant itself, all are factors influencing the distance covered by the fumigants in the soil (Knight et al. 1997).

15.10.3.3.3 Disinfestations of Warehouses

Storage houses should be cleaned and debris should be removed or burned to prevent the stored stocks from infection (that remained in the storage house from the infected crop of the previous year). Sodiūm hypochlorite (generally known as bleach) or any other disinfectant is used to wash the floors and walls. Tear gas (commonly known as chloropicrin) is efficiently used for fumigation of storage houses. The doors should be tightly closed. Temperature and relative humidity plays an important role in fumigation storage house temperature at 25–30 °C with relative humidity 100% must be maintained. The doors of storage house are then opened for ventilation at least after 24 h (Champ et al. 1984).

15.10.3.3.4 Control of Insect Vectors

Insect vectors are a major threat as they play a key role in the dispersal of infection from crop to crop or from field to field. It is very important to control these vectors to reduce a disease spread. Insecticides can effectively control almost all insects that are vectors of bacteria and fungi. Insects that carry fastidious bacteria, viruses and mollicutes must be controlled effectively in case the vector overwinters on another crop or feed on other plants, before dispersing the disease to a particular crop. The disease will not be controlled effectively if the insect vectors are controlled after their arrival on the crop to protect. Even after efficient control of vectors, insects may persist for enough time to disseminate infection. However, losses due to such pathogens can be reduced significantly by killing the vectors. Disturbance in the aphids' ability to acquire and transfer viruses leads to a more significant reduction in virus dissemination, as compared to the reduction of virus dispersal by insects killing. Application of well graded mineral oil for a number of times on plants disturbs the aphids ability. Such oil plays a role in disturbing the dissemination ability of virus acquiring aphids. However, such oil has no toxic effect on aphids nor a significant effect on feeding and penetration behaviour. Aphid-borne viruses such as *Potato virus V* on pepper and *Cucumber mosaic virus* on pepper and cucumber have been controlled effectively by oil applications on plants (Hadidi et al. 1998; Harris and Maramorosch 1982).

15.11 Resistance or Immunity: Way to Go

One of the best way to manage any disease is resistance. Resistant plants can totally overcome the effect of pathogens. If this strategy is not possible, immunity should be elicited so that plants can withstand the effect of diseases. In both cases disease induced losses are minimum. As antibiotic production is not present in plants, scientist were able to induce plants to produce antibiotics called plantibodies, through genetic engineering. Genes which encode the plantibodies production in mice were engineered against specific plant pathogens (Buddenhagen 1977).

Plant immunization can be achieved by cross protection. Plant are inoculated with pathogens, mostly mild strains of pathogens, which induces the activation of the defence system. As a result, when a virulent strain attacks the plant, an immune reaction is already in place against such pathogen. This process is known as induced plant resistance. Viruses are the target of most treatments that induce plant resistance or cross protection, as they can only be controlled through resistance. Similarly, there are few chemicals available which can be applied to plants, i.e. salicylic acid, dichloroisonicotinic acid and a few benthiazoles. Their application results in the activation of plant resistance known as systemic acquired resistance. Systemic acquired resistance can also be activated by inoculation with certain pathogens.

Another type of resistance in plants is called pathogen-derived. In such resistance, particular segments of the pathogen DNA or specific genes are identified and incorporated in the plant, through genetic engineering. Upon expression these genes express plant resistance against one or more pathogens. Resistance of most plants has been improved through pathogens-derived resistance, mostly against viruses.

The activation of genes or set of genes for resistance is very important, as resistance can be activated by simply improving the plant growing condition. Proper irrigation, balance nutrition, weeding or soil fertility have prominent effects on the health and productivity of plants. Once a vigorous growth is achieved, plants are automatically able to withstand most of the problems (Buddenhagen 1977).

Activation of resistance or introduction of genes or DNA segments for resistance through genetic engineering is by far the best methods to improve plant health. Most of resistance in plants, especially against viruses, has been achieved through genetic engineering. The combination of conventional breeding programs with these advance molecular techniques are effective ways of controlling plant pathogens.

15.11.1 Cross Protection

The concept of cross protection relies on infecting a plant with a mild strain of specific pathogen so that when severe strain attack, he plant is already been immune to that. Success through cross protection has been achieved against viruses. The attack of mild strain of viruses mostly results in activation of defence system (Beachy et al.

Table 15.1 Few examples of cross protection

Host	Mild Strain inoculated
Tomato	Mild strain of TMV
Citrus	Mild strain of CTV
Papaya	Mild strain of PRSV (Papaya ring spot virus)

Table 15.2 Systemic acquired resistance against different pathogens

Treatment	Resistance induced against
Tobacco inoculated with tobacco mosaic virus	TMV, other viruses, *Phytophthora parasitica* var. *nicotianae, Pseudomonas tabaci*
Tobacco inoculated with *Pseudomonas syringae*	Tobacco mosaic virus
Bean inoculated with *Collectotrichum lagenarium*	*Collectotrichum lagenarium*

1990; Lomonossoff 1995). Some of the example of cross protection are given in Table 15.1.

Application of cross protection is still an issue, in particular when it comes to field application, as failures to produce expected results have been reported. Most important of all is the chance of mutation, which can turn the mild strain towards a new, virulent strain. The cross protection ability of mild strain may be lost after few years in perennial crops. Trees show a limited spread of less severe strains, so after a few years the most severe strain can became established (Gal-On and Shiboleth 2006).

15.11.1.1 Systemic Acquired Resistance

In Systemic acquired resistance (SAR) the infection with a necrotizing pathogen (producing a hypersensitivity response) will spread the resistant reaction to the whole plant (Table 15.2). This is a sort of physiological immunity. SAR is effective against most but not all pathogens, (Baker et al. 1997).

15.11.1.2 Pathogenesis Related Proteins (PR Proteins)

These are the first component induced by necrotizing pathogens (producing a hypersensitive response). The mild strains produce wounds on infection which result in secretion of PR proteins, intercellularly. PR proteins are soluble at pH 3. Basic homologs are also found in vacuoles. They cannot inhibit proteinase but can induce resistance against proteinase. PR proteins secretion is part of a developmental process, as PR can be produced without any wound or infections, e.g. at flowering (Baker et al. 1997).

15.11.1.3 Induced Systemic Resistance (ISR)

This type of resistance is more related to a biotic stress, especially plant growth promoting rhizobacteria (PGFR) (Pieterse et al. 2014), or to herbivore-induced resistance. Both SAR and ISR are forms of induced resistance. SAR is more inclined toward mild pathogens and chemicals induced resistance, while ISR is linked to beneficial microorganisms, more specifically PGPR. Activation time of both resistance is important. ISR mode of action is based on barriers for pathogens, that may be physical or chemical. ISR do not directly kill or inhibit the pathogens. As compared to SAR, ISR activates some signal transduction pathways. Negative impact of ISR on the plant-insect interaction has also been reported. ISR applications on melon, bean, potatoes, rice and tobacco has shown a significant success. Research on ISR is very much active, especially on artificially activation of its pathways. However, there is still no major success and it does not appear as the major way for management of plant diseases (Choudhary et al. 2007).

15.11.1.4 Synthetic Compounds for Resistance Activation

Several types of synthetic compounds have shown very useful to activate plant defences against tobacco mosaic virus, *Peronospora tabacina* and *Pseudomonas syringae* (Sreeja 2014). These compounds can be applied by injection, spraying or absorption depending on the plant part to be treated (see Table 15.3).

Treatments with UV-C light (254 nm) at low dosages, immediately activate expression of certain defence-related-genes. These control the induction of enzymes such as chitinases 3-1, 2-glucanase and phenylalanine lyase. This indicates that UV treatment may induce systemic resistance. Silicon, a plant activator, applied as nutrient solution, also showed protective effects on plant against powdery mildews, by induction of the papilla formation, callose production, and phenolics accumulation.

Table 15.3 Some plant defence activator

Compound name	Common name	Disease control
Benzothiadiazole	Actigard	?
Acibenzolar-S-methyl (derivative of benzothiadiazole)	Blockade	Downy mildews of leaf vegetables
DL-β-aminobutyric acid	BABA	Effective against biotic and abiotic stresses
3-allyloxy-1,2-benzisothiazole-1,1-oxide	Probenazole	Effective against rice blast

15.11.1.5 Growth Conditions

Vigorous plant growth is very important, as it improves resistance to pathogens' attack. It can be maintained by improving cultural conditions. All cultural condition such as balance nutrition (fertilization), drainage, irrigation at recommended times, maintenance of plant-plant and row-row distance and weeding (to reduce competition), not only increase growth and production but also have an impact on disease resistance. For example, balancing nutrition and irrigation can help plants to withstand canker of fruits and trees. Resistance against soil borne pathogens such as *Pythium*, *Phytophthora* etc. can be increased by seed priming, that controls the hydration of seeds. Upon germination such seeds produce healthy and vigorous seedlings, avoiding the attack of damping off and other diseases (Agrios 2005).

15.11.1.6 Genetic Resistance

Resistant varieties provide a method that is economic, effective, environment friendly, with board spectrum, offering an easy way out of a disease problems (Wolfe 1985). When a resistant variety is cultivated, the increase in production due to the disease eradication of diseases is flanked by savings on the chemical sprays application. Environment is also protected because the dispersal of hazardous chemicals is avoided. For viruses and other obligate pathogens such as rust, smut, powdery and downy mildew this is the only mean of control. It is likely the most acceptable way of controlling such diseases, without application of pesticides, with satisfactory yields (Buddenhagen 1977).

More than 85% of total cultivated varieties of crop plants are resistant. Different organizations such as federal, state level seed certification and commercial seed companies, have resistant seed available to farmers. These varieties helped to control some important diseases which otherwise were impacting farmers very badly (Buddenhagen 1977). Most important diseases being managed through resistant varieties are rusts, smuts, powdery mildew and wilts caused by fungi, with also many diseases caused by bacteria, nematodes and viruses. The application of resistance in case of forest and fruit trees is still limited. There are only few cases in which attempts to manage a disease through genetic resistance Were successful, i.e. apple scab, for which resistant varieties are now available and cultivated. Another example is the management of white pine blister rust and fusiform rust (Hodson and Nazari 2010).

Genetic resistance is of two types i.e. vertical and horizontal. Vertical resistance is specific and usually known as single gene resistance, and is also called as "initial inoculum limiting" resistance. Horizontal resistance is characterized by a broad spectrum, multigene resistance. It is also known as "rate limiting" resistance.

Work carried out for genetic resistance focuses on both types. In a resistant variety both vertical and horizontal resistance genes may be expressed. It is always batter that either one or few specific genes are expressed, with an unspecified number of other genes. Vertical resistance is easy to break down, as shown by obligate parasites (rust, smuts, powdery and downy mildew), mostly with an airborne inoculum, allowing genetic recombination or mutations, that very commonly establish new races or stains. Vertical resistance is then no more effective against these new races, as they result from mutations of genetic recombinations. These new virulent then become widespread as resistance has been broken. The varieties with only one type of resistance (vertical) need to be replaced sporadically, many after 5 to 10 years of cropping (Buddenhagen 1977; Agrios 2005; Leung et al. 2003). How long a single variety will survive in the field depends upon multiple factors, such as:

1. Pathogen genetic pliability,
2. Gene or set of genes involved in resistance,
3. Environmental conditions which favour the disease development,
4. Degree of activation and arrangement of resistance genes.

Innovation in science and technology promises quick ways available to develop resistant varieties. Genetic engineering is one of them, which has facilitated researchers to speed up the process of variety development by transferring single genes, or group of genes. After a new variety has been developed, its field testing is far more important to increase its life span. It has firstly to be tested against as many as pathogens and races as possible. The second step is its cultivation in many locations, with different climatic conditions and in some case, in different states or countries. After this long testing period, the varieties showing resistance to many pathogens or races are released for commercial cultivation (Buddenhagen 1977; Agrios 2005; Leung et al. 2003).

Once the resistant variety is released, proper management plans should be followed to increase its life span in the field. Cultural practices such as field sanitation, seed treatment with fungicides or application of fungicides are expected to reduce the disease pressure. This may ultimately increase the life span and productivity of a given variety. Mono-culturing, i.e. growing a single variety with a single source of resistance should not be followed, as it will provide favourable conditions for pathogens which require time for dispersion and attack, e.g. soil borne pathogens. Varieties with different resistance sources should be rotated each other. This practice not only will reduce the population density of pathogens specific to a single variety, but will also increases all varieties' life span (Buddenhagen 1977; Agrios 2005; Leung et al. 2003).

In case of cereals, regional research centres should find the set of genes that may differ from one region to another. Because of the widespread cultivation of cereals, each region is characterized by a different variety with different genes as resistance source, that will last longer. If a new virulent pathogen race develops in a particular region, it will not spread to other regions because the sources of resistance locally adopted will be different. That particular race will hence remain confined in the region where it first appeared (Buddenhagen 1977; Agrios 2005; Leung et al. 2003).

Another concept is multiline breeding. These program involves isogenic, meticulously related or unrelated lines, each one containing different genes for resistance. Multiline breeding means developing a variety by mixing all the resistance genes into a single one. It will therefore be resistant to larger number of pathogens.

Benefits include pathogen population reduction, reduction in disease development, tolerance to salt, drought, temperature and other abiotic stresses. For example, tolerance of rice against dehydration or salts stresses was increased by incorporation of two genes from wheat. Similarly, yeast genes coding for a trehalose-6-phosphate synthase were transferred to tobacco to increase its drought tolerance (Buddenhagen 1977; Agrios 2005; Leung et al. 2003).

15.12 Biological Control Agents: Management Plus Environmental Protection

Biocontrol means life against life. Population management of severe pathogen strains may be achieved by application of avirulent strains or natural antagonist (Lumsden and Walter 1995). In other words it is also called as cross protection or hypovirulence. The mechanisms of action of biological control agents (BCA) are still under debate. Biological control is a long lasting and economic method once it is established. Main concern for use of BCA in management of diseases regards the limited success sometimes observed in the field (Thurston 1990).

Antagonist can affect pathogens in following ways:

1. Direct parasitism,
2. Competition for food, space, water etc.,
3. Production of antibiotics, toxins etc.,
4. Production of enzymes that can attack structural components,
5. Indirect toxic effect of volatile substance released by antagonists' metabolic activities,
6. Combination of above methods.

Some important examples of antagonistic microorganism capable to reduce the amount of pathogens are shown in Table 15.4.

Presence of antagonistic microorganism in crop soil is well reported. They exert biological control also I natural microcosms, regardless of human activities. Farmers may attempt to increase the population of these antagonist either by providing favourable conditions or by artificially increasing their population. However, BCA regulate but rarely extinguish their hosts, thus arising some question marks on their effective success. Promising results of laboratory and greenhouse experiments encourage the use of BCA, but little success under field conditions hits back. Problems may be the BCA establishment, their inability to compete with natural resident microflora, the ability to withstand changing soil conditions and cultivation practices and the effect of changing environmental conditions. Most of time these factor are not suitable nor maintained to enhance or sustain the BCA population, Resulting in a limited success and disease control. However, a number of cases for direct plant protection by BCA have been reported (Table 15.5) (Kerr 1980; Harman and Bjorkman 1998).

Table 15.4 Some antagonistic microorganisms of economically important pathogens

Antagonist	Pathogens
Trichoderma harzianum	*Rhizoctonia* spp., *Sclerotium* sp., *Pythium* sp., *Fusarium* sp. and *Fomes* sp.
Laetisaria arvalis	*Rhizoctonia* and *Pythium* spp.
Sporidesmium sclerotivorum	*Sclerotinia sclerotiorum*
Gliocladium virus	*Sclerotinia sclerotiorum*
Coniothyrium minitants	*Sclerotinia sclerotiorum*
Chaetomium sp.	*Venturia inaequalis*
Tuberculina maxima	*Cronartium ribicola*
Darluca filum	*Rust fungi*
Verticillium lecanii	*Rust fungi*
Ampelomyces quisqualis	*Powdery mildew fungi*
Tilletiopsis sp.	*Sphaerotheca fuligena*
Nectria inventa	*Alternaria* sp.
Gonatobotrys simplex	*Alternaria* sp.
Streptomyces sp.	*Pythium* sp.
Pseudomonas sp.	*Pythium* sp.
Aphelenchus avenae	*Rhizoctonia* sp. *Fusarium* sp.
Vampyrella	*Cochliobolus sativus, Gaeumannomyces grammis*
Catenaria auxiliaris	*Heterodera* and *Globodera* sp.
Nematophthora gynophila	*Heterodera* and *Globodera* sp.
Pochonia chlamydosporia	*Heterodera*, *Globodera*, *Meloidogyne* spp.
Verticillium lecanii	*Heterodera* sp.
Dactylella oviparasitica	*Meloidogyne* sp.
Pasturia penetrans	*Meloidogyne javanica*

Table 15.5 Important examples of direct protection through biological control

Disease	Control agent
Root and butt rot of conifers (*Heterobasidion annosum*)	*Peniophora gigantea*
Chestnut blight (*Endothia parasitica*)	Hypovirulent strain of the pathogen
Fusarium wilt of sweet potato (*Fusarium oxysporium* f.sp. *batatas*)	Non-pathogenic strain of the same fungus
Botrytis rot of grapes and strawberries (*Botrytis cinerea*)	*Trichoderma* sp.
Cucumber powdery mildew (*Sphaerotheca fuliginea*)	*Ampelomyces quisqualis*
Wheat leaf rust (*Puccinia recondita*)	*Darlucaa filum*
Citrus green mould (*Penicillium digitatum*)	*Trichoderma viride*
Tomato wilt (*Fusarium oxysporium* f.sp. *lycopersici*)	*Mycorrhizae*
Crown gall of pome, stone and small fruits (*Agrobacterium tumefaciens*)	Strain K_{84} of *Agrobacterium radiobactor*
Brown rot of peach (*Monilinia fructicola*)	*Bacillus subtilis*
Fireblight of apple (*Erwinia amylovora*)	*Erwinia herbicola*
Bacterial leaf streak of rice (*Xanthomonas translucens* pv. *oryzicola*)	Isolates of *Erwinia* and *Pseudomonas*

15.13 Integrated Disease Management-Examples

In some situations, the IDM objective is to manage crop and many diseases rather than a single one. Some examples of integrated control measures employed to improve yield of both perennial and annual crops are summarized in Table 15.6, 15.7, and 15.8.

15.14 Conclusion

Any effective IDM strategy demands that the crop and its environment should be considered as an ecosystem. It requires realistic assessment of the economic significance of disease and the estimation of the ecological, sociological and financial aspects affecting agricultural practices and disease insurgence. In absence of these information, any attempt to devise an IDM system will remain only partly successful.

IDM has to be considered as part of the agro-ecosystem. It has to provide bases to ensure health of the entire cropping programme in a given ecosystem. However, this comprehensive approach has not been practically demonstrated so far. In Pakistan, we have very little information on the effect of agricultural practices on disease management. Due to this lack of information, devising IDM for different crops will work partially. With the increasing world population, any success in crop production is important. Environmental stresses play an important role in plant growth. To increase crop production, focus should be on developing varieties that perform better or are less affected by environmental stress. Suitable techniques need to be developed by utilising the information of weather forecast and ecological

Table 15.6 Integrated measures applied to maintain high yields of sugar beet

Measures	Disease affected
Crop hygiene	Aphids- yellow viruses
	Aphids-mosaic virus
	Downy mildew
	Powdery mildew
	Rust
Crop protection	Cyst nematode
	Pygmy beetle
Sowing date	Aphids-viruses
	Downy mildew
Plant spacing and cover crop	Aphid-viruses
	Ramularia leaf spot
	Curly top virus
Cultivar	Downy mildew
Pesticides	Aphid-viruses
Predator and parasite	Aphids-viruses

Table 15.7 Integrated disease management of perennial fruit crops (apple, peach or citrus)

Management Strategy	Purpose
Plant pathogen-free nursery stocks. Establishment of certified and properly inspected nurseries for stocks.	To prevent occurrence and spread of virus, crown gall, fungi, bacteria, nematode diseases in new, susceptible fruit trees.
The location must be free from *Phytophthora, Armillaria* or nematode pests. Fumigation of planting material before plantation. Use of resistant rootstock for grafting varieties against these pathogens.	To prevent occurrence of soil borne pathogens, especially nematodes.
Drainage should be properly maintained at the location.	To prevent occurrence of soil borne diseases and nematodes.
Plantation patterns: Planting young healthy plants close to old plant infected with transmissible diseases (i.e. canker, pollen or insect transmitted viruses, mycoplasma etc.) should be avoided.	To prevent the spread of diseases,
Plant protection: Balance nutrition, proper irrigation, off season pruning, use of pesticides against diseases and insects.	To maintain tree vigour and reduce occurrence of insect and disease causal agents.
Uprooting of infected plants: Especially infected by viruses and mycoplasma (systemic pathogens)	To minimize or eliminate the chances of a disease spread.
Field sanitation: Burning of plant diseased and debris after pruning	Elimination of primary inoculum to avoid infection in subsequent growing season (spring)
Farm equipment disinfection: Use disinfected equipment for pruning especially when moving from infected to healthy plants for pruning.	To prevent spread of the pathogens from tree to tree.
Chemical treatment (fungicides, bactericides or mixture of fungicides and insecticides) of plants before bud breaks.	To prevent the attack of fungi, bacteria and insects activated during spring.
Chemical protection of leaves and blossoms (fungicides, bactericides and insecticides)	To prevent infection of blossoms and young leaves which are usually highly susceptible.
In rainy areas, plant leaves and blossoms should be protected continuously by the application of chemical sprays (fungicides, bactericides and insecticides). Chemical spray applications on growing tissues shall continue until there is a chance of spore release, oozing (bacteria) and wet conditions. Weather forecasting will be helpful in this regard.	To prevent infection of flowers appearing and of rapidly broadening leaves by fungal spores and bacteria present in abundance in wet weather.
Repetition of spray scheduled for infection spreads to the new growth (fruits, leaves etc). In case of new pathogens or insects, chemical sprays should be replaced accordingly.	To prevent infection of young fruits by pathogens and insect attack on fruits. The systemic fungicides spray followed by protectant to prevent fungicide resistance strains.
Fungicides to avoid fruit rotting should be sprayed until harvesting. Insect control be continued.	To protect fruits from pre- and post-harvest pathogens.
Avoid fruit injuries during pre and post-harvest handling.	To avoid fruit infection during harvesting and handling

(continued)

Table 15.7 (continued)

Management Strategy	Purpose
Harvesting and picking containers should be clean. Fumigation (formaldehyde and Sulphur dioxide) of packing house and warehouse.	To reduce primary inoculum of certain fruit rotting fungi.
Washing with water containing a to avoid post-harvest rotting pathogens	To protect fruit during storage and transport
Discard infected fruits	To reduce secondary inoculum of fruit-rotting pathogens.
The fruits should be stored and transported in vehicles with low temperature facilities.	To slow down infection process in attacked fruits and avoid infection of healthy fruits.

Table 15.8 Integrated disease management in potato crops

Management Strategy	Purpose
Use healthy tubers in disease-free fields.	To prevent occurrence of viruses, late blight, ring rot and several fungi, bacteria and nematodes carried by potato seeds. The field may have pathogens such as *Verticillium, Fusarium* and cyst or root-knot nematodes. Using a field free of these pathogens would reduce chances of infection through soil.
Crop rotation with unrelated crops.	To reduce pathogens' population build-up.
Cull piles of potato should be properly managed or eliminated	To prevent the spread of *Phytophthora* sporangia to potato plants.
Cut tubers with disinfected knives and treat the seed pieces with chemicals (i.e. fungicides, bactericides or insecticides). Soil treatment.	Protection against ring rot of potato. Protection of seed against post-harvest decay. Protection against *Verticillium*, nematodes and *Fusarium*.
Sowing seed at proper time for vigorous growth.	Slow-growing or unhealthy growth is prone to the attack by *Rhizoctonia* in low temperature conditions.
Proper field drainage.	Protection against damping off, seed and root rots.
Use resistant varieties: Schedule spraying to protect or control early and late blight using weather forecast. Insecticide sprays may be given to reduce virus attacks. Using weather forecast can help in spraying at right time.	To prevent occurrence and spread of blights and transmission of viruses.
Destruction of diseased plants in the field.	To avoid the contact of *Phytophthora*-inoculum to tubers.
Avoid wounding.	To prevent entry of storage rot fungi (*Fusarium, Pythium* and others) into the tuber.
Discard damaged tubers	To reduce chance of tuber infection.
Store tubers at about 15 °C then at about 2 °C. keep storage room clean.	Wound healing occurs at 15 °C. development of fungus rot in storage is prevented at about 2 °C. keeping storage rooms clean will reduce chances of tuber infection.
Potato cull piles be destroyed as soon as possible.	To prevent the chance of tuber infection and spread of various diseases.

models. Last but not the least, both developed and especially developing world should emphasize the development of models for better crop production that are practically applicable and successful in the long run.

References

Agrios, G. N. (2005). *Plant pathology* (p. 348). New York: Elsevier Academic Press.

Akoroda, M. (2010). *Global perspectives on cassava production*. In: FAO African Regional Conference, Luanda, Angola, 2010.

Baker, B., Zambryski, P., Staskawicz, B., & Dinesh-Kumar, S. P. (1997). Signalling in plant-microbe interactions. *Science, 276*, 726–733.

Beachy, R. N., Loesch-Fries, S., & Tumer, N. E. (1990). Coat protein-mediated resistance against virus infection. *Annual Review of Phytopathology, 28*, 451–474.

Beattie, B., McGlasson, W. B., & Wade, N. L. (1989). *Postharvest diseases of horticultural produce, volume 1: Temperate fruit*. Melbourne: CSIRO.

Buddenhagen, I. W. (1977). Resistance and vulnerability of tropical crops in relation to their evolution and breeding. *Annals of the New York Academy of Science, 287*, 309–326.

Champ, B. R., Highley, E., & Johnson, G. I. (1984). *Postharvest handling of tropical Fruits*. In Proceedings of an international conference, Chiang Mai, Thailand, 19–23 July 1993, Australian Centre for International Agricultural Research, Canberra, ACT.

Chandrashekara, C., Bhatt, J. C., Kumar R., & Chandrashekara, K. N. (2012). Supressive soils in plant disease management, chapter 14. In *Eco-friendly innovative approaches in plant disease management*. https://www.researchgate.net/publication/232957603

Chaube, H. S., & Singh, R. (2004). *Introductory plant pathology* (1st ed.). Charbagh: International Book Distribution Company.

Choudhary, D. K., Prakash, A., & Johri, B. N. (2007). Induced systemic resistance (ISR) in plants: Mechanism of action. *Indian Journal of Microbiology, 47*, 289–297.

Ciancio, A., & Mukerji, K. G. (2007). *General concepts in integrated pest and disease management*. Dordrecht: Springer.

Dhaliwal, G. S., Dhawan, A. K., & Singh, R. (2007). Biodiversity and ecological agriculture: Issues and perspectives. *Indian Journal of Ecology, 34*(2), 100–109.

Fry, W. E. (1982). *Principles of plant disease management* (1st ed.). New York: Academic Press, a subsidiary of Harcourt brace Jovanovich, Publishers.

Gal-On, A., & Shiboleth, Y. M. (2006). *Cross-protection*. https://doi.org/10.1007/1–4020-3780-5_12. https://www.researchgate.net/publication/227091176

Hadidi, A., Khetarpal, R. K., & Koganezawa, H. (Eds.). (1998). *Plant virus disease control*. St. Paul: APS Press. 684 pp.

Hall, R. (1995). Challenges and prospects of integrated pest management. In R. Reuveni (Ed.), *Novel approaches to integrated pest management* (pp. 1–19). Boca Raton: Lewis Publishers.

Harman, G. E., & Bjorkman, T. (1998). Potential and existing uses of Trichoderma and Gliocladium for plant disease control and growth enhancement. In G. E. Harmon & C. P. Kubicek (Eds.), *Trichoderma and gliocladium, vol. 2: Enzymes, biological control and commercial applications* (pp. 229–265). London: Taylor and Francis.

Harris, K. F., & Maramorosch, K. (Eds.). (1982). *Pathogens, vectors, and plant diseases: Approaches to control*. New York: Academic Press. 310 pp.

Hewitt, H. G. (1998). *Fungicides in crop protection*. Wallingford: CAB International.

Hodson, D., & Nazari, K. (2010). Serious outbreaks of wheat stripe or yellow rust in central and West Asia and North Africa – March/April 2010. *Borlaug Global Rust Initiative, Newsroom, Rust in the news, April 2010 News Items*. http://globalrust.org/traction/permalink/Pathogen206

Kasirajan, S., & Ngouajio, M. (2012). Polyethylene and biodegradable mulches for agricultural applications: A review. *Agronomy for Sustainable Development, 32*, 501–529.
Kerr, A. (1980). Biological control of crown gall through production of agrocin 84. *Plant Disease, 64*, 25–30.
Khetarpal, R. K., & Gupta, K. (2005). Status of plant protection in India in the wake of international agreements. *Indian Journal of Plant Protection, 33*, 153–163.
Khoury, W. E., & Makkouk, K. (2010). Integrated plant disease management in developing countries. *Journal of Plant Pathology, 92*(4, Supplement), S4.35–S4.42.
Knight, S. C., Anthony, V. M., Brady, A. M., Greenland, A. J., Heaney, S. P., Murray, D. C., et al. (1997). Rationale and perspectives on the development of fungicides. *Annual Review of Phytopathology, 35*, 349–372.
Kumar D., & Shivay, Y. S. (2008). *Modern concept of Agriculture: Integrated crop management*. https://www.researchgate.net/publication/237569012
Leung, H., Zhu, Y., Revilla-Molina, I., Fan, J. X., Chen, H., Pangga, I., Vera-Cruz, C. M., & Mew, T. W. (2003). Using genetic diversity to achieve sustainable rice disease management. *Plant Disease, 87*, 1156–1169.
Lomonossoff, G. P. (1995). Pathogen-derived resistance to plant viruses. *Annual Review of Phytopathology, 33*, 323–343.
Lumsden, R. D., & Walter, J. F. (1995). Development of biocontrol fungus Gliocladium virens: Risk assessment and approval for horticultural use. In M. T. Hokkanen & J. M. Lynch (Eds.), *Biological control: Benefits and risks* (pp. 263–269). Cambridge: Cambridge University Press.
Mathys, G. (1975). Thoughts on quarantine problems. *EPPO European and Mediterranean Plant Protection Organization Bulletin, 5*(2), 55.
Mills, F. D., & Nutter, F. W., Jr. (1991). Late leaf spot control: Can the costs be reduced? *The Peanut Grower, 3*, 36.
Mishra, R. K., Ojha, M. D., Pandey, V. S., & Verma, R. B. (2016). In G. Chand & S. Kumar (Eds.), *Recent advances in the diagnosis and management of major diseases of onion and garlic in India, crop diseases and their management: Integrated approach*. Boca Raton: CRC Press. https://www.taylorfrancis.com/books/e/9780429156038/chapters/10.1201/b19891-18
Murga Gutierrez, S. N., Rosso, L. C., Colagiero, M., Finetti Sialer, M. M., & Ciancio, A. (2012). Root-knot nematodes and associated biological antagonists from asparagus crops in northern Peru. *Nematropica, 42*, 57–62.
Nelson, D. L. (2004). Plant pathology and managing wildland plant disease systems. Chapter 15, USDA Forest Service Gen. Tech. Rep. RMRS-GTR-136 2004. https://www.fs.fed.us/rm/pubs/rmrs_gtr136_1/rmrs_gtr136_1_187_192.pdf
Nutter, F. W. (2007). In A. Ciancio & K. G. Mukerji (Eds.), *The role of plant disease epidemiology in developing successful integrated disease management programs in general concepts in integrated pest and disease management*. Dordrecht: Springer.
Nutter, F. W., & Guan, J. (2001). Disease losses. In O. C. Maloy & T. D. Murray (Eds.), *Encyclopedia of plant pathology* (pp. 340–351). New York: John Wiley and Sons.
Nutter, F. W., Jr., & Mills, F. D. (1990). Cost/benefit comparison of a weather-based fungicide scheduling program versus a calendar spray program to control late leafspot of peanut. *Phytopathology, 80*, 989.
Overton, J. (1996). *Ecologically based Pest management-new solutions for a new century*. Washington DC: National Academy Press.
Palti, J. (1981). *Cultural practices and infectious crop diseases*. Berlin: Springer.
Pathak, V. N., Khatri, N. K., & Pathak, M. (2006). *Fundamental of plant pathology* (3rd ed.). Jodhpur: Updesh Purohit for AGROBIOS.
Pieterse, C. M. J., Zamioudis, C., Berendsen, R. L., Weller, D. M., Wees, S. C. M. V., & Bakker, P. A. H. M. (2014). Induced systemic resistance by beneficial microbes. *Annual Review of Phytopathology, 52*, 347–375.
Razdan, V. K., & Sabitha, M. (2009). *Integrated disease management: Concepts and practices*. Dordrecht: Springer.

Salamanca, L. R. (2015). *Michigan state university extension.* https://www.canr.msu.edu/news/sanitation_is_critical_to_prevent_plant_diseases_part_2_field_sanitation
Shtienberg, D., Elad, Y., Bornstein, M., Ziv, G., Grava, A., & Cohen, S. (2010). Polyethylene mulch modifies greenhouse microclimate and reduces infection of Phytophthora infestans in tomato and Pseudoperonospora cubensis in cucumber. *Phytopathology, 100*(1), 97–104.
Sreeja, S. J. (2014). Synthetic plant activators for crop disease management–a review. *International Journal of Thesis Projects and Dissertations, 2*(1), 19–28.
Strange, R. N. (2003). *Introduction to plant pathology* (1st ed.). Southern Gate Chichester: John Wiley and Sons Ltd.
Thomas, M. B., & Waage, J. K. (1996). *Integration of biological control and host plant resistance breeding: A scientific and literature review*. Wageningen: Technology Centre for Agricultural and Rural Cooperation. European Union.
Thurston, H. D. (1990). Plant disease management practices of traditional farmers. *Plant Disease, 74*, 96–101.
Thurston, H. D. (1992). *Sustainable practices for plant disease management in traditional farming systems*. Boulder: Western Press.
Whetzel, H. H. (1929). The terminology of plant pathology. In *Proceedings of International Congress, Plant Science, Ithaca, NY* (Vol. 1926, pp. 1204–1215).
Wood, R. K. S., & Tveit, M. (1955). Control of plant diseases by use of antagonistic organisms; Botanical review, Vol. 21, No. 8, *Control of plant diseases by use of antagonistic organisms* (pp. 441–492). Springer on behalf of New York Botanical Garden Press Stable. URL: https://www.jstor.org/stable/4353536
Wolfe, M. S. (1985). The current status and prospects of multi- line cultivars and variety mixtures for disease resistance. *Annual Review of Phytopathology, 23*, 251–273.
Zadoks, J. C. (1985). On the conceptual basis of crop loss assessment: The threshold theory. *Annual Review of Phytopathology, 23*, 455–473.
Zadoks, J. C., & Schein, R. D. (1979). *Epidemiology and plant disease management*. New York: Oxford University Press, 427 pp.

Index

© Springer Nature Switzerland AG 2020
I. Ul Haq, S. Ijaz (eds.), *Plant Disease Management Strategies for Sustainable Agriculture through Traditional and Modern Approaches*, Sustainability in Plant and Crop Protection 13, https://doi.org/10.1007/978-3-030-35955-3

MIX
Papier aus verantwortungsvollen Quellen
Paper from responsible sources
FSC® C105338

If you have any concerns about our products,
you can contact us on
ProductSafety@springernature.com

In case Publisher is established outside the EU,
the EU authorized representative is:
Springer Nature Customer Service Center GmbH
Europaplatz 3, 69115 Heidelberg, Germany

Printed by Libri Plureos GmbH
in Hamburg, Germany